ALIEN
BASE

Other Books by
Timothy Good

ALIEN UPDATE
THE UFO REPORT
ALIEN CONTACT

ALIEN BASE

The Evidence for Extraterrestrial Colonization of Earth

TIMOTHY GOOD

AVON BOOKS ◆ NEW YORK

To Gordon Creighton

AVON BOOKS, INC.
1350 Avenue of the Americas
New York, New York 10019

Copyright © 1998 by Timothy Good
Cover photograph by NASA
Back cover author photograph by Dorothee Walter
Published by arrangement with Century/Random House UK Limited
ISBN: 0-380-80449-2
Library of Congress Catalog Card Number: 99-94870
www.avonbooks.com

First Avon Books Trade Paperback Printing: August 1999

AVON TRADEMARK REG. U.S. PAT. OFF. AND IN OTHER COUNTRIES, MARCA REGISTRADA, HECHO EN U.S.A.

Printed in the U.S.A.

OPM 10 9 8 7 6 5 4 3 2 1

Contents

Acknowledgements

It would be impracticable to acknowledge here all those who have contributed directly or indirectly to the production of *Alien Base*, but I would like to record my thanks in particular to the following individuals and organizations:

The George Adamski Foundation; Walter Andrus and the Mutual UFO Network; Rafael Baerga; Leslie Banks, DFC, AFC; Margaret Barling; Juan José Benítez; Ted Bloecher; Ralph and Judy Blum; Jonathan Caplan QC; Filiberto Caponi; Gerry Casey and the *Western Flyer* (Tacoma, Washington); Ronald Caswell; the Central Intelligence Agency; Paul Cerny; Antonio Chiumiento; Jerome Clark and the J. Allen Hynek Center for UFO Studies; Sir Arthur C. Clarke; Terence Collins; Stephen Darbishire; *Domenica del Corriere*; Britt and Lee Elders; Lucius Farish and the *UFO Newsclipping Service*; the Federal Bureau of Investigation, Ugo Furlan; Bruno Ghibaudi; Dr Daniel Rebisso Giese; Horacio Gonzales; the Göteborg (Sweden) Information Centre on UFOs; Charles Gourain; Antonio Giudici; Jane Thomas Guma; Bill Gunston, OBE; Dr James Harder; Carol Honey; William E. Jones; Peter Jordan; Tony Kimery; Kevin McNeil; Carlos Mañuel Mercado; Lawrence Moore and Central Independent Television; James W. Moseley; the National Aeronautics and Space Administration (NASA); Héctor Antônio Picco and *Crónica* (Buenos Aires); *Norddeutscher Rundfunk* (*NDR*); Lieutenant Commander Rolan D. Powell, US Navy (Retired); the Public Record Office; Madeleine Rodeffer; Pedro Romaniuk; Herbert Schirmer; Dr Berthold E. Schwarz; Dr Irena Scott; William T. Sherwood; Warren Smith; Dr Leo Sprinkle; Lieutenant General Thomas P. Stafford, US Air Force (Retired); Ray Stanford; Hal Starr; Bill Steele; General Boris Surikov, Soviet Air Force (Retired); Neil Thomas and the *Staffordshire Newsletter*; Donald R. Todd; Marc Tolosano; the United States Departments of the Air Force, Army and Navy; Dr Jacques Vallée; Walter N. Webb; Don Worley; the Wroclaw (Poland) Club for UFO Popularization and Exploration.

I am especially indebted to: Warren Aston, for his research material relating to the case of Udo Wartena; Jeannie Belleau, for terrific transportation around New Mexico in 1997; Mark Ian Birdsall, for research material from his forthcoming book *Flying Saucers of the Third Reich* and for his report on the Jan Siedlecki case; Mark Booth, my editor at Century, for his unswerving loyalty; Lieutenant Colonel Philip Corso, US Army (Retired), William J. Birnes and Pocket Books (Simon & Schuster), for extracts from *The Day After Roswell*; Gordon Creighton, to whom *Alien Base* is dedicated, for huge chunks of material from *Flying Saucer Review*, of which he is editor; Frédérique, for advice and help relating to her mother Joëlle's contact story; Rachael Healey and her team of publicists at Century, for first-rate promotion; Air Marshal Sir Peter Horsley and his publisher Leo Cooper, for lengthy extracts from Sir Peter's autobiography, *Sounds From Another Room*; Desmond Leslie, for much material from *Flying Saucers Have Landed* (which he co-authored with George Adamski), a great deal of additional information, and for some unforgettable visits to Castle Leslie; Andrew Lownie, my agent, for his loyalty, support and sensible advice; Jorge Martín and his wife Marleen, for a great deal of material from their magazine *Evidencia OVNI*, for additional information, and for driving me around Puerto Rico to meet witnesses. Their courage and dedication, often in the face of adversity, are admirable; Howard Menger and his wife Connie, for much material from their books, in particular Howard's *From Outer Space to You*; Joël Mesnard, editor of *Lumières Dans La Nuit*, for articles and drawings; Carlos L. Moreno, for translations, and for acting as interpreter and providing transportation in Puerto Rico; Ludwig F. Pallmann – wherever he may be – for the story of his remarkable encounters with extraterrestrials; Sue Phillpott, for superb proof-reading; Bob Pratt, for extracts from his book *UFO Danger Zone*; Walter Rizzi, for the report and sketches of his encounter; Liz Rowlinson, Mark Booth's assistant at Century, for her patience and professionalism; Jane Selley, for superb copy-editing; Captain Graham Sheppard, for his transcription of the voice tapes relating to the America West encounter over New Mexico, and for additional advice relating to matters aeronautical; Jean Sider, for the use of extracts from his book *Ultra Top-Secret* and for additional help and advice; Neville Spearman, publisher of *Flying Saucers Have Landed* (Leslie and Adamski) and *Inside the Space Ships* (Adamski), for lengthy extracts from these books; Lieutenant Colonel Wendelle C. Stevens, US Air Force (Retired), for accounts from several of his books; Mrs Leonard Stringfield, for reports from her late husband's books; Sir Mark Thomson, for his contribution and support, including the funding of a reconnaissance flight to Dulce, New Mexico; Dorothee Walter, for help with word-processing and translations; Carroll and Rosemary Watts – wherever they may be – for their report and photographs of Carroll's encounters; Haroldo Westendorff, Michael Wysmierski of *The Brazilian UFO Report* and the Grupo de Pesquisas Científico-Ufológicas (GPCU, Pelotas, Brazil), for material relating to the Lagoa de los Patos incident.

Finally, I am indebted to those who have contributed or helped but are not named.

ALIEN
BASE

Introduction

It seems that scarcely a week goes by without some story in the sensationalist media about little bug-eyed aliens abducting a hapless victim and performing all manner of sinister experiments upon his or her person. A
proliferation of films, television documentaries, books and magazines on
the subject has contributed not only to growing interest in the UFO
phenomenon but also to an acceleration of wild claims.

Following publication of best-selling books on alien abduction, such as
those by Budd Hopkins, Whitley Strieber, Dr David Jacobs and Dr John
Mack, many people began claiming that they, too, had been abducted by
small, grey, bug-eyed creatures (the so-called 'Greys'), sometimes even
taken from their beds, beamed aboard spaceships and subjected to
physical procedures such as the extraction of ova, for the aliens' declared
purpose of creating a hybrid race.

Having met many abductees, I have no doubt that some of them have
interacted with alien life-forms; however, while it is feasible that an
explanation for the proliferation of such claims might be simply that
witnesses feel more inclined to come forward due to increased public
awareness, a corollary is that many such stories may be contaminated. Is
it coincidental that in the United Kingdom, for example, abduction
reports increased with the landing of *The X Files* television series on
British shores?

Another problem with the abduction phenomenon is the alarming
increase in the number of 'backstreet abductionists'; unqualified
hypnotherapists with little or no knowledge of general medicine and
psychiatry, who conduct hypnotic regression sessions with the thousands
of people now claiming to have been abducted. Most leading investigators urge caution in this undefined area. 'I feel that the present fad of
hypnotizing "abductees", which is being engaged in by untrained investigators, will inevitably lead to suffering,' wrote Whitley Strieber. 'These
investigators usually make the devastating error of assuming that they
understand this immense mystery.'[1] In 1991, Strieber went further: 'The
"abduction reports" that they generate are not real,' he wrote. 'They are
artifacts of hypnosis and cultural conditioning.'[2]

Stories of alien abduction have become fashionable. This has led, in my opinion, to an unbalanced perspective. One of the principal reasons for writing this book is to redress the balance, by recounting many stories of the so-called 'contactees', some dating back as far as 1920; accounts now either forgotten, confined to early literature collections, privately published, or taken from my own hitherto unpublished files.

It is now mandatory to scorn contactees, due in part to sometimes banal and evangelical messages imparted to them by the 'space brothers', and because the extraterrestrials they have encountered do not conform to preconceived notions of alien appearance, behaviour, or purported planetary origin. Budd Hopkins, whom I admire for his pioneering work and intelligent approach to the abduction phenomenon, encapsulates this attitude succinctly in his important book, *Witnessed*:

> In the 1950s and early 1960s a number of so-called contactees claimed to have ridden in flying saucers to Venus or Mars or elsewhere in our solar system, and there to have received from beautiful Space Brothers and Sisters antiwar messages and warnings about the environment . . . virtually all of their accounts were designed to make them seem special – honoured Earthlings, proud recipients of the Space Brothers' flattering attentions and intergalactic wisdom.[3]

This overstates the case. First, claims of non-abducting contacts with extraterrestrials not only pre-date the 1950s but, though increasingly rare, have continued to the present time. Secondly, most of the lesser-known contactees did not seek attention in any way; rather, the contrary. Thirdly, with respect to 'antiwar messages and warnings about the environment', Hopkins overlooks the fact that this is precisely what some abductees report having heard. Equally banal messages have sometimes been imparted to the abductees. In these and other respects, therefore, differences as between abductees and contactees can be marginal and confusing.

It is not enough arbitrarily to dismiss reports by contactees in terms of ego-gratification, hallucinations, an overly vivid imagination, and so forth. There are too many such accounts from too wide a variety of sources and countries, all of which contain scientifically interesting data elements in common. These beg for analysis, interpretation and explication. 'The statements of those who purportedly have had actual contact with "space people" should not be dismissed offhand as mere romance,' declared Rear Admiral Herbert Knowles, a graduate of the US Naval Academy, in 1957. 'Perhaps there is some real information here. One cannot afford to be dogmatic in this matter if the truth is to be found.'[4]

As I have stated in my previous books, it is my opinion that a secondary reason behind the reluctance of governments to acknowledge the alien presence is fear of ridicule. No one likes to look silly, least of all political and military leaders. Those few who have admitted to such a presence – such as President Ronald Reagan, who hinted at it during an important speech before the United Nations General Assembly in 1987,[5] and Javier Pérez de Cuellar, former Secretary-General of the United Nations, who reportedly was abducted together with his security guards and others in 1989[6] – invariably are mocked by the media, thus effectively stifling serious discussion on these matters and discouraging others in high office from coming forward. In 1997, Air Marshal Sir Peter Horsley, former Deputy Commander-in-Chief of Strike Command, and for seven years in the personal service of Her Majesty the Queen and HRH Prince Philip as Equerry, revealed in his autobiography[7] that in 1954 he had had a meeting with an apparently extraterrestrial being (see Chapter 10). The reaction was predictable. 'Oh God,' commented a former Ministry of Defence senior officer. 'How unfortunate that the public will learn that the man who had his finger on the button at Strike Command was seeing little green men.'[8] In a patronizing article in *The Times*, entitled 'Air marshal's flight of fancy', Dr Thomas Stuttaford suggested that the air marshal was either deluded or had suffered an hallucination,[9] overlooking the fact that the meeting had been arranged via a British Army general and that a witness had been present throughout the two-hour discourse.

My own investigations show that delusion (as well as deception) does indeed feature in a number of reports by both abductees and contactees. One example will suffice. In 1981, I interviewed a South African contactee, the late Elizabeth Klarer, a cultured, striking lady who claimed to have given birth to a child fathered by a spaceman. Klarer's initial encounter with 'Akon', her lover from Venus, supposedly occurred in the 1950s. In her fascinating 1980 book, however, Akon's origin changed to 'Meton, one of the planets of Proxima Centauri'[10] (4.26 light years from Earth). During my interview with Klarer in Johannesburg, she mentioned casually that Akon still visited her from time to time. 'Do you have any kind of evidence to show me?' I asked. 'Oh yes,' she replied, 'he brought me this beautiful plant!' I took some photographs. On my return to London, I soon discovered that the plant was the very terrestrial maidenhair fern (of the genus *Adiantum*), and I duly informed Elizabeth. I never heard from her again. Perhaps, like other claimants, she did have a genuine experience of some sort but later began to fantasize and embellish it. I believe that a number of well-known abductees and contactees have inflated their claims, either to nourish their egos or retain a following, or both. These motives alone do not negate the validity of all their claims.

Another reason for this book is to show that alien species are more varied than we commonly suppose, and that the little grey men appeared on the scene relatively recently. From gorgeous girls, through to grotesque goblins – and yes, even little green men – most types of alien species that have been reported are represented in this book.

Objections are frequently raised regarding the anthropomorphic nature of reported aliens. 'One of the chief reasons I have never been able to take reports of alien contact seriously,' wrote the great space pioneer and science-fiction writer, Arthur C. Clarke, in 1997, 'is that no spaceship ever contains aliens – the occupants are always human! Oh, yes, they *do* show a few minor variations such as large eyes, or pointed ears . . . but otherwise they are based on the same general design as you and I. Genuine extraterrestrials would be *really* alien . . .'[11] Aside from the fact that some alien creatures, such as the so-called 'chupacabras', look like nothing on Earth, we cannot disregard the testimony of thousands of men and women from all walks of life who, throughout the twentieth century at least, have testified that the aliens they have encountered are indeed generally humanoid, though sometimes with unusual differences. 'You certainly make a good case,' Clarke wrote to me, in response to my critique of his article. But, he added: 'My main argument against ETs is that none have yet called on *me*, which seems a deplorable oversight.'[12]

As to the similarities, one reason could be that *Homo sapiens* is related to an extraterrestrial species. The idea may seem far-fetched, yet the probability that our galaxy has been colonized by other, more advanced races, cannot be dismissed out of hand.

Much of the material in this book will be new, even to seasoned investigators. Interspersed with the contact encounters are new or little-known reports of sightings by military and civilian pilots as well as sightings of unidentified submergible objects reported by naval observers. I have also included some cases involving alien vehicles apparently grounded for repairs, as well as alleged retrievals, by military forces, of alien vehicles and bodies, occurring well before the so-called Roswell Incident. With all these varied claims, proof remains elusive. I believe, nonetheless, that the evidence is mounting.

In 1997, the United States Air Force issued its third 'final' explanation for the Roswell, New Mexico, incident of July 1947, when an alien spacecraft and its occupants were recovered by military forces. The bodies, stated the Air Force in its 230-page report (timed to coincide with the 50th anniversary of the Roswell Incident), were really dummies used in experiments in the 1950s to test the effects of high-altitude parachute drops on human bodies.[13] This 'down-to-earth' explanation totally fails to account for the fact that no such dummies were dropped until several years after the Roswell Incident; moreover, only a dummy could confuse dummies with alien bodies. US Air Force 'special planners' in the Pentagon seem to be expending an extraordinary 'public perception management' effort aimed at destroying the central icon of ufology, the Roswell Incident, with the goal that, if successful, the whole fabric of ufology will come apart. Similarly, in August 1997, the Central Intelligence Agency released an article on its role in the study of UFOs (1947–90) in which it is stated that many UFO sightings in the 1950s and 1960s actually were of Lockheed U-2 and SR-71 spy planes.[14] The CIA's 'explanation' is so flawed that even hard-core UFO sceptics are sceptical. With such shifting explanations, it is little wonder the public is becoming even more aware that it is being misled. 'The old cover-up stories just don't wash any more,' said John E. Pike, director of space policy at the Washington-based Federation of American Scientists. 'The UFO community is definitely on to something.'[15]

Perhaps those few in authority who are aware of the alien presence will be more forthcoming in the near future: indeed, there are encouraging developments. In June 1997, a remarkable book was published which, despite its egregious hyperbole and factual errors, seems to lift the lid off much that has been concealed by certain elements of the US intelligence community since 1947. Lieutenant Colonel Philip J. Corso, US Army (Retired), who served on President Dwight Eisenhower's National Security Council staff[16] and the Army staff's Research and Development Foreign Technology desk at the Pentagon, claims not only that he saw the alien bodies recovered from the Roswell crash, but also that, from 1961 to 1963, he was steward of an Army project that 'seeded' alien technology at American companies such as IBM, Hughes Aircraft, Bell Labs and Dow Corning – without their knowledge of where the technology originated. Corso alleges that the materials found aboard the Roswell craft were precursors for today's integrated circuit chips, fibre optics, lasers and super-tenacity fibres. He also discloses the role that alien technology played in shaping geopolitical policy and events; how it allowed the United States to surpass the Russians in the space programme, led to the Strategic Defense Initiative ('Star Wars'), and ultimately brought about the end of the Cold War. Corso also claims that certain intelligence analysts were aware at that time that Earth was being 'probed' by one or more alien cultures, and that they posed a serious threat. 'Maybe, in the 1960s' he points out, 'we didn't have the technology we have now to intercept their ships, but by using new satellite surveillance techniques we believed we'd be able to pick up the signatures of an alien presence on the face of our planet. If we made it too difficult for them to set up shop with bases on Earth, military intelligence planners speculated, maybe they would simply go away . . .'[17]

They have not gone away.

Alien Base is dedicated to Gordon Creighton, the indefatigable, multi-lingual editor of Britain's *Flying Saucer Review*,[18] established in 1955, from which I have selected a lot of the material in this book. Creighton is one of the few remaining pioneers, having become interested in UFOs as far back as 1941, when he saw one while serving in the British Embassy at Chongqing, China. In the 1950s, as an intelligence officer at the Ministry of Defence in Whitehall, he became aware of top-secret investigations into the phenomenon conducted jointly by the Royal Air Force and the United States Air Force.[19] Many of those interested in this enduring enigma have benefited from his scholarship.

Finally, I must stress that, while the reports of the numerous witnesses described herein have been faithfully reproduced, the various interpretations and explanations offered by those witnesses must be viewed as subjective attempts to apply some degree of rationality to their extraordinary encounters.

NOTES

1 Strieber, Whitley, *Transformation: The Breakthrough*, Century, London, 1988, p. 254.
2 Strieber, Whitley, *The Communion Letter*, summer 1991.
3 Hopkins, Budd, *Witnessed: The True Story of the Brooklyn Bridge UFO Abductions*, Pocket Books, Simon & Schuster, New York, 1996, p. 398. Also published by Bloomsbury, London, 1997.
4 *UFO Investigator*, National Investigations Committee on Aerial Phenomena, Washington, DC, vol. 1, no. 1, July 1957.
5 Good, Timothy, *Beyond Top Secret: The Worldwide UFO Security Threat*, Sidgwick & Jackson, London, 1996, p. 539.
6 Hopkins, op. cit. (Hopkins does not reveal the name, Pérez de Cuellar, to whom he refers as 'the third man', in this book: privately, he has confirmed it, however.)
7 Horsley, Sir Peter, *Sounds From Another Room: Memories of Planes, Princes and the Paranormal*, Leo Cooper, London, 1997.
8 Barton, Fiona, 'Close encounter in a Chelsea flat', the *Mail on Sunday*, London, 10 August 1997.
9 Stuttaford, Dr Thomas, 'Air marshal's flight of fancy', *The Times*, 14 August 1997.
10 Klarer, Elizabeth, *Beyond the Light Barrier*, Howard Timmins, Cape Town, 1980.
11 Clarke, Arthur C., 'Why ET will never call home', *The Times*, London, 5 August 1997.
12 Letters to the author from Arthur C. Clarke, CBE, 29 August and 2 October 1997.
13 McAndrew, Capt. James, *The Roswell Report: Case Closed*, Superintendent of Documents, US Government Printing Office, Mail Stop SSOP, Washington, DC 20402-9328, June 1997.
14 Haines, Gerald K., 'CIA's Role in the Study of UFOs, 1947–90', *Studies in*

Intelligence, Central Intelligence Agency, spring 1997.

15 The *New York Times*, 3 August 1997, quoted by Nick Hopkins in the *Daily Mail*, London, 5 August 1997.

16 Interestingly, Corso (who died in July 1998) also served from 1954–57 on the National Security Council's Operations Coordination Board, later known as the 'Special Group' or '54/12 Committee'. This was the group which, spanning several Presidential Administrations, approved and evaluated the most sensitive covert operations ever mounted by the United States.

17 Corso, Col. Philip J., with Birnes, William J., *The Day After Roswell*, Pocket Books, New York and London, 1997, p. 129.

18 *Flying Saucer Review*, FSR Publications Ltd., PO Box 162, High Wycombe, Bucks. HP13 5DZ, England.

19 Good, op. cit., p. 25.

PART ONE

Chapter 1

Strategic Reconnaissance

It was the morning of 5 April 1943. US Army Air Forces flying instructor Gerry Casey, together with a student, had taken off in a Vultee Valiant BT-13 trainer from the USAAF Ferry Command Base at Long Beach, California. After climbing through the cloud deck they cruised back and forth at 5,000 feet for 40 minutes on the southeast-northwest legs of the Long Beach low-frequency radio range. Above the clouds, visibility was unlimited. At 09.50, looking east towards Santiago Mountain, Casey thought he saw a flash of light. 'Peering intently, I saw an aircraft in a moderate dive aimed at our BT-13 with a perfect interception angle [and] I prepared to take evasive action if needed.'

> The craft coming at us appeared to be painted an international orange and was now about to pass on our left side. Unable to determine the craft's make or model, I knew it was unlike any airplane I'd ever seen. As I studied it, I was shocked to see it make a decidedly wobbly turn that quickly aligned it off our left wing in instant and perfect formation.

Ordering his student to come out from under the hood used for practising 'blind' flying (preventing the student from seeing anything except his instruments), Casey exclaimed that he thought Lockheed's new secret plane, rumoured to be propellerless (the Lockheed P-80 Shooting Star jet, which first flew in January the following year), was flying in formation with them. Instinctively, Casey reached for his camera, but realizing he could get into serious trouble if he photographed a secret plane, he put it away. The unknown aircraft defied rational explanation:

> I'd noticed that its turn appeared totally independent of air-reaction but that when it was off our wing, the adjustment to our altitude and course was perfect and instantaneous. Its position with us was held as if an iron bar had been welded between the two . . . its color was a radiant orange, which appeared to shimmer in the bright sunlight. As we watched, its aft end made a slight adjustment and it shot away from our position, disappearing in a climbing turn toward the ocean. Later, both of us agreed that it was gone from sight in two seconds.

AN EXOTIC AIRCRAFT

After landing, Casey and his student discussed the 'exotic aircraft'. Both agreed that it was orange in colour, changing to white when it accelerated. No openings or signs of a cockpit could be detected, nor could the means of propulsion be determined. Size was difficult to estimate, but both pilots thought that if the object had been 10 feet in diameter, it would have been 35 to 50 feet off their wingtip; if 50 to 75 feet in diameter, it would have been 100 or more feet away. Casey felt certain the object was elliptical, while the student was certain it was circular. Both agreed it had a rounded hump on the top and a smaller hump on the bottom. Casey later computed its departure speed to be over 7,000 m.p.h. – a computation with which the student agreed.

'Trying to recapture the details of an event that had consumed less than 90 seconds kept my thoughts occupied,' said Casey. 'I drew a pencil sketch of the craft's profile to confirm my opinion that it had been designed and built of parabolic curves rather than compass-drawn arcs. I could not reconcile its wobbling flight nor its sudden and unbelievable acceleration.'

CONCLUSIONS

Following the war, Gerry Casey became an inspector for the Civil Aeronautics Administration (CAA – later the Federal Aviation Administration) at the Boeing Airplane Company in Seattle. He remained convinced that what he and his student had seen was a flying machine 'light years' in advance of anything on Earth. For fear of ridicule he dared not discuss the incident with his colleagues. Then, in 1948, a memo came through the CAA concerning the United States Air Force's Project Sign (later Project Grudge, still later Blue Book, USAF's official UFO investigations), urging that any personnel who had a UFO experience should report it. 'They added that the person and the date would be investigated,' said Casey. 'I did as suggested but never received any acknowledgement or contact.' An enduring curiosity led him to make his own investigations and to form his own conclusions:

Since that early time in UFO history, sightings throughout the world have been reported by too many credible witnesses to ignore . . . Airline and military pilots the world over have had similar brief encounters with exotic machines such as seen by my student and me in 1943 . . . For anyone to dismiss all sightings by professional airmen, scientists, and radar and air traffic personnel only displays the critic's closed mind . . . For any airman who has had a similar experience to mine, the conscious event cannot be erased. Nor can it be rationalized through comparisons with any known thing on Earth . . . Credible scientists have noted that many sightings have occurred in the vicinity of our atomic plants or military installations. Other viewings have indicated that close approaches were made in isolated areas.

Casey explained that 'the sorry state of mankind versus his environment and his apparent headlong flight into self-destruction' finally caused him to bare his soul by coming forward with this important report. 'If it is true that we creatures are moving headlong into a self-destructive mode,' he concluded, 'possibly the failure of our planet could upset the balance of others in our, or a nearby, planetary system. If this is true, then any other superior race of creatures would be seriously concerned.'[1]

ENCOUNTER OVER UKRAINE

On an unspecified date in 1944, during a mission to Romania to bomb oil refineries used by the Germans, Boris Surikov and his commander, Major Bajenov, were flying at an altitude of five kilometres over southwest Ukraine when they had an encounter with a highly unusual aircraft. 'In front of the plane, a large elliptical-shaped object flew towards us,' Surikov told British television producers Lawrence Moore and Livia Russell in Moscow in 1994:

> We'd read in the newspapers about new German weapons, but we'd seen nothing like this. What happened was that our heavy plane [unspecified], of 14.5 tonnes, started shaking, the oil pressure rose, and when I leaned towards the window I felt a strong electrostatic charge. I was worried that the plane would burst into flames. It passed us and disappeared, but our plane was still affected: I looked at the wings, and they were covered in electrical discharges.

Major Bajenov, equally concerned that the plane was about to catch fire, ordered the crew to jettison the bomb load. 'The whole plane was fluorescent and the wings were glowing like a rainbow,' said Surikov. 'If it had been up to me, I would have carried on and tried to fulfil the mission, and if the plane would have caught fire I would have jumped with a parachute.' The commander, having more experience, believed there was real danger of the plane catching fire and the bombs exploding, so gave orders to jettison the two-tonne bomb load in southwest Ukraine instead of in Romania.

Mentioning nothing about the incident in their report on the mission, Bajenov and Surikov stated merely that they had successfully bombed Romanian oil refineries. 'If we had said we had not carried out our mission,' Surikov explained, 'we could have been taken to court as cowards.'

Surikov described the unknown object as similar in some respects to the Russian and American space shuttles. 'It lit up the air around it. It looked like a localized sunset, but in the centre was a strange-looking flying object. It didn't look at all like the burst of an anti-aircraft shell, which is about 10 metres in diameter. It was larger and longer than our Buran space shuttle – I think about twice as long.'

Years later, Surikov asked scientists for their opinion as to what he had seen. 'I was told that one could not rule out the possibility that the electrification of the plane was due to the close proximity of a UFO with a new type of propulsion system which ionized the atmosphere.'

Surikov later became a specialist in rockets and nuclear weapons. For a long time he worked at Soviet Army headquarters as its chief authority on weapons of mass destruction. 'But I am proud,' he points out, 'to be one of those who developed a treaty on the restriction of anti-missile defence systems. For a long time I worked as an expert in Geneva, where we were trying to promote the disbanding of certain types of weapons of mass destruction – nuclear, radiological, and so on.'

Now retired from the Soviet Armed Forces with the rank of general-major, Surikov specializes in environment problems. He has also pondered the significance of the UFO phenomenon:

> We cannot rule out the possibility that creatures who may well be superior to us are interested in what is happening to our Earth ... Scientists with whom I have discussed these matters think that in those civilizations new types of energy have been discovered which allow them to fly very far at great speed, so it is very important for us to study them in order to make use of these discoveries and to improve life on Earth.[2]

AN AERIAL MERRY-GO-ROUND

During the German occupation of France, Daniel Léger, 19 years old at the time, lived in a little village in the department of Sarthe, 15 kilometres to the south of Le Mans, working at a butcher's shop. On a hot afternoon in the summer of 1941, the exact date not recalled, Léger was in the street when he noticed a kind of 'merry-go-round' in the sky, beneath cirrus clouds and above the airfield occupied by the Germans at Raineries. About a dozen German aircraft were circling around a large, aluminium-coloured 'cloud', shaped like the handle of a frying pan and surrounded by puffs of cloud. It was moving slowly and horizontally, though somewhat tilted. Above all this hovered another, larger and luminous 'cloud', moving at the same speed.

Deciding to take a closer look at this spectacle, Léger jumped on his bike and raced towards the scene. A mile further on, together with many other curious people, he was stopped by a wall of soldiers. The German aircraft continued circling the 'cloud' and manoeuvred as if they were attacking, but without opening fire. When the aircraft came close to the cloud they appeared to fall down like leaves, only to rise again seconds later, as if their engines had been stopped momentarily. The witness observed this merry-go-round for an hour as it moved from west to east, then he returned home.

After the war, Léger visited the municipal library in Le Mans to search for press commentary on this extraordinary event: he found none. Perhaps Luftwaffe records pertaining to this incident, censored at the time, may one day be located, as well as testimony from other witnesses. Léger had insisted on anonymity, so in his book, *Ultra Top-Secret*, researcher and author Jean Sider used the pseudonym 'S. Théau'.[3] Following the death of the witness in 1993 or 1994, Sider kindly provided me with the source's real name, which appears here for the first time. Important though this case is, it pales into relative insignificance when compared to an extraordinary encounter which befell Léger in 1943, to be described further on.

Regarding the aircraft losing altitude when they approached the 'cloud', it is interesting to note that, according to information disclosed to the well-known author, Jacques Vallée, by a former engineer with US intelligence in Germany, Americans were already aware by 1943 that UFOs (or 'foo-fighters', as they were then dubbed by USAAF air crews)[4] could interfere at a distance with internal combustion engines. Investigators at the time suspected that electrostatic effects were the cause. A secret investigation into the phenomenon, including an investigation into German research on jet aircraft, was conducted in 1943 by the then US National Bureau of Standards (now the National Institute of Standards and Technology), under the direction of Professor Dryden.[5] A distinguished aerodynamicist, Dr Hugh L. Dryden developed America's first successful radar-guided missile. He was chief of the aerodynamics section of the National Bureau of Standards, later chief of the National Advisory Committee for Aeronautics, then the first deputy administrator of NASA.

In 1944, American planes returning to England from bombing missions over the Continent were plagued repeatedly by engine cut-outs. 'The engines would suddenly become rough, cutting in and out,' a former US intelligence officer explained to author Ralph Blum. 'There was considerable discussion among intelligence people as to what should be done. The general feeling – that some new German device was causing the electrical problems – presented one major difficulty: the amount of electricity required to short out a [bomber] engine was calculated as greater than all the known electrical energy output of Europe!'[6]

FOO-FIGHTERS OVER ENGLAND

In June 1944 the Germans began launching against London the first of their two *Vergeltungswaffen* (retaliation weapons) – the V-1. Dubbed 'doodle-bugs' or 'buzz bombs' by the British, the V-1s, powered by a pulse-jet, were launched from ramps in northwestern France and directed to their target by a pre-set guidance system. Over 8,000 V-1s were launched against London alone, while thousands of others were launched against Allied-held targets on the Continent. Owing to their unreliability and relatively low speed (350 m.p.h.), about a quarter of them failed and about half were destroyed by countermeasures: aircraft interception, anti-aircraft fire and barrage balloons. Only a quarter reached their target areas and even then, some failed to explode.[7]

In the summer of 1944, Bill Steele, who frequently observed V-1s being attacked by Royal Air Force fighters, was operating an excavator at Fairlight Quarry near Hastings, Sussex. On clear days, he could see the aircraft patrolling, the V-1s approaching and the Observer Corps firing rocket signals to alert pilots. On two occasions, he told me, he noticed something very unusual:

> It was while watching one of these that I saw these curious discs. The impression was of hub-caps, though a little larger, circling the V-1 and running alongside with ease. I put them down to something the RAF were using. They had no effect whatsoever upon the V-1s. The odd thing was the curious melodious whistle-like noise that I heard . . .[8]

HUMANOIDS IN FRANCE

It was not just 'foo-fighters' that were observed during the Second World War. In the summer of 1944 a French witness claims to have observed an unknown flying machine together with its unusual occupants. Although only 13 years old at the time, Madeleine Arnoux retained a vivid recollection of the incident, which occurred while she was picking berries at a farm near the village of Le Verger, in the department of Saône et Loire.

'I was walking, slowly, looking for berries as I went,' she recalled. 'Up there ahead, something was standing beside the trees, and there were some beings quite near it. Looking back on it now, I think the machine must have been of about the size of one of our small cars today [and] it was of a dull metallic grey colour.'

> The beings that were beside it must have been less than one metre in height, and were dressed in a sort of brown-coloured overalls. They made no gesture in my direction, and I, for my part, was rooted to the spot . . . I remember the oppressive atmosphere and the thundery state of the weather, and I remember how I had the feeling that I was unable to move. Then suddenly I was able to do so. I wanted to get my bike, which was lying a few metres from me. It took just the time needed for me to bend down as I got it, and then, when I looked up again towards the strange apparition, there was nothing there . . . All there was to be seen, at the spot where it had been, was a violent wind blowing the trees about. I didn't think of looking up in the air, where, if I had done, I should no doubt have still been able to see the machine as it was flying away.

Terrified, the witness made off at full speed to the farm-house, where she mentioned nothing about her experience. Nor did she tell anyone when she arrived home, fearing to be branded a hoaxer. For a very long time, Madeleine Arnoux thought about her weird encounter, then forgot about it. It was only when people began to talk about *soucoupes volantes* (flying saucers) in the 1950s that she began to put two and two together.

> After all these years, the picture is still very clear in my memory and I know perfectly well that I wasn't dreaming and that what I saw that day in the woods wasn't anything that is 'known' . . . It was 1944, and the Maquis people of the [French] Resistance were quite plentiful in the area, but it couldn't have been any of them. Nor could it have been German soldiers, and had it been either Maquis or Germans, unquestionably they would have challenged me. So one can only think that I must have been a witness of one of the earliest UFO visits.[9]

This fascinating report was sent to the editor of the magazine *Lumières Dans La Nuit* in 1972. It comes across as an account by a sincere, puzzled witness. Of particular significance is Madeleine Arnoux's reference to feeling paralysed, a feature increasingly evident in close encounters in the years following the Second World War. The 'violent wind blowing the trees about' during take-off is also reported in many other such cases.

RETRIEVALS OF CRAFT AND BODIES
While it is widely believed that the first recovery by US military forces of a crashed UFO took place in New Mexico in July 1947 (the so-called Roswell Incident),[10] there are intriguing stories of UFO recovery operations some years earlier.

MISSOURI, USA
Raymond Fowler, a UFO researcher who once served in the United States Air Force (USAF) Security Service, learned of the recovery of a crashed UFO near Cape Girardeau, Missouri, in the spring of 1941. Fowler's source was Charlotte Mann, granddaughter of a witness, Reverend William Huffman, a Baptist. In a letter to the late Leonard Stringfield, a former USAF intelligence officer and the leading specialist in what he called 'UFO crash/retrievals', Mann shared her knowledge of the case:

> About 9 to 9:30 one evening, granddad got a telephone call from the police department, saying they had received reports that a plane had crashed outside of town and would he go in case someone needed him . . . A car was sent to get him, but grandmother said it wasn't a police car. After grandfather returned that night, he explained what he had seen to my grandmother, my

father, Guy, and Uncle Wayne, but they were never to speak of it again as he had given his word. Grandmother said he never did talk about it after that.

He said they drove out of town 13–15 miles or so, then parked the cars on the side of the road and had to walk a quarter of a mile or so into a field where he could see fire burning. Grandfather said it wasn't an airplane or like any craft he'd ever seen. It was broken and scattered all around, but one large piece was still together and it appeared to have a rounded shape with no edges or seams. It had a very shiny metallic finish. You could see inside one section and see what looked like a metal chair with a panel with many dials and gauges – none familiar-looking to him.

He said when he got there, men were already sifting through things. There were some police officers, plain-clothes people and military men. There were three bodies, not human, that had been taken from the wreckage and laid on the ground. Grandfather said prayers over them so he got a close look but didn't touch them. He didn't know what had killed them because they didn't appear to have any injuries and they weren't burnt. It was hard for him to tell if they had on suits or if it was their skin but they were covered head to foot in what looked like wrinkled aluminum foil . . .

There were several people with cameras taking pictures of everything. Two of the plain-clothes men picked up one of the little men, held it under its arms. A picture was taken. That was the picture I later saw. Then, one of the military officers talked to granddad and told him he was not to talk about or repeat anything that had taken place for security reasons and so as not to alarm the people. Granddad returned home, told his family. That was it. About two weeks after it happened, he came home with a picture of the two men holding the little man . . .

My recollection from what I saw in the picture was a small man about 4 feet tall with a large head and long arms. He was thin and no bone structure was apparent; kind of soft-looking. He had no hair on his head or body, with large, oval, slightly slanted eyes but not like an oriental from left to right, more up and down. He had no ears at all and no nose like ours. There appeared to be only a couple of small holes where his nose should have been. His mouth was as if you had just cut a small straight line where it should have been. His skin or suit looked like crinkled-up tin foil and it covered all of him . . . I believe he had three fingers, all quite long, but I can't be sure on this.

One of the first checks Stringfield made was to establish whether the photograph seen by Charlotte Mann was identical to a bogus photograph released as an April Fool's Day joke in Germany in 1950, showing some men in hats and coats holding a little silver man. It was in no way similar, as Mann confirmed. (The photo was lent by her grandfather to a friend, who never returned it.)

If this event really occurred, one is left to wonder why no more witnesses have come forward. Were they all intimidated? Whatever the case, Stringfield was impressed. 'After discussing the incident several times with Charlotte by phone, I felt increasingly comfortable with her manner of response to my questions,' he concluded. 'To me, she sounded sincere.'[11]

SONORA, MEXICO

Dr James Harder, a professor of civil engineering and a UFO investigator, believes that a crashed spacecraft may have been recovered from the Sonoran desert in Mexico (south of Arizona) late in 1941. Reportedly, recovery was effected by a team from the US Office of Naval Intelligence (ONI). One member of the team, unable to contain the importance of what had been discovered, brought home to share with his immediate family some photographs supposedly showing an unusual craft and small bodies.[12]

Dr Harder told me that the location of the incident was determined by one of his co-investigators, the late Jim Lorenzen, co-founder of the Aerial Phenomena Research Organization (APRO), to which Harder was a consultant. Together with another APRO consultant, Dr Leo Sprinkle, Harder interviewed a relative of the witness. He informed me that although the relative was only 10 years old or so at the time of the alleged incident, she seemed reliable and convincing. 'She had been sitting on the information for a long time, and had decided that APRO was the organization to talk to,' Harder explained. He continued:

> She was looking from a stair landing when she observed her uncle showing her family a sheaf of 8 x 10 prints of a UFO crashed horizontal on level ground. In one of the pictures, he himself was holding up a small spindly dead body, about 3.5 feet tall – the picture had been taken by a friend. A small pile of other small bodies was over to one side. The UFO took up nearly the entire frame. By various deductions from the witness's age, etc., we estimated the time of her seeing the prints was about Halloween, 1941 . . . Jim Lorenzen did send, against my advice, some local investigator around to question the retired officer [name supplied], who naturally clammed up, so at least he knows we know, if he is still alive.[13]

'NORTH OF GEORGIA', USA

According to another of Leonard Stringfield's sources, a crash of an unknown craft took place in a state 'north of Georgia' during the summer of 1942. The source, given the pseudonym Mary Nunn, served in a key civilian capacity in one of the two armed forces at the time of the incident.

Nunn claimed that a spacecraft had crashed at an Army base, causing damage to a building and minor ruptures to the side of the craft. Described as generally round in shape, the craft was 15 feet wide and 10 feet high, divided into three main sections: a control room, a compartment with four seats and a bottom bay equipped with a trap-door exit. In the control section was one large window and a number of smaller windows around the sides. The craft was silver in colour with markings on the inside as well as the outside. There also were crew members, Stringfield was told:

> The four crew members, taken alive, died about two weeks later of apparent starvation. Described as five feet tall, weighing about 90 to 100 pounds, the skin . . . was a milky white, smooth like a baby's, and without hair. Facial features were generalized; the eyes large and black like bug eyes, ears were small, lips thin slits. The fingers, numbering five, were long, bony; the feet flat, about size 6 with half-inch bony toes. The female had small breasts and, according to the source, the race could reproduce but there was no hint as to their genitalia . . . They had teeth, very white, wide and short. I asked about nourishment, how it was digested and eliminated. No answers . . .
>
> Communications were telepathic but when I asked about other details, anatomical, organic or emotional or about their craft, propulsion and all the other usual questions, came silence; that was it . . .[14]

ENCOUNTER WITH A GROUNDED CRAFT AND AVIATRIX

In 1943 Daniel Léger (whose 1941 observation of German aircraft in pursuit of unknown objects is described earlier) was conscripted by the Germans for *Service du Travail Obligatoire* (compulsory working service) at a labour camp in Gdynia (renamed Gotenhafen by the Germans), north of Gdansk, on the Baltic coast of Poland. Léger had obtained permission from the head of camp to visit Exelgroud, near Gdynia. Because transportation was not available, he set out on the sunny afternoon of 18 July on foot, by way of a short cut over the sand dunes running along the beach.

Reaching the top of a dune, Léger observed a peculiar metallic object, of a greyish aluminium hue, embedded in the sand. Approaching the device, he saw a human figure crouched on the ground, attempting manually to remove the sand that covered the lower part of the object. Although only the back of the figure was visible, Léger noted that it was a woman, with long blond hair, slim waist and broad hips. He assumed that she was a German Air Force pilot, because at that time there were a number of female Luftwaffe pilots in Exelgroud, as well as female technicians who loaded and transported torpedoes at the local German naval base. Léger was reluctant to make his presence known, but the woman seemed to have been aware of him. She turned around and stood up, revealing her height to be about 1.75 metres – above average for her sex.

The aviatrix wore a tight-fitting one-piece suit of dark-brown cloth, without pockets or fasteners, emphasizing her feminine form. The witness also observed a pair of pads on each calf, the upper part of which appeared like boots, of the same colour, forming an integral part of the suit. A four-inch-wide belt encircled her waist, the same colour as the suit, with the exception of a square silvery buckle. Her features were regular, with white skin, devoid of any kind of cosmetics, but with slightly slanted, Asian-like eyes. Her hair, parted equally, fell freely down her back. The only other visible part of her body was her long, slim hands, with short-cut nails, 'like a pianist', lacking any nail varnish.

The craft, embedded in the sand, looked like a 'colonial hat'. Later, as it took off, Léger could see that it was constructed like two plates joined together, separated by a middle section consisting of two rings with a black line between them. The craft was estimated to be about six metres in diameter and two metres in height. Several square portholes with rounded edges were spaced on the upper section, the exact number being indeterminable. No insignia, seams, weldings or connections were apparent on the upper part, which seemed to be made in one piece.

The aviatrix began talking to Léger in a language he could not interpret. Although it sounded quite guttural, it corresponded with neither German nor Polish. Because the most common sounds were vowels and diphthongs, he assumed it was not Russian. (Years after the war, Léger met Tahitians, whose native tongue sounded similar, though by no means identical, to that of the aviatrix.) In any event, Léger had the impression that she understood his French (this was just an impression, not a certainty). Gesturing animatedly with her hands while talking, she gave him to understand from this that she wanted him to continue the work she had left off, removing sand from the craft. Léger, being accustomed to obeying orders from the Germans, went about this task, and after some 10 minutes succeeded in freeing the 'new fighter plane' from the sand.

The woman appeared to be happy about this and, smiling contentedly, continued talking with Léger. Suddenly realizing with some surprise that he did not understand a word, she finally pointed to the sky, tapped her chest with the palm of her hand two or three times, and did the same to him. She then placed her hand on her buckle, whereupon a rectangular opening immediately became visible on the lower part of the craft's hull. First, a panel appeared on the hitherto seamless hull, then the panel was withdrawn several centimetres into the object and slid aside. The woman entered her craft after indicating to Léger that he should move away. The panel closed, leaving the hull looking as if the door did not exist.

Through one of the portholes, Léger observed that the interior was devoid of instruments. He saw the aviatrix sprawling 'on all fours' – or rather, in a stretched position – in the middle of the floor, as if she was driving a motorbike in a competition. (Investigator Jean Sider remarks that although this seemingly ludicrous detail tends to minimize credibility in the report, it was precisely this detail that led him to believe in the reality of the incident, based on comparison with a little-known 1954 case – see p. 170.) A slight rumbling sound could be heard and two rings on the craft began to rotate at an ever-increasing speed; the lower one clockwise and the upper one counterclockwise. The dark stripe separating the rings became luminous and began to vibrate, at which point the craft rose from the ground, slowly at first, then suddenly accelerated and disappeared in a northerly direction, at a speed far in excess of any German aircraft with which the witness was familiar.

Although Léger had touched the hull of the craft a few times while removing the sand, he did not notice any untoward physiological effects, during neither the hours nor the days following his adventure. Convinced that he had just witnessed the landing of an experimental aircraft, he quietly continued on his walk to Exelgroud, deciding not to discuss the experience with others.

LATER DEVELOPMENTS

After the Liberation, Léger got a job at an American Army Air Forces unit, where some USAAF gunners told him that during several missions they were accompanied by luminous disc-shaped objects – nicknamed 'foo-fighters' – which they assumed were German or Russian in origin. This conversation later inspired Léger to discuss the 1943 report with his workshop leader, one Sergeant Chappedelaine. The latter informed the colonel commanding the unit, who asked Léger to relate the incident to him in person.

At the end of 1945 Léger began to work with the Renault company in Le Mans, a job he held until his retirement in March 1982. After the war, he followed closely developments in aeronautics, hoping to find information about the unusual aircraft he had witnessed at Gdynia. He studied everything he could about the Peenemünde experimental centre on the Baltic coast, 130 miles to the west of Gdynia, where the Germans developed the V-1 flying bomb, the V-2 rocket and other weapons under the direction of Dr Wernher von Braun. Of particular interest to him was Hanna Reitsch, the legendary German female test-pilot who had flown aircraft such as the V-1e (a piloted version of the V-1 flying bomb, developed to test control problems with the V-1) and the Messerschmitt Me-163 rocket-powered fighter. Facts Léger gathered about research by German aeronautical experts supported a growing conviction: he had seen a unique fighter aircraft, tested not by Hanna Reitsch, but by a colleague of hers. (Reitsch's features, which he had seen in various magazines, did not match those of 'his' aviatrix.)

Because of the V-7 project – allegedly involving several conventionally propelled German aircraft or helicopters, some with lenticular-shaped wings, which had been developed towards the end of the war –

rumours circulated in the sensationalist press that Nazis exiled in South America after the war had developed these 'flying saucers', with which they were planning to avenge the Nazi defeat. Such rumours further confused Léger, who remained uncertain how to explain his own astonishing encounter.[15]

Interestingly, there is circumstantial evidence that at least one of the V-7 project aircraft was prototyped. According to the researcher and author Mark Ian Birdsall, several projects involving a circular-wing aircraft were conceived during the war, the most elaborate of which was constructed by Dr Richard Miethe at facilities in Breslau (Wroclaw), Poland, and in Prague. A small prototype was rumoured to have flown over the Baltic Sea in January 1943, and two full-scale aircraft with a diameter of 135 feet were eventually built. Also, reports Birdsall, another V-7 project was a 'spinning saucer', based on helicopter principles, about 35 feet in diameter, designed by Rudolf Schriever, a small prototype of which was allegedly first flown in 1943.[16] Could either of these small prototype aircraft have been the one seen by Léger in July 1943? It is unlikely. Apart from being of differing designs, small prototypes would have been unmanned. This is not to rule out the possibility that the Germans actually produced a number of circular flying machines at this time. The question remains: Did they actually fly? Though the majority of aviation experts are completely sceptical, it is difficult to disregard altogether the testimony of Professor Hermann Oberth, one of the great pioneers in astronautics and the teacher of Wernher von Braun, who claimed, rather extravagantly, that the V-7 and various modifications were responsible for many UFO reports during the closing stages of the war:

> At the end of the war we developed, first in Prague, then in Vienna, the V-7 helicopter – this could easily have been mistaken for a flying saucer. Instead of having rotor blades like an ordinary helicopter, the V-7 had rotating tubes which released an 'exhaust' of flame. As the tubes rotated, the helicopter appeared to have a circle of flame round it, and at a distance it looked like a shining disc. When it hovered, the flame was dark-red and dim. At higher speeds the disc appeared lighter and the flame looked yellowish, then white. At its highest speed . . . the V-7 tipped over and flew on its side. A significant feature is that it was extremely noisy in flight, and produced a thick trail in the stratosphere.

Dr Oberth (who was firmly convinced that unexplained UFOs were extraterrestrial in origin) added that: 'The V-7 certainly does not explain UFO reports before the end of the war.'[17] Certainly too, from information provided by Oberth, the V-7 helicopter was nothing like the grounded disc reported by Léger, and a great deal noisier. Furthermore, despite Oberth's claim, I doubt that a full-scale version of the V-7 ever flew. Science writer Brian Ford, author of a book on German secret weapons, believes that although some progress may have been made towards the construction of a small disc-like aircraft, the results were destroyed, before they could fall into enemy hands.[18] Ronald Humble, an aerospace and defence expert, also concludes that there is no hard evidence for German disc-aircraft having actually flown.[19]

It was not until the late 1950s, when Léger read about flying saucer sightings as well as the first artificial satellites, that he became divided in his opinion as to the origin of the craft he had seen, though he remained convinced that it was most likely a prototype of a revolutionary aircraft that the Germans were unable to develop in large numbers, owing to the intensive Allied bombing missions at that time.

In spite of his many contacts with first-generation French UFO researchers, such as René Fouéré and Marc Thirouin, Léger's natural reserve prevented him from discussing his experiences with them. Furthermore, he had read about ridicule heaped on many UFO witnesses. Then, in June 1989, he approached Jean Sider. This led to several meetings, and so impressed was Sider that he published the accounts in *Ultra Top-Secret*.

COMPARISONS WITH ADAMSKI'S INITIAL ENCOUNTER

The description by George Adamski of the 'man from Venus' he claims to have encountered in the Californian desert in November 1952 (Chapter 6) closely resembles Léger's account. In his first book, *Flying Saucers Have Landed*, co-authored with Desmond Leslie, Adamski described the 'Venusian' as follows:

> His hands were slender, with long tapering fingers like the beautiful hands of an artistic woman. In fact, in different clothing he could easily have passed for an unusually beautiful woman; yet he definitely was a man.
>
> He was about five feet, six inches in height . . . and I would estimate him to be about twenty-eight years of age, although he could have been much older. He was round faced with an extremely high forehead; large, but calm, grey-green eyes, slightly aslant at the outer corners; with slightly higher cheek bones than an Occidental, but not so high as an Indian or an Oriental . . . As nearly as I can describe his skin the colouring would be an even, medium-coloured suntan. And it did not look to me as if he had ever had to shave, for there was no more hair on his face than a child's. His hair was sandy in colour and hung in beautiful waves to his shoulders . . .
>
> His clothing was a one-piece garment . . . Its colour was chocolate brown . . . A band about eight inches in width circled his waist . . . I saw no zippers, buttons, buckles, fasteners or pockets of any kind, nor did I notice seams as our garments show . . . He wore no ring, watch, or any other ornament of any kind.[20]

In spite of some differences in attire (the man's, for example, was not tight-fitting) and footwear (he wore unusual shoes, as opposed to 'boots') from that described by Léger, there are some extraordinary parallels. It is of course possible that Léger fabricated his story, based on Adamski's account: Léger was very well versed in UFO literature, having 90 books on the subject (in French). From the end of the war, he was driven by an irresistible desire to read anything concerning aerial mysteries. Asked for his opinion of Adamski, he replied immediately that he thought Adamski's first meeting was authentic, but that since then he had gone on to tell fanciful stories. This may well be the case, as I shall discuss later.

Léger's experience, like so many other contact stories, contains apparent absurdities. Yet Jean Sider is convinced by the witness's account. 'I would like to stress the fact that during our first meeting on June 25, 1989, I got a very strong impression that I could rely on him,' he reports. 'At no time did he give me the impression that he was fabricating . . . His account is very hard to believe, I must confess. But it was given by a man who appeared to me to be very down-to-earth. He was sixty-seven years old in 1989, but still very active, both physically and mentally. He was quite frank and possessed an intellectual honesty, and he escaped from a few small traps I set for him to see if he would contradict himself at some point. Furthermore, I doubt that his intellectual faculties – and I must stress that they are really modest – would have allowed him to fabricate such a complicated case . . .'[21]

AN ENCOUNTER ABOVE THE WORLD'S FIRST ATOMIC PLANT

Rolan D. Powell was serving at the US Naval Air Station, Pasco, Washington, in the summer of 1945, training new pilots in preparation for aircraft carrier operations in the Pacific. In addition, he and other pilots were detailed to protect the top-secret Hanford Engineering Works, the large plutonium-production facility, located 60 miles away from Pasco. Although few of the pilots expected a Japanese attack on the plant, aircraft were kept in a state of constant readiness.

At noon on a certain date, estimated by Powell to have been about six weeks before the Japanese surrender on 2 September, Pasco radar detected a fast-moving object that assumed a holding pattern directly above the Hanford plant. Six Grumman F6F Hellcat fighters were scrambled to intercept. In an interview nearly fifty years later with Walter Andrus, a US Navy electronics technician programme instructor during the Second World War and currently International Director of the Mutual UFO Network (MUFON), Powell, a war hero who later became a test pilot for McDonnell Douglas, described the object as at an estimated altitude of 65,000 feet and 'the size of three aircraft carriers side by side, very streamlined like a stretched-out egg and pinkish in colour'. Powell also reported that some kind of vapour was emitted around the edges from portholes or vents, which he speculated was for the purpose of camouflage.

The Navy pilots were unable to believe what they were seeing. Under orders, they forced the Hellcats to 42,000 feet – well above their rated ceiling of 37,000 feet – but were unable to reach the unknown object's altitude and so returned to base. After hovering in a fixed position above the Hanford plant for twenty minutes longer, the object disappeared vertically. [22] [23]

THREE MYSTERIOUS STRANGERS

During an air-raid warning a few days before the heavy bombardment of a central German city in January 1945, citizens gathered in an air-raid shelter were visited by three young men who asked to inspect the shelter. The visitors wore peculiar dark, tight-fitting, high-necked one-piece suits, and despite an outside temperature of eight degrees below zero, they wore neither headgear nor scarves. Their footwear reminded one witness of gym shoes, which hardly suited the snowy conditions. All present were struck by the fact that the shoes made no noise on the ground. A further striking factor was that the men did not speak to one another, and only one of them spoke at all, with an odd accent. All three looked similar, with 'beautiful, symmetric features' and dark hair.

The mysterious strangers looked around the shelter briefly. Their 'spokesman' remarked merely that, in the event of another air-raid warning, everyone should be inside the shelter. They then left the building. Investigations later revealed that these three men had been seen only by those present in the shelter. In a nearby clearing, about 135 feet square, footprints were found in the snow, but no trails of cars or other vehicles led to the spot. Furthermore, officials stated that no inspectors had been sent to the air-raid shelter on the night in question. Suspicions arose that the three men might have been English spies, but would spies have made themselves so conspicuous?

For fear of ridicule, the person who provided this story remains anonymous. Years later, when reading *Flying Saucers Have Landed*, this person noticed that in a drawing of Adamski's 'Venusian', the visitor's footwear looked exactly like that of the three strangers in the shelter. Interestingly, during heavy bombardment a few days afterwards, the witness's apartment block was completely destroyed, but all its occupants, heeding the strangers' warning, were inside the air-raid shelter. [24]

Alternative interpretations might explain this event in prosaic terms, of course; yet there are parallels here with those reports of encounters with 'angelic' human beings dating back for thousands of years; parallels which become even more apparent when we examine encounters reported, increasingly, in the twentieth century.

NOTES

1 Casey, Gerry A., 'UFO: The time for telling has come', *Western Flyer*, Tacoma, Washington, 7 July 1989.
2 Interview with General-Major Boris Surikov by Lawrence Moore and Livia Russell, Moscow, February 1994. Part of this interview was shown in the documentary *Network First: UFO*, produced, written and directed by Lawrence Moore for Central Productions, 1994.
3 Sider, Jean, *Ultra Top-Secret: Ces ovnis qui font peur*, Axis Mundi, Avenue Calizzi, 20220 Ile-Rousse, France, 1990, pp. 369–70.
4 For further information on 'foo-fighters' and UFOs reported during the Second World War, see *Beyond Top Secret: The Worldwide UFO Security Threat* by Timothy Good (Sidgwick & Jackson, London, 1996).
5 Vallée, Jacques, *Forbidden Science: Journals 1957–1969*, North Atlantic Books, Berkeley, California, 1992, p. 309.
6 Blum, Ralph, with Blum, Judy, *Beyond Earth: Man's Contact with UFOs*, Corgi Books, London, 1974, p. 67.
7 Von Braun, Wernher, and Ordway III, Frederick I., *History of Rocketry and Space Travel*, Thomas Nelson & Sons, London, 1966, pp. 104–5.
8 Letter to the author from Bill Steele, 19 August 1992.
9 Lagarde, F., 'A French Landing in 1944', *Lumières Dans La Nuit*, no. 118, June 1972, translated by Gordon Creighton and published in *Flying Saucer Review Case Histories*, supplement no. 12, December 1972, p. 8.
10 Good, op. cit.
11 Stringfield, Leonard H., *UFO Crash/Retrievals: The Inner Sanctum*, Status Report VI, July 1991, pp. 69–71. Published by Stringfield, 4412 Grove Avenue, Cincinnati, Ohio 45227.
12 Harder, Dr James A., 'The Ins and Outs of UFOs and Secrecy since 1940', *The APRO Bulletin*, Aerial Phenomena Research Organization, Tucson, Arizona, vol. 32, no. 2, 1984, p. 6.
13 Letter to the author from Dr James Harder, 15 January 1997.
14 Stringfield, Leonard H., *UFO Crash/Retrievals: Search for Proof in a Hall of Mirrors*, Status Report VIII, Stringfield, February 1994, pp. 5–6.
15 Sider, op. cit., pp. 370–9.
16 Birdsall, Mark Ian, *Flying Saucers of the Third Reich: The Legacy of Prague-Kbely* (pending publication).
17 Oberth, Professor Hermann, 'They Come from Outer Space', *Flying Saucer Review*, vol. 1, no. 2, May–June 1955, pp. 12–15.

18 Ford, Brian, *German Secret Weapons: Blueprint for Mars*, Ballantine Books, New York, 1969.

19 Humble, Ronald D., 'The German Secret Weapon/UFO Connection', *UFO*, vol. 10, no. 4, 1995, pp. 21–5.

20 Leslie, Desmond, and Adamski, George, *Flying Saucers Have Landed*, Werner Laurie, London, 1953, pp. 195–6.

21 Sider, op. cit.

22 Powell, Rolan D., Varner, Byron D., and Andrus, Walter, 'UFO Sighting over Hanford Nuclear Reactor in 1945', *MUFON UFO Journal*, no. 344, December 1996, pp. 13–14.

23 Varner, Byron D., *Living on the Edge: An American War Hero's Daring Feats as a Navy Fighter Pilot, Civilian Test Pilot, and CIA Mercenary*, available from Rolan Powell, PO Box 1307, Round Rock, Texas 18680.

24 *UFO-Nachrichten*, Wiesbaden, Germany, January 1961.

Chapter 2

A Pantomime of Unrealities

For a few hundred years, rumours persisted that a strange group of people, who kept their distance from the local populace, resided in the vicinity of California's majestic Mount Shasta. In the late nineteenth century, there were sporadic reports of individuals seen emerging from the forests in the vicinity of Shasta to visit local towns and trade nuggets and gold dust in exchange for basic commodities. Described as tall, graceful and agile, with distinctive features such as large foreheads and long curly hair, the strangers wore unusual clothes, including head-dresses with a special decoration that came down from the forehead to the bridge of the nose.

On some occasions, powerful illuminations were observed in the forests, and strangely beautiful music could be heard. Invariably, when an investigator approached the area, he would be met by a 'heavily covered and concealed person of a large size who would lift him up and turn him away' from the area. Other intruders reportedly were affected by some invisible influence, causing them to become temporarily paralysed.[1]

All attempts by the local community to get close to or photograph these mysterious individuals proved fruitless. On some such occasions, it was alleged, the strangers would either run away or suddenly vanish into thin air. 'Those who have come to stores in nearby cities, especially at Weed,' reported author Wishar Cervé, 'have spoken English in a perfect manner with perhaps a tinge of the British accent, and have been reluctant to answer questions or give any information about themselves. The goods they have purchased have always been paid for in gold nuggets of far greater value than the article purchased, and they have refused to accept any change, indicating that to them gold was of no value and that they had no need for money of any kind.'[2]

Not only were powerful lights often seen emanating from certain areas – years before electricity was in use – but there were also reports, in the early twentieth century, of cars that stalled on approaching the remote area apparently inhabited by these beings; a curious circumstance that was to become common during close encounters with UFOs reported years later. 'At an unexpected point where a light flashed before them the automobile refused to function properly,' commented Cervé in 1931, 'for

the electric circuit seemed to lose its power and not until the passengers
emerged from the car and backed it on the road for a hundred feet and
turned it in the opposite direction, would the electric power give any
manifestation and the engine function properly.'

Still others reported encountering strange cattle, 'unlike anything seen
in America', which would run back towards the area inhabited by the
mysterious group. Of particular relevance are the (undated) sightings of
peculiar aerial vessels:

> There are hundreds of others who have testified to having seen peculiarly
> shaped boats which have flown out of this region high in the air over the hills
> and valleys of California and have been seen by others to come on to the
> waters of the Pacific Ocean at the shore and then continue out on the seas as
> vessels . . . and others have seen these boats rise again in the air and go upon
> the land of some of the islands of the Pacific . . . Only recently a group of per-
> sons playing golf on one of the golf links of California near the foothills of the
> Sierra Nevada range saw a peculiar, silver-like vessel rise in the air and float
> over the mountaintops and disappear. It was unlike any airship that has ever
> been seen and there was absolutely no noise emanating from it to indicate that
> it was moved by a motor of any kind.[3]

Were these mysterious people the survivors of the mythical lost continent of Lemuria – a theory espoused by Rosicrucian mystics, including
Wishar Cervé – or might they have been of extraterrestrial origin? I do
not have the answer to these questions, but evidently there are parallels
with other, later accounts.

ZRET AND THE NORCANS

'Oh help me, help me!' It was June 1920, and 16-year-old Albert Coe was
on a canoeing vacation in Ontario with his companion Rod. Alone at the
time, Coe heard the muffled cry while clambering to the top of an out-
cropping of rocks in remote and rough terrain on the Mattawa River.
Looking around, Coe could see no one, so he let out a yell. Slightly to his
right and ahead came an answer. 'Oh help me, I'm down here.'

'I still couldn't see anyone,' said Coe, 'and had walked about 25 feet in
the direction of the voice when I came to a five-foot-wide cleft in the base
rock that ran diagonally toward the river. Wedged down this narrowing
crevice was a young man with his blond head some two and a half feet
from the surface. He only had one arm free, so I reached over and
grabbed his wrist, but could not budge him. We always carried a coil of
rope and a hunting knife, so I cut down a sapling about 10 or 15 feet long
to use as a lever, and working my rope under the pit of his pinned arm,
circled it around his back and chest, bringing a loop to ground level, at
the same time telling him I would try to pry him out. If I failed, I told him

not to worry, for my pal was somewhere on the other side of the river, and between the two of us we would free him.'

Slipping the pole through the loop and using the opposite edge as a fulcrum point, Coe gave a heave and felt the stranger move. Raising the lever end higher he propped it on a tree branch, jumped the crevice and pulled the man out. His legs were so numb that he was unable to stand, and the left hip, knee and shin were badly lacerated. The first thing he asked for was water, so Coe clambered back down the rocks to the river and fetched some in his felt hat. Slitting two of his bandannas, Coe also bathed and dressed the wounds. Then some oddities became apparent:

As I was helping him my curiosity was rising as to the identity of my 'patient'. I told him of our trip and that I was searching for a way to open water, at the same time noticing he was wearing an odd silver-gray, tight jumper-type garment that had a sheen of silk to it. It had a leathery feeling without a belt or visible fasteners attached, but just under the chest was a small instrument panel. Several of the knobs and dials were broken, from being jammed against the rock in his fall. Being so many miles from any form of civilization, I pointedly asked where he was from, if he was on a canoe trip, also when and what had happened to cause his misfortune.

He said that he was not canoeing, but had a plane parked in a clearing, three or four hundred yards downstream, and had started out early the previous morning to do some fishing. In attempting to jump over the crevice, the loose earth and moss had given way underfoot and he had just about given up all thought of ever getting out alive when he heard some of the stones, loosened in my ascent, bouncing down the rock . . . he decided to cry out and said that my answering yell was like a miracle.

'Well, planes were very primitive in those days,' Coe went on, 'and if you can imagine the side of the mountain, coming down, and all those rocks and branches, how the heck did he manage to land a plane? So I didn't say anything to him – I was thinking the guy was nuts. I thought maybe he'd banged his head and was having hallucinations.'

The stranger requested Coe's name and address, expressing his eternal gratitude for having been saved. He asked Coe to look for his small tackle-box and fishing-rod which he had dropped when he fell down the crevice. Coe was unable to find the tackle-box, but he did locate the fishing-rod. 'The mystery of this strange person deepened within me,' said Coe, 'the peculiar outfit, a plane landing in this rocky forest and now a fishing-rod, the likes of which I had never seen.'

The butt was about three-quarters of an inch in diameter and had the same leathery touch as his suit, but bright blue and formed in a slight rounded protuberance just above it. It had a tiny slot in either side and continuance in a slender aluminum shaft. It had no guides or reel, for the fine line came directly out from the inside at its tip, as a fine filament, to which was attached a conventional dry fly. I asked where he had purchased such a rod and the question was partially parried with a reply that his father was a research engineer and it was one of his own design.

By now, the circulation was beginning to return to the stranger's numbed limbs. Although occasionally grimacing from pain, the man's overall composure, without apparent reaction to the stress or shock from such a torturous ordeal, was astonishing. An offer to help the man back to his plane was at first declined.

Coe and his companion, Rod, had come up against a lot of logs and other flotsam in the river and were anxious to find passage through or around it. The stranger said that, observed from his aircraft, five to six tough miles lay ahead, though he thought the teenagers could perhaps pull their canoe through some of the shallow, swampy water. 'He did not want to impose upon me any further and said I had better think of starting back, for he had already been quite a burden,' explained Coe.

From the condition of his leg I doubted that he could even walk, but made no comment as I helped him up. He took two steps, swayed and grabbed a tree to keep from going down. I threw one arm around his waist, lifted his left arm over my shoulder and insisted again that he accept my aid . . . He finally gave in, but on condition of a promise; asking for my solemn word that I would not divulge to anyone, not even my partner, anything that had taken place today, or what I may see. He then told me that his father had developed a new type plane that was still in an experimental stage and highly secret, but he often helped in the lab when home from school. As sort of a test, his father had permitted him to use the plane for this fishing trip. In the future he would fully explain the reason for his request that I keep my promise.

Agreeing to this request, Coe supported and half-carried the man downstream to his aircraft.

THE CRAFT

In a clearing beside the river, no more than 70 or 80 feet wide, stood the aircraft. Fully expecting it to be some type of conventional plane, Coe was astounded by what he saw.

A round silver disc, about 20 feet in diameter, was standing on three legs in the form of a tripod, without propeller, engine, wings or fuselage. As we approached, I noticed a number of small slots around the rim, and it sloped up to a rounded central dome. I had to duck to walk with him underneath, between the legs, although it was slightly concave and only about four and a half feet from the ground.

He said, 'Surprised?' That wasn't actually the word for it, but I did not press him with questions, realizing he was suffering a great deal of pain. The only thing I was trying to figure was, how the hell does the damn thing fly?

'I grabbed hold of him and he said, "Take me toward the centre of the craft." He reached into the end of one of three recessed panels in its bottom that fanned out centerwise from the base of each leg, pressed a button, and a door swung down with two ladder rungs moulded on its inner surface. I clasped my hands under his good foot and boosted him in. He peered down at me over the rim of the opening, and said, "I will never forget you for this day. Remember to keep your promise, and stand clear when I take off."' Coe retraced his steps to just within the trees at the edge of the clearing and turned round to watch.

I was musing over its lack of windows or portholes and wondered how he could see out, unless they were on the other side. Just then, the perimeter edge began to revolve. At first it gave off a low whistling sound, picked up speed mounting to a high-pitched whine, finally going above the audible capabilities of the ear. At that time I experienced a throbbing sensation, which was felt rather than heard. It seemed to compress me within myself. As it lifted a few feet above the ground, it paused with a slight fluttering, the legs folded into the recesses as it swiftly rose with the effortless ease of thistle-down caught in an updraft of air, and was gone.

Coe set off back towards camp in a state of bewilderment. 'It all seemed like a pantomime of unrealities,' he commented. 'It was an episode lasting not much more than an hour that may have carried me a thousand years into the future, and yet left an uneasy feeling of witnessing something that did not actually exist, an impression of disconnected sequences only found in dreams.'

He ran back to hunt for the tackle-box, without success, but part of a blood-stained bandanna, the lever pole, its stump and branches were all still there.

Coe and Rod prepared for their journey along the Mattawa River, the conditions of which turned out exactly as the stranger had indicated. Eventually, they joined the Ottawa River and spent the next two weeks enjoying their vacation in the wilderness.

One night, less than a day's paddle from Ottawa, while Rod was inside the tent, Coe relaxed outside beside their camp fire. 'My musing was interrupted as I caught a glint of silver over the tree-darkened outline of the hills across the river that disappeared for a few seconds, and then I was sure,' said Coe. 'Framed in a background of stars was my strange friend's stranger plane. He hovered motionless, not more than 70 feet above me and just off the shoreline, then dipped from side to side in an unmistakable gesture of hello . . . I knew that it was his way of telling me he was well again, and I made a mental note, if ever I did meet him, to surely question [him] as to how he could know my exact location in the darkness of the night.'

THE RETURN

Almost six months after this initial encounter, Coe received a note, signed 'Xretsim', requesting a meeting over lunch at the McAlpine Hotel, Ottawa. Coe felt certain that this was indeed the mysterious stranger whom he had befriended earlier in the year.

'I did have a few "butterflies" wondering if I would remember his face,' said Coe. 'I entered the lobby as he came toward me with out-stretched hand and the greeting of, "You surely look a lot different than when we first met," which echoed my own thought, doubting very much if I would have recognized him in the conventional suit, white shirt and tie.'

There was something odd about the handshake on this occasion. The man held a small gadget as the two shook hands, which Coe learned later was a device that registered the 'vibrational frequency' of his body, the data from which could be shown on a television-like screen elsewhere. Once this 'vibrational frequency' was registered, Coe claimed, his every move could be monitored. 'They did this to make sure I'd keep my promise,' he explained.

'I first asked the pronunciation of his name and inquired about his injuries. With a mischievous chuckle he replied, "Just call me Zret for now. In the future you will figure it out. Thanks to your timely intervention with first aid, the leg and I are in good shape."' (Xretsim was simply Mister X spelled backwards.)

'There were a million questions on the tip of my tongue,' Coe continued, 'but most remained unuttered as he carried a good part of the conversation, regarding my trip, my school work, activities, ambitions, etc. He told me that he had spot-checked our progress, as far as Ottawa, to be sure we were OK [and] cleared up the mystery of the night I saw his plane, explaining that he was fishing [!] on the opposite bank when we set up camp, and could see my outline by the embers' glow . . .'

Following lunch, Zret explained that Coe would not be hearing from him for two or three months, but promised that they would take a fishing trip together in late spring. In early May, the two met at Hastings Station in Ottawa, and drove in Zret's (conventional) car to Lake Mahopac. During the drive, the alien angler gave Coe much of the information he had hoped and longed for. It was one of many such meetings, which were to span six decades.

THE MISSION

Zret began the conversation by asking Coe if he had told his parents about the encounter. Coe replied that he had not done so and would never betray the secret. Zret continued:

You probably already have an inkling that I'm a stranger to your modern world. This decision of explanation is a personal responsibility. Our mission here will forever be cloaked in the tightest secrecy. If the events that we foresee do not come to pass, our presence will not become known. The great depth of gratitude that I feel toward you, coupled with the things that you have seen and know exist, has influenced a violation of [a] law of disclosure . . . I am sure, if you can be as tight-lipped in the future as you have been in the past, that I will have nothing to fear, but a breach of this trust could result in the direst of consequences.

Zret went on to say that his true identity and address and details of his personal life had to remain secret, though he did explain that he was one of a group which had come to monitor Earth's scientific advances. While on Earth, he 'doubled' as a student majoring in electronic engineering. Man's capacity for developing weapons of ever-increasing power was the prime motive for the visits to Earth. 'Only recently,' he continued, 'many of the more "intelligent" and "cultural" nations of Earth have concluded a long, bloody war, and during its progress several innovations, designed specifically for the mass slaughter of humanity, were introduced . . . As each new invention was applied to a military potential, its horizon broadened to the eventual horror, brutality and devastation that emerged as a "world war". This conversion of inventive genius from the brain of Earth's inhabitants, to ever greater devices of destruction, was the prime factor that motivated our mission . . .'

Later, Coe was to learn that in 1904, Zret's people had paved the way for a hundred of them to infiltrate every major nation of our planet – as small groups of technicians – to observe and evaluate every step of our scientific advancement. Their main concern was that we were on the verge of discovering secrets of the atom which could have disastrous consequences for our planet.

A Dweller on Three Planets

Coe was anxious to learn about his friend's origin. Here, as usual, we run into difficulties. Contactees frequently are given seemingly ludicrous points of alien origin which tend to devalue their accounts. Zret replied that his present homes were on two planets: '. . . one, the planet Mars, nearing the end of an evolutionary life, and the other, planet Venus, younger in evolutionary processes than Earth, but its higher regions are not too drastically different from the environment here'. Evidently, he also spent a considerable amount of time on planet Earth.

In later meetings, Zret explained that his race had originated on a planet called 'Norca', slightly smaller than Earth, with four moons, orbiting 85 million miles around Tau Ceti (a star about 11 light years from ours, similar in age and type to our own). Fourteen thousand years ago,

Norca began dehydrating slowly, inexorably, to the extent that drastic action was necessary to preserve the race. Everything was tried to counteract the effects of dehydration, but nothing worked. The only solution was to migrate to another solar system. Ours – having a similar sun – was chosen. Eventually, following a successful exploratory mission to Earth, during which contact was established briefly with Cro-Magnon humans, the expedition returned to Norca. It was decided that Norcans would colonize Earth. Supposedly, 243,000 Norcans eventually left their planet in sixty-two huge spacecraft, together with various species of animals, plants and insects. Owing to unforeseen and tragic circumstances whereby nearly all of the ships were drawn into our Sun, only one 'Norcans' Ark' made it; even then, it crash-landed on Mars, killing many on board. Nonetheless, 3,700 out of the 5,000 or so on board survived.

The Norcans, claimed Zret, overcame the challenge of Mars' relatively hostile environment and spent about 900 years on the planet. 'Succeeding generations,' he explained, 'once again advanced to the scientific potential of launching twin probes, to Venus and Earth, both of which were subsequently colonized. In the primary stages of this expansion, bases of research were established on Venus to study its peculiar atmosphere, [but] the main colonization was concentrated on Earth.' If Coe – and Zret – are to be believed, these colonization areas were, in chronological order: the mythological continent of Atlantis; the Cuzco Valley in the Andes; the legendary continent of Lemuria (at a point about 1,000 miles east of what is now known as the Marshall Islands); northern Tibet; and, finally, Lebanon. Norcans reproduced with native inhabitants. Irrespective of skin pigmentation, Zret explained, the indigenous Earth people at that time had black or brown hair and eyes, and the interbreeding led to a blond-haired, fair-skinned people.

'On Venus, the man form had not appeared,' said Zret. 'Today our basic home is the high land of Venus, although a good part of our research is still conducted on Mars, especially electronic probe[s], for its thin atmosphere and peculiarity of magnetic fields lends itself, as an ideal laboratory, to almost distortion-free reception.'

The atmospheres of both Mars and Venus are known to be far too inhospitable for *unprotected* human existence. On Mars, the atmosphere is far too thin and cold, while Venus's atmospheric pressure is said to be about 90 times that of Earth, with a temperature averaging around 470 degrees Celsius and a massive carbon-dioxide atmosphere (97 per cent), with no water. Ten per cent of Venus's terrain is highland, and the highest point on the planet is the mountain known as Maxwell Montes, towering 35,400 feet above Venus's 'sea level' and 27,000 feet above a huge highland region the size of Australia, known as Ishtar Terra. Because Venus has often been named as an abode of certain aliens – Coe supposedly being the first to be told this – we are left with a paradox. Assuming that neither Coe nor Zret was lying, could it be that the Norcans, utilizing highly advanced technology, were able to convert the hostile environment – which in any case may be less extreme in the highland regions – to suit their requirements? This idea is not wholly fanciful – even with terrestrial technology. The late Carl Sagan, a leading authority on planetary sciences, hypothesized that injection of appropriately grown algae into the Venusian atmosphere 'would in time convert the present extremely hostile environment of Venus into one much more pleasant for human beings'.[4]

FURTHER EXPLANATIONS

Zret explained that at around the time of his initial encounter with Coe he had been on summer vacation, and had taken advantage of this period to rejoin some of his own people 'who operated one of their established bases off the planet'. With his craft, he was then able to 'enjoy the wonderful fishing of the otherwise inaccessible rivers and lakes of Canada'. On leaving the base, he told his colleagues not to worry if they did not hear from him for a week, as long as the 'all-clear' radio signal emitted from the craft in its regular 20-minute cycle. This was an inexcusable error, admitted Zret:

> You probably remember the little control panel that was attached to the front of my flying suit. Well, attached within the suit are a series of what I will simply call electrodes, that come in contact with various nerve centres of the body. At the back of my neck, under the base of my brain, are two more, the left one receiving brain impulses and the right one receiving all signals from the pituitary gland, the 'master switch' of the body. All these comparatively weak waves feed into a section of that panel below my chest and any impulse of stress or emergency thought should have been transposed and amplified through it, to automatically record in the craft's control and change the all-clear signal to a rapid tonal wave of distress. Help would have arrived in three or four hours. The manual controls of this panel also activate many of the functional duties of the craft, even to an unmanned flight back to its base . . .

The control knobs on Zret's suit panel were severely damaged when he fell into the crevice, and he was unable to operate them. He explained that, once on board his craft following the rescue by Coe, he immediately switched the ship's transmitter to the emergency mode. Not many minutes after setting the craft's automatic 'homing' device, he collapsed. Later, he learned, 'I was "picked up" by one of our larger ships that had intercepted the distress call and taken me aboard, craft and all.'

Coe asked about Zret's age. 'My age is going to surprise you,' replied the very youthful-looking Zret. 'I am exactly 304 years older than you.' Decades later, many contactees claimed that aliens enjoyed phenomenal longevity, so it is interesting that Coe was told (or said he was told) this in 1921. Zret explained that such longevity was achieved through a rejuvenation process. 'We have to go through this system every 105 years. We have a life potential of about 630 years, but we must go through this rejuvenation process. If we don't, we die as you do. We go through this five different times, and then the internal mechanism, chemical decomposition, and so on, wears out.'

In addition to their extraordinary technology, the Norcans apparently were equally advanced in mental skills, such as telepathy, and seemed to be highly advanced ethically.

LATER DEVELOPMENTS

Albert Coe's meetings with Zret – and others of his group – continued into the late 1970s at the rate of 10–12 times a year. He kept his promise and told no one about these meetings; until 1958, that is, when, with the go-ahead from Zret, he told his wife. 'She thought I was kidding at first,' Coe told Dr Berthold Schwarz, a psychiatrist and UFO researcher, in 1977. 'And then she wanted to meet him. Of course, that's a no-no. You see, these people are very secretive. They have very good reasons to be, and I wouldn't want to be the one that let that secret out.'

Coe was also given permission to bring some of his story into the public arena. He began to give interviews on various radio and television stations in Washington, DC, and wrote a book detailing some of his experiences, privately published in 1969. He also claimed to have had several meetings with US Government officials in Washington. 'And they'd pump me,' he said, 'always trying to break down my story.'

In 1958 Coe began to receive the attention of what he assumed were federal agents. 'For one year, they followed me all over,' he claimed. 'I used to live in Beverly, New Jersey, and I had a little apartment next to a barber shop. The barber was a very good friend to me, and said these men used to pump him. They wanted to know where I went, what I did, who my friends were, and so on.' Many contactees in the 1950s, such as George Adamski, were investigated by federal agents, as numerous FBI files – released under the Freedom of Information Act – show. Interestingly, when asked by Dr Schwarz what he thought of Adamski – whose much later encounters contain many parallels – Coe replied simply that he was a 'faker'!

What are we to make of Albert Coe's outrageous story? Coe had an excellent work record as a mechanical engineer, and, so he told Dr Schwarz, had not suffered from delusions, encephalitis, hallucinations or paranoia, nor had he spent time in a mental hospital. Having listened carefully to the 90-minute interview with him conducted by Dr Schwarz,[5] and studying his book, *The Shocking Truth*[6] – both of which are used here as references – I conclude that he told the truth; at least, the truth as he believed it.

GROUNDED FOR REPAIRS

During the summer of 1933, numerous people in Nipawin, Saskatchewan, Canada, reported strange lights in the sky and at ground level. One night, two men and a woman (whose names are known to the investigator, John Musgrave), determined to get to the bottom of the mystery, drove in a pick-up truck to the area. After driving as close as was practicable, the trio headed on foot towards the direction of a glow in the woods. Although a stretch of muskeg (bog) prevented them from getting closer than about a quarter of a mile away, the witnesses were able to observe that the light came from a large oval-shaped craft, supported on landing legs, in the middle of a marsh.

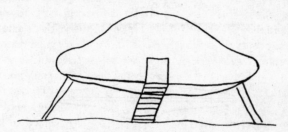

Fig. 1. Sketch by one of the witnesses of the landed craft in Nipawin, Saskatchewan, Canada, in 1933.

The craft was domed on top and slightly rounded on the bottom (see Fig. 1). From a central doorway or hatch, about a dozen figures could be seen going up and down a ladder. Appearing to be slightly shorter than the average-sized man, wearing silver-coloured suits and helmets or 'ski' caps, the figures seemed to be busily running around 'repairing' their craft. As reported in numerous cases years later, a strange silence pervaded the immediate area. A bright-orange glow emanated from the craft, lighting up the surrounding area. After about half an hour, the witnesses headed back to town, hoping to find a way around the marshy area to get a better look. Regrettably, when finally they did find a way through, they realized they did not have sufficient gasoline to take them there and back.

A few nights later, the witnesses returned to the site. There was no sign of the craft itself, but traces could clearly be seen at the landing site. As John Musgrave reported:

Six large square imprints that must have been the bases of the legs that supported the craft proved that there indeed had been something there that night. Each imprint was the same size – 2 to 2½ feet square, and approximately 8 to 10 feet apart. The imprints were 2 to 3 inches deep, and reminded the three of them of a kind of mark that would be made by a boiler plate stomped into the ground. They could also see markings where the base of the stairway met ground. As if this wasn't remarkable enough, a great burn mark in the centre of the area covered a circle approximately 12 feet in diameter. They looked for footprints but found none though there was some scuffing of the vegetation surrounding the spot where the craft had been.[7]

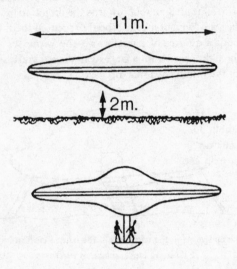

Fig. 2. The flying disc observed by military witnesses at Guadalajara, Spain, in 1938. (*Terence Collins/FSR Publications*)

HUMANOIDS IN SPAIN

Throughout the years, a number of cases have been reported of aliens being lowered from their craft on a platform (for example, the incident investigated by Lord Mountbatten on his estate in Hampshire, England, in 1955, described in my book, *Beyond Top Secret*). Additional cases will be cited later, but here follows the earliest example of which I am aware, given to the investigator Oscar Rey.

At 23.30 on 25 July 1938, a military man and his assistant came across a dark, lenticular object, 11 metres in diameter, hovering about two metres above the ground, and only 60 metres away from them, at Guadalajara, Spain. Two humanoid figures descended silently on a platform at the bottom of a column (see Fig. 2). A circle of blue light focused on the witnesses, who felt chilled by it. The platform ascended, the two sections of the object began to spin in opposite directions and, glowing with an intense white light, the object took off.[8]

In the majority of such landing cases, the aliens beat a hasty retreat when humans appear on the scene. Nearly two years later, though, contact was allegedly established with some of the more approachable species, remarkably similar to those encountered by Albert Coe.

A STRONG BOND OF FRIENDSHIP

On a clear day in early May 1940, in remote country near the village of Townsend, southeast of Helena, Montana, 37-year-old Udo Wartena was moving some boulders at his gold-mining claim. Suddenly there came a loud, turbine-like droning sound, which at first he assumed to be an Army plane that flew periodically in the vicinity. Climbing to higher ground, he was astonished by the sight of an unusual craft, shaped like two soup plates of a stainless-steel colour. It was hovering a short distance away, above a meadow where he had built a small dam using water diverted from a nearby stream. The craft measured over 100 feet in diameter and about 35 feet in height.

A circular stairway with a solid base, forming part of the craft's hull, lowered, and a man came down and approached the miner. 'I went to meet him,' said Wartena. 'He stopped when we were about 10 or 12 feet apart. He wore a light-grey pair of coveralls, a [circular] cap of the same material on his head, and on his feet were slippers or moccasins.' The stranger shook hands with Wartena, explaining apologetically that it was not their normal policy to intrude or to allow themselves to be seen.

'He then asked me if it would be all right if they took some water from the stream,' said Wartena, 'and as I could not see why not, I said sure. He then gave a signal and a hose or pipe was let down. His English was like mine, but he spoke slowly, as if he were a linguist and had to pick his way.'

The stranger politely invited Wartena to board the craft. As the bemused witness came directly underneath it, he noticed that the droning sound, though quieter now, seemed to go through him. Once inside, though, the noise was barely perceptible.

> We entered into a room about 12 feet by 15 feet, with a close-fitting sliding door on the farther end, indirect lighting near the ceiling, and nice upholstered benches around the sides. There was an older man already in the room, plainly dressed, but his hair was snow-white . . . the younger man's hair was also white.

Both men were extremely good-looking, with perfect, almost translucent skin, and appeared to be very youthful and strong. Becoming sus-

picious about their origin, the witness asked where they came from. 'We
live on a distant planet,' they replied, pointing to the sky.

Asked why they wanted to take water from the stream, rather than
from a nearby lake, the younger man responded that stream water was
purer, as it contained no algae, and that it was 'convenient' – which pre-
supposes that they had taken water from there before.

The younger man explained some details of the craft's propulsion
system:

> As you noticed, we are floating above the ground, and though the ground
> slopes, the ship is level. There are in the outside rim two flywheels, one turn-
> ing one way and the other in the opposite direction . . . this gives the ship its
> own gravitation; or rather, it overcomes the gravitational pull of the Earth and
> other planets, the Sun and stars. Though this pull is but light, we use [it] to
> ride on, like you do when you sail on ice.

The 'flywheels' or rings, about three feet wide and several inches thick,
were separated by rods turned by motors and powered by 'battery- or
transformer-like' units positioned around the inside perimeter of the
craft. For interstellar travel, the craft supposedly could be 'focused' on a
star and its energy, drawing itself through space at speeds faster than that
of light (186,000 miles per second) – 'skipping upon the light waves', as
they put it. 'They *use* gravity,' Wartena declared, 'they don't just overcome it.' He then asked where they got the energy to run such a large
ship. 'They said from the Sun and other stars, and [they] could store this
in [the] batteries, though this was for emergency use only. They carried
another source, but did not explain this to me.' Many years later,
Wartena indicated to a family member that hydrogen extracted from the
water provided the craft with its fuel source.

As with many such accounts where extraterrestrials explain the capabilities of their crafts, it is not readily apparent how nullifying gravity in
itself induces propulsion. In many cases, however, they have made clear
their reluctance to elaborate, for the perfectly valid reason that Earthlings
would misuse such revolutionary technology. Even so, in my estimation,
Wartena seems to have obtained some important clues.

Feeling completely at ease with the cosmonauts, Wartena accepted an
invitation to be 'monitored for impurities', and a type of 'X-ray' machine
was passed over his body. Regrettably, no details as to the purpose or
result of this physical examination are available. It is worth noting that
other contactees, such as Carroll Watts (Chapter 14), have reported being
physically examined.

The witness asked his hosts their age. One claimed to be about 600
years of age, and the other, who looked 'slightly older' – as well he might
– said he was over 900 years old, as *we* measure time – impressive even by

Norcan standards. They said they spoke many of our languages, and were continually improving their knowledge of them.

But why were they coming to Earth? 'As you have noticed, we look pretty much as you do,' they told Wartena, 'so we mingle with you people, gather information, leave instructions, or give help where needed.' They further explained that they were monitoring the 'progression or retrogression' of our society. A deeply committed Christian, Wartena asked if they knew about Jesus and religion. 'We would like to speak of these things but are unable to do so,' came the response. 'We cannot interfere in any way.'

The friendly spacemen invited the witness to come with them, but he declined, explaining that it would inconvenience too many people. 'Later, I wondered why I said that,' he remarked. Perhaps his reluctance to accompany them was related to an incident two years earlier, when a young man vanished without trace in the vicinity. Had he gone with the visitors?

Wartena felt that it was now time to take his leave. The men advised him not to discuss his encounter with anyone, as no one would believe him, but said that he could do so in years to come.

As in other such contact encounters, the witness was cautioned to stay well clear of the craft as it took off. 'When I walked away from the ship, they raised the stairway, and when I got a couple of hundred feet away, I turned round. A number more portholes had opened up and though I could not see anyone, I felt sure they could see me; anyway I waved at them.' Again came the loud, turbine-like noise, and he watched as the craft – which had remained hovering throughout the two-hour encounter – lifted, wobbled briefly, then shot off at high speed.

Some type of 'energy field' permeated the area, preventing the witness from walking for several hours.

This remarkable story came to me by way of the Australian investigator Warren Aston, who learnt about it from an American source who had known the witness. Udo Wartena, of Dutch descent, was described as unsophisticated and honest by all who knew him, including his wife, whom Aston met. He wrote a full account of the experience but kept the story secret for nearly 30 years before confiding in his closest friend. He died in 1989.

Perhaps the most impressive aspect of the encounter for the witness was the strong bond of friendship he felt towards the cosmonauts. They were men, he stressed, 'just like us and very nice chaps', and he felt even 'love, or comfort' in their presence. Alas, some reported encounters with extraterrestrials around the world have proven to be less beneficial.

NOTES
1 Cervé, Wishar S., *Lemuria: The Lost Continent of the Pacific*, The Rosicrucian Press, San Jose, California, 1980, pp. 250–2.
2 Ibid., pp. 256–7.
3 Ibid., pp. 259–61.
4 Sagan, Carl, *The Cosmic Connection: An Extraterrestrial Perspective*, Coronet Books, London, 1975, pp. 151–2.
5 Interview with Albert Coe by Dr Berthold Schwarz, Philadelphia, 8 May 1977.
6 Coe, H. Albert, *The Shocking Truth*, The Book Fund, Beverly, New Jersey, 1969.
7 Musgrave, John Brent, 'Saskatchewan, 1933: UFO Stops for "Repairs"', *Flying Saucer Review*, vol. 22, no. 6, November–December 1976, pp. 16–17.
8 Ballester Olmos, Vicente-Juan, 'Survey of Iberian Landings: A Preliminary Catalogue of 100 Cases', *Flying Saucer Review*, Special Issue no. 4, August 1971, p. 46.

Chapter 3

A Festival of Absurdities

It was the late, great French researcher Aimé Michel who described the UFO phenomenon as 'a festival of absurdities' – which it often seems to be. No matter how diligently we try to define the phenomenon, it resists any single, adequate hypothesis to account for the plethora of craft and their occupants, as well as for their sometimes peculiar behaviour. Yet there are often parallels; concatenations that provide clues and patterns as to the nature and purpose of some of the encounters, ranging as they do from the deadly serious, to the silly, to the sublime. With few exceptions, the varied encounters selected for this chapter took place during a sixteen-month period in 1946–47.

A GHASTLY DEATH
Of all countries, perhaps none has experienced as many disturbing encounters with extraterrestrials as Brazil, particularly in its more remote regions. One of the earliest known cases dates back to 1946. It is a story which reads like something from a 1950s Hollywood B-movie – or the wildest of *The X Files* – yet the incident was witnessed by a number of credible people, who were interviewed by several equally credible investigators.

It was Shrove Tuesday, 5 March, in the little town of Araçariguama, in the administrative region of São Roque, State of São Paulo. At that time, the town lacked electricity and telephones, neither were there any physicians. For some time, strange lights had been seen which darted around in irregular manoeuvres above the mountains and forests of the region. At about 20.00, João Prestes Filho, a 40-year-old married farmer and businessman, returned home from a day's fishing trip on the Rio Tieté. His wife was not in the house: she and the children were attending carnival celebrations in town, and Prestes had arranged for a window to be left slightly ajar so he could get in. Although there was a light mist, the sky was clear.

Suddenly, as Prestes was lifting the window, he was struck by a beam of light coming from some outside source. He put up his hands to protect his head and eyes, then fell to the ground, stunned, for a few moments. Picking himself up, he fled into the centre of town to seek help. He arrived in a state of terror at his sister Maria's house and repeatedly explained what had happened. Neighbours were summoned, including Aracy Gomide, fiscal inspector of the Prefecture of São Roque, who was the principal witness to the events which ensued. Though not a physician, Gomide was medically knowledgeable; thus he was charged with the job of caring for sick or injured patients in the region.

Gomide noticed that although Prestes's eyes were dilated and his voice distraught with terror, there were no traces of burn marks anywhere on his body. Then began a scene of incredible horror. According to the witnesses, who were interviewed by Dr Irineu José da Silveira, a dental surgeon, this is what happened:

> Prestes' insides began to show, and the flesh started to look as though it had been cooked for many hours . . . The flesh began to come away from the bones, falling in lumps from his jaws, his chest, his arms, his hands, his fingers, from the lower parts of his legs, and from his feet and toes. Some scraps of flesh remained hanging to the tendons . . . Soon every part of Prestes had reached a state of deterioration beyond imagination. His teeth and his bones now stood revealed, utterly bare of flesh.

Amazingly, while all this was going on, Prestes gave no signs of feeling any pain. Then his nose and ears fell off and slid down his body on to the floor. What remained of him was literally carted off to the nearest hospital, Santa Casa, at Santana de Parnaíba. Six hours later, Prestes' body was brought back to Araçariguama. He had died without reaching the hospital. Right up to the end, said witnesses, guttural sounds continued to come from what was left of his mouth.

The death certificate merely recorded that Prestes had died of 'generalized burns'; a less than satisfactory explanation. Nothing was ever found outside or inside the victim's house which might have yielded some clues as to the nature of the beam of light, nor were atmospheric conditions present at the time that might have accounted for the tragedy, such as 'ball-lightning'.[1] According to a report published in a Brazilian newspaper many years later, the French Government requested the bones of Prestes for test purposes.[2]

It is, of course, impossible to establish that Prestes was killed by aliens; nevertheless, circumstantial evidence suggests that he was struck by a beam of light from an unknown source, causing a type of rapid, awesome deterioration. Furthermore, unusual flying lights had been seen in the vicinity at the time.

SPACEMEN AND WOMEN IN SWEDEN
Several months prior to the famous wave of sightings of the so-called 'ghost rockets' – reported mostly from Scandinavia[3] – a most unusual

flying craft is said to have landed in Angelholm, north of Helsingborg, Sweden. The witness was Gösta Carlsson, now famous for his pioneering research into bee pollen extracts.

One evening in May 1946, Carlsson was walking home along the shores of the Kattegat. Owing to the gathering dark he used a forehead lamp. Spotting a light in a nearby clearing in the forest, he approached the area and was surprised to see a disc-shaped contraption on the ground. The object had a cupola that looked like a cabin with oval windows. Above the cabin was a 'mast', and beneath the disc could be seen a large oblong 'fin' stretching from the centre to the underside which, together with a small ladder and two landing legs, rested on the ground. Carlsson went on to describe further details:

The object was approximately 16 metres in diameter and 4 metres from top to bottom at the middle. I know this because I measured the marks on the following day. There were a lot of holes around the edge of the disc, like those of a turbine, and it was from these that jet-beams came which burned the grass when the object departed. The light came from the mast. It was about 5 metres in height, and three antennae were suspended from its top. Lower down something like a lampshade was hanging. It was shining with a strange purple light which covered not only the whole object but also the ground a couple of metres beyond it. The light was flowing and pulsating from the 'lampshade' like water from a fountain. Where the light hit the ground I could see a sparkling effect.

On the ground, beyond the area of light, stood a man in closely fitting white overalls. He seemed to be some sort of guard, said Carlsson:

He raised his hand towards me: it was a gesture that could not be misunderstood, so I stopped. I was less than 10 metres from him. He was approximately as tall as I am, maybe a few centimetres shorter, but he was thinner than me. There were others like him, but the strange thing was that nobody said a word. It seemed as if they had just finished repairing a window, because they put their tools away and looked at me. Everything was silent. The only thing I heard was the sound from the guard when he walked on the grass. There were three men working at the window, and two more were standing alongside. There were three women as well, and one more came out of the object later. On the far side there was another guard. In all I saw eleven persons.

They wore short black boots and gloves, a black belt around the waist, and a transparent helmet. The women had ashen-coloured hair, but I could not see the hair of the men as they wore black caps. They were all brown-coloured, as if sunburned.

I went a few steps closer, but then the guard raised his hand again. After that I stood still. The guard had a black box on his chest which was suspended by a chain around his neck. It looked like an old black camera. He turned it towards me and I thought he was going to take a picture of me, but nothing happened, except that I thought I heard a click from my forehead lamp. The lamp did not work after that, but that may have been purely coincidental. When I returned home I found that the battery had run out, although it was a new one.

It seemed as if the 'cheese-dish cover' of light stood like a wall between us. I think it was created to isolate them from our world and atmosphere. One of the women came out of the cabin with an object in her hand. She went to the edge of the wall of light and threw the object beyond the area of light. At the same time I heard her laugh.

Carlsson later retrieved the object. (Analysis conducted in 1971 revealed nothing out of the ordinary, the object consisting, among other things, of silicon. Its shape had been changed by the witness, and it looked like a staff.) Carlsson then left the scene.

. . . it is difficult to explain what one does, and why one does it. I thought the disc-like object could be some sort of military device. The whole scene seemed so strange. I never take alcohol, and I knew it was not an hallucination, but nevertheless I decided to go back to the seashore, and from there return to the opening to see if the object was still present. I was aware of a smell like that from ozone (O_3) following an electrical discharge.

Thirty minutes later Carlsson arrived back at the site, returning via another route so he could observe the object from the other side. Before he had time to leave the shoreline, he suddenly saw a bright-red light rising slowly with a whining sound above the tree-tops. The contraption ascended with a corona of red lights streaming from the 'turbine holes'. At about 400–500 metres it slowed down and wobbled momentarily, the red light became brighter then turned to purple, and it accelerated into the distance at tremendous speed.

FURTHER INVESTIGATIONS

The foregoing account was given by Gösta Carlsson to Eugen Semitjov, a prominent space-science journalist who wrote an article about the incident in a Swedish magazine.[4] Semitjov believed the witness was telling the truth as well as he could remember it, 25 years after the event. Alerted to the case, further investigations were carried out by the Göteborg Information Centre on UFOs (GICOFF). At the alleged landing site – about 30 to 40 metres in diameter – the investigators noted a large outer ring, a smaller ring, two small circles (presumably from the landing legs) and a straight line, almost half the diameter of the larger ring (perhaps from the 'fin'). But something was wrong, explained investigator Sven-Olof Fredrickson:

What was strange, however, was that the marks had been made recently: someone had dug a circle 10 centimetres deep, 10 centimetres wide and 16 metres in diameter, and then filled it in with sand. The same had happened with the marks of the supposed 'landing legs' and 'fin'. Mr Semitjov assured us that this had not been done by him, and that there were no signs of digging when he was there two months earlier. The original marks were still visible without digging them up, he said, so who had done it and why?

The GICOFF investigators learned from a reliable witness, who had been at the landing site a day after the Swedish magazine article was published, that the marks had already been dug up, by person or persons unknown. My feeling is that Carlsson, who in 1998 showed me the landing site (now a tourist attraction), may have done this himself. In any event, GICOFF discovered that some of the original marks (the outer and the inner rings) appear in aerial photographs taken in 1947 and 1963. Photos taken in 1939 do not appear to show any such marks, which gives at least a modicum of support for Carlsson's claims.[5]

By extraterrestrial standards, Carlsson's contraption seems relatively primitive. We are told of 'holes around the edge of the disc, like those of a turbine [from which red] jet-beams came which burned the grass'; a 'mast'; the occupants 'repairing a window'; and what looked like 'an old black camera' suspended by a chain around the guard's neck. One could argue that the craft was a post-war Russian variant of Germany's V-7 jet-turbine-powered helicopter, reportedly developed towards the end of the war and described by the pioneering astrophysicist Dr Hermann Oberth as having 'rotating tubes which released an "exhaust" of flame . . . When it hovered, the flame was dark-red' (see p. 23). It is a tempting argument, but I believe it to be a specious one, for the following reasons:

First, though Oberth claimed in 1955 that 'Russia has now obtained the plans and a model of the V-7, and has built some models of her own which could account for some UFO reports',[6] he implies not only that this was a relatively recent development (that is, after the date of the Angelholm landing) but also that models, rather than full-scale craft, were test-flown. Secondly, I find it impossible to believe that the Soviets, presumably building the hypothetical V-7 variant with the aid of captured German aeronautical engineers, could by 1946 have perfected the design to such an extent that it would have been able to fly with a crew of eleven, rather than a test-pilot or two. Thirdly, conventional jet-engines would leave a smell of kerosene, not ozone. Fourthly, Carlsson's account refers to several peculiarities of design and performance which are not consistent with what we know about the V-7 and which simply do not correspond with contemporaneous technology. Finally, Bill Gunston, one of the world's leading aviation authorities, told me: 'I have asked two men who would have known, and I think I am on good ground in stating that nothing like the V-7 was ever tested in the Soviet Union.'[7] The most likely hypothesis, therefore, is that the craft came from somewhere else. As Oberth himself put it:

My own explanation of the unsolved percentage of UFOs is that they are machines built in some place other than Russia and countries on the Earth ... I do not, in fact, think that Russia is building any UFOs at all; on the contrary, I believe they originate exclusively from outside the Earth.[8]

Interestingly, according to Lieutenant Colonel Philip Corso, the former head of the US Army staff's Foreign Technology desk at the Pentagon who claims to have stewarded an Army project that seeded alien technology at various American companies (see p. 5), both Dr Oberth and Dr Wernher von Braun were consulted on the nature of the recovered alien materials.[9] So, presumably, both men were convinced of extraterrestrial visitation. And in an early 1960s report, Corso noted that: 'Dr Hermann Oberth suggests we consider the Roswell craft from the New Mexico desert not a spacecraft but a time machine . . .'[10]

ANGELS UNAWARES?

Shortly before his discharge from the United States Army in June 1946, Allan Edwards was admitted to an Army hospital at Camp Lee, Virginia, suffering from a minor ailment. The main ward being full, he was given a private room. After a night's sleep he went down to the ward seeking someone to chat with. Of the patients, some of whom were sitting beside their beds, one stood out. Even from 50 feet away, there was something unusual about this man, noted Edwards who, as an academically trained portrait painter, had made a thorough study of anatomy.

'I arose and walked the length of the ward to get a closer look,' said Edwards. 'The bed at the end of the room on my left was vacant and, assuming it to be his, I asked him if he minded if I sat on it. He smiled and said, "Go right ahead." I perched on the edge and, trying not to be too obvious, I studied this amazing man.' He continued:

Never in my life had I seen such beauty, yet there was absolutely nothing feminine about his features. They were perfectly formed. His forehead was extremely high, the fine veins showing faintly through the transparency of the skin at the temples. His blond hair seemed to glow with an inner light of its own; in fact, his entire head seemed to be radiant, whether from the beauty of his complexion or some mysterious factor I did not know. His eyes were softly blue beneath the pure whiteness of his brow and seemed, to me, to be filled with great compassion. His nose was perfectly shaped and the colouring of his cheeks had a freshness and purity that I had never beheld in any human being.

The extraordinary height of his forehead amazed me but the physical characteristic that I found even more unusual was the depth of his head from the forehead to the back. This was definitely an abnormality according to all rules of skeletal structure and yet, as I continued to stare, I realized that for the first time I was seeing perfection.

Edwards then became aware of a strange sensation, a sensation reported by a number of contactees, such as George Adamski, in later years. 'Somehow I seemed to be in two places at once, as though I were raised up into another dimension,' he tried to explain. 'The feeling of well-being was beyond description, almost in the nature of a spiritual experience, and I felt that in some way it was connected with the man seated near me.'

Hesitating to question the stranger, Edwards glanced at the bedside table, hoping to get some clue as to his identity. On it was a pitcher of water, a tumbler and a copy of a pocket magazine called *Pageant*. The magazine was opened at an article entitled 'Easter in Oklahoma', an article later to assume significance.

Edwards was called away for a medical examination, then returned to his room. At around midnight, he heard a commotion in the corridor outside. Peering out, he saw two attendants struggling with a young man who, they explained to Edwards, had been brought into the hospital by the military police, having been badly beaten up during a drunken brawl. Both eyes were black and swollen shut, his forehead was badly bruised, and his nose was a bleeding pulp. The young man was put into the room next to Edwards, who was unable to sleep for hours owing to the groans from the suffering man.

At breakfast the following morning, Edwards sat near the strange man, on his right, then looked around the table at the other men.

My eyes rested on a young lad seated at my left at the end of the table. This was the same boy who had been brought in the night before, the one who had been so badly beaten – yet it couldn't be, there was not a blemish on his face! . . . I felt strongly that the man seated on my right had been responsible for this miraculous transfiguration. Again I felt the odd sensation of being in two places at the same time. Was no one else at the table aware of what was taking place? I looked about me at the others. Then I realized that an amazing thing had happened. Each one of the men seated at that table was changed . . . It was as though a grey veil had been lifted off my eyes and for the first time I saw true beauty of colour. I wondered if they were aware of their transfiguration or whether it was some strange trick of my own vision.

A TOWN CALLED CEMENT

Edwards was profoundly moved by these experiences. He also began to have repetitive dreams about cement and concrete. 'I seemed to be involved in mixing it, pouring it and even being buried in it,' he wrote.

Six weeks later, immediately after his discharge from the Army, he bought a copy of *Pageant* magazine with the article 'Easter in Oklahoma'. This told about the annual pilgrimage to Lawton, a small town in the southern part of the state. Every Easter a pageant, depicting the last days of Christ, was held by the townspeople and attended by thousands from all over the country. Edwards wrote to the Lawton Chamber of Commerce to learn more about the town, and soon received a package of brochures extolling its virtues. Because Edwards and his wife had no plans for the future, they decided to move to Lawton. It was, he said, 'a decision based on nothing more than an article in a pocket magazine'.

A few days before they were due to leave, Edwards was walking down the main street of Petersburg, Virginia, where he had been living during his Army service. Walking in the same direction on the other side of the street was the extraordinary man whom he had encountered at Camp Lee, accompanied by another man. Both were wearing US Army uniforms. Edwards crossed the street and followed them. The men parted company when they reached the corner, and Edwards made his approach.

> I caught up with him, a thousand questions on my tongue. I found myself looking up at him: I am fairly tall but he towered above me. I managed to stammer out, 'Do you remember me?' He smiled and said, 'Yes, you were in the hospital at Camp Lee.' All the questions that I wanted to ask him suddenly disappeared and I found myself saying, 'We are going to move to Lawton, Oklahoma,' . . . he said, 'I come from a small town near Lawton called Cement.'

Edwards introduced himself, and the stranger gave his name as Suder. 'I cannot recall any further conversation [and] he walked away until I lost sight of him when he turned a corner.'

Within a few days, Edwards and his wife realized that Lawton was not the place they wanted to live in, so they packed their belongings and moved to Los Angeles. Several years later, Edwards returned to Oklahoma and made a point of visiting Cement. No one there had ever heard of a man called Suder, and there the matter ended.

FURTHER ENCOUNTERS

The following year, in Seattle, Washington, Edwards was once again hospitalized, this time for a respiratory ailment. During his stay in hospital, he became increasingly curious about two attendants.

> Both were quite young, one tall and blond, the other short and dark. Physically there was nothing extraordinary about them but I quickly discovered that both had the ability to read every thought that passed through my mind, an ability which proved quite disconcerting to me. To have a

question answered before it is put into words is an intriguing experience and I must admit that I was quite awed by it.

At times, Edwards reported, he felt impelled, against his own will, to do things that would ameliorate the circumstances of those sicker than himself. 'I had not had much experience with those who were in pain and yet I found myself administering to them with expert hands which did not somehow seem to belong to me. I regret, now, my reticence about questioning the two attendants, although I truly do not believe that I was meant to.'

Not long after his dismissal from hospital, Edwards felt impelled to go down to the street from his hotel room in Seattle. It was a clear, moonless night. 'My gaze was drawn to one star which was unusually large . . . it started to expand and gradually grew brighter and brighter until it was about half the size of a full moon [and] after a few moments it decreased in brilliance until it was back to its original size. I was startled and perplexed. I had never heard of space ships or flying saucers and could only suppose that it was some strange phenomenon upon a distant planet or star.' It was not until several years later, when Edwards was given a copy of George Adamski's book *Inside the Space Ships*, that he began to wonder if there was a connection between this sighting and the strange encounters in Camp Lee, Petersburg and Seattle.

Other sightings followed. One night in Virginia Beach, Virginia, Edwards and several others watched nine UFOs as they manoeuvred over the coast, describing sharp right-angled turns. Several years later, following sightings in Texas and Hawaii, Edwards saw a large glowing object in Puerto Rico. Numerous sightings and close encounters have been reported in Puerto Rico over the years; per capita, possibly more than in any other locale. A number of local people have told me they are convinced that an alien base has been established in the area (see Chapter 19).

While in Puerto Rico, Edwards spoke with a man who, one evening, with several others, had seen a large fluorescent object, the size of a DC-6 airliner, on the landing strip of an abandoned military airfield on the southern side of the island. 'It took off while they watched, accelerating quickly, and disappeared,' Edwards reported. 'They were particularly impressed by the strange vibration which they claimed they felt within their own bodies. They were only about 300 feet away and yet heard no sound. The sugar cane beside the road trembled with the same vibration and they made note of the fact that there was no wind.'

SPIRITUAL MOTIVATIONS?
In his attempt to rationalize earlier encounters, Allan Edwards remained objective.

I wish to make it clear that I do not claim to have been in contact with beings from outer space, since the people . . . did not so identify themselves. I can only present the facts as I remember them and let the reader form his own conclusions. I have placed these people in the category of 'extraordinary' due to the unusual powers that they possessed. It is quite possible that they are of this earth but have reached a higher state of development than the average . . . I cannot help, however, but hope that my intuition is right and that there are those from other planets who are anxious to assist us in our present predicament here on this earth and who, being more spiritually inclined, may be able to guide us through the darkness that seems to lie ahead. In reference to this, a verse from Hebrews in the New Testament seems pertinent: 'Let brotherly love continue. Be not forgetful to entertain strangers; for thereby some have entertained angels unawares.'[11]

ALIEN ANTICS

On 23 July 1947, José Higgins, a Brazilian topographer, together with a team of workers, was surveying an area of land northeast of Pitanga, State of São Paulo, Brazil, when a piercing, high-pitched whistling sound was heard coming from the sky. Looking up, the men watched as a large disc appeared about to land. The workers fled, leaving Higgins to fend for himself.

A 'strange, circular airship with protruding edges', as Higgins described it, landed about 150 feet away. It appeared to be made of a grey-white metal, was about 150 feet wide (not including the edges, which were about three feet in width) and was about 15 feet in height. 'It was crossed by tubes in several directions,' said Higgins, 'but there was no smoke or fire, only that odd sound coming from the tubes.' The object landed on curved metallic-looking poles which bent even more as it touched the ground. Higgins approached, and as he came closer he noticed a kind of thick window or porthole, through which he could see two strange-looking 'people' who were eyeing him closely. One turned his back as though talking to another in the craft. Suddenly Higgins heard a noise from inside and a hatch on the underside of the craft opened and three strange beings emerged.

About seven feet tall, the beings were enclosed in a kind of inflated, transparent suit which enveloped them from head to foot. On their backs were metal boxes that seemed to be part of the suit. Through the transparent covering, Higgins saw that they were wearing shirts, shorts and sandals! The clothes appeared to be made not of cloth, but of a brilliantly coloured paper. They had large, round, almost hairless heads, as well as large, round, browless eyes, and appeared identical to each other – 'like twins or brothers'. The length of their legs was greater in proportion to their bodies than those of normal human beings.

One of the beings carried a metal tube, which was pointed at Higgins, with no apparent effect. They moved around him with amazing agility for their size, speaking among themselves in an unknown tongue, then the one with the tube motioned for him to approach the entrance of the craft. He saw a small chamber with another 'door' on the inside, and the end of a pipe. He also noticed several round 'beams' on the protruding edge. Presuming that he was being invited for a trip, Higgins began questioning them (in Portuguese), using many gestures, asking where they intended to take him. Seeming to understand, one of them made a drawing on the ground with a round spot, encircled by seven circles, pointed to the Sun in the sky (uttering a word that sounded like 'Alamo'), then to the central spot on the ground, then to their craft and finally to the seventh orbit ('Orque'), repeating the latter gesture several times.

At this point Higgins became frightened and began to plan his escape. He knew it would be useless to fight them but, noticing that the beings avoided direct sunlight for more than a few seconds, he walked towards the shadow. Taking out a photo of his wife from his wallet, he showed it to them, explaining by gestures that he wanted to bring his wife on the trip. They seemed to understand, so he casually walked away into the nearby forest. Having found a safe place from which he could observe the beings without being seen, Higgins was astonished to see them frolicking about like children, jumping in the air and throwing huge stones around! After about half an hour and a careful examination of their surroundings, they returned to their craft, which took off with a whistling sound.

The size and apparent strength of the beings, plus their throwing of large stones, could indicate that they originated on a planet with a greater gravitational pull than that of Earth; their gambolling antics possibly being experimentation with the novelty of the Earth's lesser gravitational pull – which reminds one of the Apollo astronauts who performed similar antics when on the Moon.

'I will never know if they were men or women,' Higgins told the press later. 'Despite the characteristics I described, they were somehow beautiful and appeared in excellent health.' In company with many contactees, Higgins was thoroughly bemused by the experience. 'Was it a dream? Was it real?' he asked. 'Sometimes I doubt that these things can happen, and then I think that if it was not for the workers together with me in the beginning, it might have been a strange and fascinating dream . . .'[12]

NOTES

1 Machado Carríon, Professor Felipe, 'The Terrible Death of João Prestes at Araçariguama', translated from the French by Gordon Creighton, *Flying Saucer Review*, vol. 19, no. 2, March–April 1973, pp. 14–15. This article first appeared in *Phénomènes Spatiaux*, no. 30, December 1971.

2 *Notícias Populares*, 23 October 1972.

3 For further details of the 'ghost rocket' sightings, see *Beyond Top Secret: The Worldwide UFO Security Threat* by Timothy Good (Sidgwick & Jackson, London, 1996).

4 *Allers*, Sweden, no. 44, 30 October 1971.

5 Fredrickson, Sven-Olof, 'The Angelholm Landing Report', *Flying Saucer Review*, vol. 18, no. 2, March–April 1972. pp. 15–17. See also *Mötet i gläntan* by Clas Svahn and Gösta Carlsson (NTB/Parthenon, Sweden, 1995). Svahn's investigation into this case covered a ten-year period.

6 Oberth, Professor Hermann, 'They Come from Outer Space', *Flying Saucer Review*, vol. 1, no. 2, May–June 1955, p. 13.

7 Letter to the author from Bill Gunston, OBE, 24 August 1996.

8 Oberth, op. cit., p. 14.

9 Corso, Col. Philip J., with Birnes, William J., *The Day After Roswell*, Pocket Books, New York and London, 1997, p. 160.

10 Ibid., p. 90.

11 Edwards, Allan W., 'An Angel Unawares?', *Flying Saucer Review*, vol. 7, no. 1, January–February 1961, pp. 7–10.

12 Lorenzen, Coral, and Lorenzen, Jim, 'The Spacemen Threw Stones: Another Contact Story from Brazil', *The APRO Bulletin*, May 1961, as republished in *Flying Saucer Review*, vol. 7, no. 6, November–December 1961. Higgins's story was first published in *Diário da Tarde*, Curitiba, Brazil, 8 August 1947.

Chapter 4

Steps to the Stars

It was the evening of 4 July 1949. For Daniel Fry, a rocket test technician employed by Aerojet General Corporation at the vast and remote White Sands Proving Grounds in New Mexico, it was to be an Independence Day with a difference.

Fry had planned on going into the town of Las Cruces to celebrate with colleagues, but having missed the last motor pool bus, he went back to his room and began reading a textbook on heat transfer, a problem of considerable relevance to the design of rocket motors. 'I was soon to learn however,' said Fry, 'that the problems of heat transfer can become as uncomfortable physically as they are interesting academically.' At 20.00 the building's air-conditioning system apparently broke down and it became unbearably hot. He decided to go for a walk, hoping it would be cooler outside.

Heading in the direction of an old static test stand, where Aerojet at the time were mounting their largest rocket engine for tests, Fry then decided to take a different route that led off towards the base of the Organ Mountains. Scanning the evening sky, he was surprised to notice one star, then another, then two more, 'going out'. Suddenly, the outline of an object of some sort, blending with the dark blue of the sky, appeared to be headed in his direction. 'As it continued to come toward me, I felt a strong inclination to run,' he explained, 'but experience in rocketry had taught me that it is foolish to run from an approaching missile until you are sure of its trajectory, since there is no way to judge the trajectory of an approaching object if you are running.'

The object was now less than a few hundred feet away, moving more slowly and seeming to decelerate. Its shape was an oblate spheroid with a diameter at its largest part of about 30 feet. 'Somewhat reassured by its rapid deceleration,' continued Fry, 'I remained where I was and watched it glide, as lightly as a bit of thistledown floating in the breeze. About 70 feet away from where I was standing, it settled to the ground without the slightest bump or jar. Except for the crackling of the brush upon which it had settled, there had been no sound at all. For what seemed a long time . . . I stared at the now motionless object as a child might stare at the rabbit which a stage magician has just pulled from his hat. I knew it was impossible, but there it was!'

Fry had been employed for many years in the burgeoning field of astronautics, yet never before had he seen such a device. As Vice-President of Crescent Engineering and Research Company in California, for example, he developed a number of parts for the guidance system of the Atlas missile, while at Aerojet he was in charge of installation of instruments for missile control and guidance systems. 'Obviously, the intelligence and the technology that had designed and built this vehicle had found the answers to a number of questions which even our most advanced physicists have not yet learned to ask,' Fry commented. He then cautiously approached to within a few feet of the landed craft and listened for any sign of life or sound from within. There was none.

I began to circle slowly about the craft so that I could examine it more completely. It was . . . a spheroid, considerably flattened at the top and bottom. The vertical dimension was about 16 feet, and the horizontal dimension about 30 feet at the widest point, which was about seven feet from the ground. Its curvature was such that, if viewed from directly below, it might appear to be saucer-shaped, but actually it was more nearly like a soup bowl inverted over a sauce dish.

The dark blue tint which it had seemed to have when in the air was gone now, and the surface appeared to be of highly polished metal, silvery in color, but with a slight violet sheen. I walked completely around the thing without seeing any sign of doors, windows or even seams . . . I stepped forward and cautiously extended my index finger until it touched the metal surface. It was only a few degrees above the air temperature, but it had a quality of smoothness that seized my attention at once. It was simply impossible to produce any friction between my fingertip and the metal. No matter how firmly I pressed my finger on the metal, it drifted around on the surface as though there were a million tiny ball-bearings between my finger and the metal. I then began to stroke the metal with the palm of my hand, and could feel a slight but definite tingling in the tips of my fingers and the heel of my palm.

THE VOICE

Suddenly, a loud male voice came out of nowhere: 'Better not touch the hull, pal, it's still hot!' Shocked, Fry leaped backwards several feet, catching his heel in a low bush and sprawling full length in the sand. Something like a low chuckle was heard, then the voice came back: 'Take it easy, pal, you're among friends.' Recovering himself, Fry looked around for some person or gadget from which the voice came, but could see none. He complained to the voice that it should turn the volume down. 'Sorry, buddy, but you were in the process of killing yourself and there wasn't time to diddle with controls,' came the response. 'Do you mean the hull is highly radioactive?' asked Fry. 'If so, I am much too close.'

'It isn't radioactive in the sense that you use the word,' replied the voice. 'I used the term "hot" because it was the only one I could think of in your language to explain the condition. The hull has a field about it which repels all other matter. Your physicists would describe the force involved as the "anti" particle of the binding energy of the atom.' The 'voice' expounded at length, and one must wonder at how Fry could have recalled so much detail:

When certain elements such as platinum are properly prepared and treated with a saturation exposure to a beam of very high-energy photons, the anti-binding energy particle will be generated outside the nucleus. Since these particles tend to repel each other, as well as all other matter, they, like the electron, tend to migrate to the surface of the metal where they manifest as a repellent force. The particles have a fairly long half-life, so that the normal cosmic radiation received by the craft when in space is sufficient to maintain an effective charge. The field is very powerful at molecular distances but, like the binding energy, it follows the seventh power law so that the force becomes negligible a few microns away from the surface of the hull.

Perhaps you noticed that the hull seemed unusually smooth and slippery. That is because your flesh did not actually come into contact with the metal but was held a short distance from it by the repulsion of the field. We use the field to protect the hull from being scratched or damaged during landings. It also lowers air friction greatly when it becomes necessary to travel at high speed through any atmosphere. The field produces an almost perfect laminar flow of air or any gas about the craft, and little heat is generated or transmitted to the hull.

'But how would this kill me?' asked Fry. 'I did touch the hull and felt only a slight tingle in my hand. And what did you mean by that remark about *my* language? You sound pretty much American to me.'

Replying to the first question, the voice explained that Fry would have probably died within a few months from exposure to the 'force field', which produces 'what you would describe as "antibodies"' in the blood stream that are absorbed by the liver, causing the latter to become greatly enlarged and congested. 'In your case,' continued the voice, 'the exposure was so short and over such a small area that you are not in any great danger, although you will probably feel some effects sooner or later, provided, of course, that your biological functions are similar to ours, and we have good reason to believe they are.' The voice continued:

As to your second question, I am not an American such as you, nor even an 'Earthian', although my present assignment requires me to become both. The fact that you believed me to be one of your countrymen is a testimonial to the effort I have expended to learn and to practise your language. If you talked with me for any length of time, however, you would begin to notice that my vocabulary is far from complete, and many of my words would seem outdated and perhaps obsolete. As a matter of fact, I have never yet set foot upon your planet. It will require at least four more of your years for me to become adapted to your environment, including your stronger gravity, [and] your atmosphere . . . I will also require the complete co-operation of someone like yourself who is already a resident of the planet.

Fry stood silently for what seemed a long time, attempting to come to terms with the profound implications of what he had seen and heard. The conversation then continued, with Fry asking a variety of questions, the first dealing with his reactions to the experience. The voice was encouraging:

One of the purposes of this visit is to determine the basic adaptability of the Earth's peoples, particularly your ability to adjust your minds quickly to conditions and concepts completely foreign to your customary modes of thought. Previous expeditions by our ancestors, over a period of many centuries, met with almost total failure in this respect. This time there is hope that we may find minds somewhat more receptive so that we may assist you in the progress, or at least in the continued existence of your race . . . The fact that, in spite of being in circumstances completely unique in your experience, you are listening calmly to my voice and making logical replies is the best evidence that your mind is of the type we hoped to find.

Fry thanked the voice, but pointed out that this statement implied that he was to be used in some project involving the advancement of the people of Earth. 'Why me?' he asked. 'Is it just because I accidentally happened to be here when you landed? I could easily put you in touch with a number of men right here at the test base who could be of far more value to you than I.'

'Perhaps they could,' came the reply, 'but would they?'

If you think you are here by accident, you greatly underestimate our abilities. Why do you think the dispatcher at your motor pool gave you incorrect information? Why did you think your air-conditioning system had failed tonight when, as a matter of fact, it was functioning perfectly? Why do you think you turned off on this small road, when your intention had been to go to your static test stand? And finally, why do you think you changed your mind about going back to your base to report [as had been Fry's initial intention] the arrival of our sampling carrier? It is seldom that we superimpose our will upon that of others . . . but this is a case of such urgency for your people that we felt an exception to the rule was warranted . . .

The voice went on to request Fry's assistance in a planned programme for 'the welfare, and in fact for the preservation of' Earth's people. Several years would pass, Fry was told, before his services would be required. 'We will be glad to offer you a short test flight in the sampling craft if it will help you to decide that we are what we say, and that our technology has much for you to learn,' said the voice, and continued:

The craft is a remotely controlled sampling device, or cargo carrier, and while I am speaking through its communication system, I am not in it. I am in a much larger deep space transport ship, or what you would call a 'Mother Ship'. At present, it is some 900 of your miles above the surface of your planet, which is as close as ships of this size are permitted to approach any planet with an appreciable atmosphere. The cargo craft is being used to bring us samples of your atmosphere so that my lungs may gradually become accustomed to it . . .

All of the previous atmosphere that was in the craft was allowed to escape while it was in space, by the opening of the remotely controlled valve in the top. There is now an almost perfect vacuum inside. When the port is again opened, which I shall do now, your air will rush in to fill the craft, and we will have a large-scale sample of your atmosphere, together with any microorganisms which may be present in it. We need these also, for study and for immunization. Your breathing of the atmosphere during this short demonstration flight will, of course, distort this particular sample somewhat, but we will have ample opportunity to obtain others before my adaptation to the environment of your planet is complete.

INSIDE THE CRAFT

A sound – partly a hiss and partly a murmur – came from the top of the craft. This lasted for about 15 seconds, then all became quiet again. 'Any port large enough to have filled a ship of that size with air in 15 seconds should have produced quite a roar,' Fry reasoned. 'I realized then that the walls of the ship were almost, if not entirely, soundproof, and since most of the sound of the entering air would be produced inside, very little would be audible outside.'

A single click was then heard coming from the lower wall of the craft, and a section of the lower side of the hull moved back on itself for a few inches then moved sideways, disappearing into the wall of the hull and leaving an oval-shaped opening about five feet in height and some three feet in width at its widest point. Fry walked towards the hatch and, ducking slightly, went inside the craft. With his feet still on the ground, owing to the curvature of the craft, he looked around.

The compartment into which I was looking occupied only a small portion of the interior of the ship. It was a room about nine feet deep and seven feet wide, with a floor about 16 inches above the ground and a ceiling between six and seven feet above the floor. The walls were slightly curved and the intersections of the walls were bevelled so that there were no sharp angles or corners . . .

The room contained four seats; they looked much like our modern body contour chairs, except that they were somewhat smaller . . . The seats faced the opening in which I was standing, and were arranged in two rows of two each in the center of the room, leaving an aisle between the seats and either wall. In the center of the rear wall, where it joined the ceiling, there was a small box or cabinet with a tube and what appeared to be a lens arrangement. It was somewhat similar to a small motion picture camera or projector, except that no film spools or other moving parts were visible. Light was coming from the lens. It was not a beam of light such as would have come from a projector, but a diffused glow . . . it still furnished ample light for comfortable seeing in the small compartment.

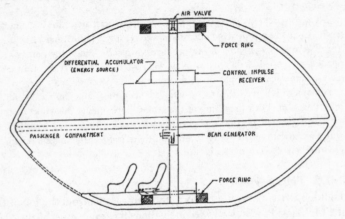

Fig. 3. A diagram of the remotely-controlled craft.

Fry noted that the seats and the light seemed to be the only furnishings in an otherwise bare metal room. 'Not a very inviting cabin,' he thought, 'looks more like a cell.'

'It's only a sampling carrier,' said the voice, in response to Fry's unspoken thought, 'and was not really designed or intended to carry passengers: the small compartment was designed for emergencies only, but you will find the seats quite comfortable. Step in and take a seat if you wish to make this test flight.'

Fry stepped up on to the floor and headed towards the nearest seat. As he did so, he heard a click as the door began to slide out of the recess in the wall behind him. 'Instinctively, I turned as though to leap out to the comparative safety of the open desert behind me, but the door was already closed. If this was a trap, I was in it now, and there was no point in struggling against the inevitable.'

'Where would you like to go?' asked the voice, now seeming to come from all around Fry, rather than beside him, as before. 'I don't know how far you can take me in whatever time you have,' he replied. 'And since this compartment has no windows, it won't matter which way we go, as I won't be able to see anything.'

'You will be able to see,' came the reply, 'at least, as much as you could see from any of your vehicles in the air at night. If you would like a suggestion, we can take you over the city of New York and return you here in about 30 minutes. At an elevation of about 20 of your miles, the light patterns of your major cities take on an especially fascinating appearance which we have never seen in connection with any other planet.'

'To New York – and back – in 30 minutes?' retorted Fry. 'Your minutes must be very different from ours. New York is 2,000 miles from here. A round trip would be 4,000 miles. To do it in half an hour would require a speed of 8,000 miles per hour! How can you produce and apply energies of that order in a craft like this, and how can I take the acceleration? You don't even have belts on these seats!'

'You won't feel the acceleration at all,' came the reassuring reply. 'Just take a seat and I will start the craft . . .'

Fry took the seat nearest to the door and found it to be quite comfortable. 'The material of which it was made felt like foam rubber with a vinylite covering. However, there were no seams or joints such as an outer covering would require, so the material, whatever it was, probably had been moulded directly into its frame in a single operation.'

ON RENDERING MATERIALS TRANSLUCENT

'I will now turn off the compartment light and activate the viewing beam,' said the voice. For a moment the room became completely dark. Then a beam of light came from the projector. 'The beam, or the part of it which was visible at all, was a deep violet, at the very top of the visible spectrum,' Fry explained. 'The beam was focused so as to exactly cover the door through which I had come, and under its influence the door became totally transparent. It was as though I were looking through the finest type of plate glass or lucite window.'

As the voice went on to say that a few of the basic technical principles would be explained, Fry began to realize that the words he had been hearing were probably not coming to his ears as sound waves but seemed to be originating directly in his brain. The voice continued:

As you see, the door has become transparent. This startles you because you are accustomed to thinking of metals as being completely opaque. However, ordinary glass is just as dense as many metals and harder than most, and yet transmits light quite readily. Most matter is opaque to light because the photons of light are captured and absorbed in the electron orbits of the atoms through which they pass. This capture will occur whenever the frequency of the photon matches one of the frequencies of the atom. The energy thus stored is soon re-emitted, but usually in the infra-red portion of the spectrum, which is below the range of visibility, and so cannot be seen as light.

There are several ways in which matter can be made transparent, or at least translucent.

One method is to create a field matrix between the atoms which will tend to prevent the photon from being absorbed. Such a matrix develops in many substances during crystalization. Another is to raise the frequency of the photon above the highest absorption frequency of the atoms. The beam of energy which is now acting upon the metal of the door is what you would call a 'frequency multiplier'. The beam penetrates the metal and acts upon any light that reaches it in such a way that the resulting frequency is raised to that between the ranges which you describe as the 'X-ray' and the 'Cosmic Ray' spectrums. At these frequencies, the waves pass through the metal quite readily. Then, when these leave the metal on the inner side of the door, they again interact with the viewing beam, producing what you would describe as 'beat frequencies' which are identical with the original frequencies of the light. As a rough analogy, the system could be compared to the carrier wave of one of your radio broadcasting stations, except that the modulation is applied 'upstream' as it were, instead of at the source of the carrier.

EARTH'S BROADCASTS MONITORED

Fry remarked that, for one who had never set foot on Earth, his unseen host seemed extraordinarily familiar with our terrestrial technology. 'You are underestimating *our* technology,' came the reply.

You have no idea of the amount of close-range observation to which your planet has been subjected by passing space craft during the past few generations. The radio messages and programs which you continually hurl into space can readily be monitored by our receiving equipment, at distances equal to several times the diameter of your solar system. Within such a volume of space there will always be at least a few ships either passing through the system or pausing to store up energy from its solar radiation. Any data received from earthly broadcasts which is considered to be of potential interest to other races will be recorded and relayed to more distant receiving points which will relay in turn, until the data is ultimately available to much of the galaxy.

LIFT-OFF

The voice announced the imminent departure of the craft. 'A moment later, the ground suddenly fell away from the ship with almost incredible rapidity,' said Fry. 'I did not feel the slightest sense of motion myself, and the ship seemed to be as steady as a rock . . . the lights of the army base at the proving ground, which had been hidden by a low hill, instantly came into sight . . . A few seconds later the lights of the town of Las Cruces came into view . . . and I knew that we must have risen at least 1,000 feet in those few seconds. The ship was rotating slightly to my left as it rose, and I was able to see the highway from Las

Cruces to El Paso [Texas] . . . I could even distinguish the very thin dark line of the Rio Grande . . . The surface of the Earth appeared to be glowing with a slightly greenish phosphorescence. The sky outside the ship had become much darker, and the stars seemed to have doubled in brilliance.' He assumed that the craft had now entered the stratosphere, in which case an altitude of more than 10 miles must have been attained in no more than 20 seconds, and without the slightest sense of acceleration.

'You are now about 13 miles above the surface,' announced the voice, 'and you are rising at about one half-mile per second. We have brought you up rather slowly so that you might have a better opportunity to view your cities from the air. We will take you up 35 miles for the horizontal flight . . .'

GRAVITATIONAL FIELDS

The voice explained the question of acceleration and why it had no effect on the occupants of their craft; a question 'which seems to have come up quite often in the minds of the men of science, and many others of your people'.

Whenever our sampling devices or landing vehicles have been observed by them, and when the velocities and acceleration are described, disbelief is always apparent . . . This has been one of the causes of disappointment to us in our evaluation of the intelligence of the people of Earth . . . The answer is simply that the force which we use to accelerate our vehicles is identical in nature to a gravitational field. It acts, not only upon every atom of the vehicle, but equally upon every atom of mass that is within it, including the mass of the pilot and any passengers. Regardless of the intensity of the field therefore, every particle of mass within the influence of the field is in a uniform state of acceleration or, as you would term it, free fall, with respect to the field. Under these circumstances acceleration has no effect upon the vehicle or anything within it.

'But in that case,' asked Fry, 'why am I not floating around in the air as things are supposed to do in a missile that is in free fall?' Back came the answer:

Before the ship's own field was generated, it was resting upon the earth, and you were resting upon the seat. There was a force of one gravity acting between your body and your seat. Since the force which accelerates both the ship and your body acts in exact proportion to the mass, and since the Earth's gravity continues to act upon both, the original force between your body and the seat will remain constant except that it will decrease as the force of gravity of the planet decreases with distance. When traveling between planetary or stellar bodies far from any natural gravity source, we find it necessary, for practical reasons, to reproduce this force artificially.

The gravity to which we are accustomed is but little more than one-half that of Earth. It is one of the reasons that it will require so much time for one of us to become completely acclimatized to your environment. If I were to land now upon your planet, I could tolerate the doubled gravitational force for a time but the double weight of all my internal organs would cause them to be displaced downward, seriously hampering their functions. The difference in blood pressure between head and feet when standing erect would be double that to which we are accustomed, and there would be several other complications . . . If, on the other hand, I remain in my own ship, the gravitational force to which I am subject can be increased by small but regular increments: the supporting tissues will gradually increase in size and strength until, eventually, your gravity will become as normal to me as my own is now.

Fry asked for an explanation of the craft's propulsion system, specifically as it related to the tremendous amount of energy presumably required to accelerate the ship to such fantastic velocities. The voice began by saying that there were several concepts and words which did not yet exist in human vocabulary, or even in human consciousness, to allow for a complete explanation. But after several analogies by way of simplification, he continued:

The large drum-like structure just above the central bulkhead is the differential accumulator. It is essentially a storage battery that is capable of being charged from a number of natural energy sources. We can charge it from the energy banks of our own ship, but this is seldom necessary. In your stratosphere, for example, there are several layers of ionized gas which, although they are rarefied, are also highly charged. By placing the ship in a planetary orbit at this level, it is able to collect, during each orbit, several times the energy required to place it in orbit. It would also, of course, collect a significant number of high-energy electrons from the Sun.

By the term 'charging the differential accumulator' I merely mean that a potential difference is created between two poles of the accumulator. The accumulator material has available free electrons in quantities beyond anything of which you could conceive. The control mechanism allows these electrons to flow through various segments of the force rings which you see at the top and bottom of the craft . . . The tremendous surge of electrons through the force rings creates a very strong magnetic field. Since the direction and amplitude of the flow can be controlled through either ring, and in several paths through a single ring, we can create a field which oscillates in a pattern of very precisely controlled modes. In this way we can create magnetic resonance between the two rings or between the several segments of a single ring.

As you know, any magnetic field which is changing in intensity will create an electric field which, at any given instant, is equal in amplitude, opposite in sign and perpendicular to the magnetic field. If the two fields become mutually resonant, a vector force will be generated. Unless the amplitude and the frequency of the resonance are quite high, the vector field will be very small, and may pass unnoticed. However, the amplitude of the vector field increases at a greater rate than the two fields which generate it and, at high resonance levels, becomes very strong. The vector field, whose direction is perpendicular to each of the other two, creates an effect similar to, and in fact identical with, a gravitational field.

If the center of the field coincides with the craft's center of mass, the only effect will be to increase the inertia, or mass, of the craft. If the center of mass does not coincide with the center of force, the craft will tend to accelerate toward that center. Since the system which creates the field is a part of the ship, it will, of course, move with the ship, and will continue constantly to generate a field whose center of attraction is just ahead of the ship's center of mass, so that the ship will continue to accelerate as long as the field is generated . . . To slow or stop the craft, the controls are adjusted so that the field is generated with its center just behind the center of mass, so that negative acceleration will result.

UNCONVENTIONAL COMMUNICATION

'Incidentally,' remarked Fry at one point, 'I don't even know your name, or do you people not have given names?'

'We have names, though there is seldom any occasion to use them among ourselves. If I become a member of your race, I shall use the name Alan, which is a common name in your country and is nearly the same as my given name, which is pronounced as though it were spelled "Ah-lahn".'

Alan said that, whenever it became necessary, Fry would be contacted again, though not necessarily by conventional means. 'We have recorded your exact frequency pattern,' he said (a method reported by Albert Coe), and went on briefly to discuss mental telepathy, or 'extra-sensory perception':

In the first place, it isn't extra-sensory at all. It is just as much a part of the body's normal perception equipment as any of the others, except that during one phase of the development of the race it falls into disuse because it is a rather public form of communication, and during this phase of development the individual requires a considerable degree of privacy in his words and thoughts. Most of your animals use the sense to a greater degree than your people, and for some of your insects, it is the only form of communication . . .

Communication could also be established by means of small, remotely controlled probes, as well as directly, using 'electronic beam modulation of the auditory nerve', Alan claimed.

FREE FALL

Prior to Alan's elaborate explanation of propulsion, the craft had flown over New York City – at a reduced height of 20 miles and a reduced velocity of 600 m.p.h., affording Fry an excellent view. 'The differing temperatures of the various air strata beneath me,' he wrote, 'caused the lights to twinkle violently, so that the entire city was a sea of pulsing, shimmering luminescence.'

Before the return trip, Alan offered Fry the chance of experiencing 'free fall'. 'To reach this condition fully under the present circumstances would be somewhat dangerous,' said Alan, 'but we can approach it closely enough so that while you will still retain some stability, you will experience the sensation of weightlessness.'

'Instantly the compartment light came on,' said Fry. 'While I was attempting to adjust my eyes to the light, my stomach suddenly leaped upward toward my chest.' Although having been through steep dives and sharp pullouts in aircraft, and ridden in many devices at amusement parks, he had never experienced anything like this.

'There was no sensation of falling. It simply felt as though my organs, having been released from a heavy strain, were springing upward like elastic bands when released from tension . . . In a few seconds I felt almost normal again.' Pushing down with his hands on the seat, he then rose slowly to the ceiling, though his body rotated and tipped forward so that he came to rest with his knees on the chair and his eyes only a few inches from the back cushion.

PLANETARY INDEPENDENCE

Fry was rather shocked to notice a very earthly symbol imprinted on the material of his seat. It was that of the tree and the serpent – the caduceus. 'It is found, in one form or another, in the legends, the inscriptions and the carvings of virtually every one of our early races,' he said. 'It has always seemed to me a peculiarly earthly symbol, and it was startling to see it appear from the depths of space, or from whatever planet you call home.' Alan responded:

> It is difficult even to outline, in a few minutes of discussion, the events of many centuries. For it has been centuries since we have called any planet home. The space ships upon which we live, and work and learn, have been our only home for generations. Like all space-dwelling races, we are now essentially independent of planets. Some of our craft are very large, judged by your standards, since they are many times larger than your largest ships. We have no personal need to approach or land upon any planet, except occasionally to obtain raw materials for new construction, and that we usually obtain from asteroids or uninhabited satellites [moons].
>
> Our ships are closed systems. That is, all matter within the craft remains there; nothing is emitted, ejected or lost from it. We have learned simple methods of reducing all compounds to their elements, and for recombining them in any form . . . For example, we breathe in the same manner as you do.

That is, our lungs take in oxygen from the air, and some of that oxygen is converted to carbon-dioxide in the body processes. Therefore, the air in our ship is constantly passed through solutions which contain plant-like organisms which absorb carbon-dioxide, use the carbon in their own growth, and return the oxygen to the air. Eventually those plants become one of our foods . . .

It may be difficult for you to conceive of a race of intelligent beings who spend all of their lives within the relatively restricted confines of a space ship. You may even be inclined to feel pity for such a race. We, on the other hand, are inclined to feel pity for the relatively primitive races which are still confined to the surface of a single planet, where they are unable to control . . . earthquakes, floods, tornadoes, tidal waves, blizzards, drought, and a dozen other hazards . . .

While our bodies seldom leave the ship, our technology has provided us with almost unlimited extensions of our senses so that . . . we can be intimately present at any time and at any place which we may choose, providing that the place is within a few thousand miles of our ship. Through a portion of our technology which your race has not yet begun to acquire, we are able to generate and apply simple forces at points quite remote from our craft. Our abilities may be somewhat startling and incredible to some of your people but they are not actually as startling and incredible as the scientific knowledge and abilities your people now have, compared to those which your ancestors possessed a few hundred years ago . . .

A COMMON ANCESTRY

Alan explained that the symbol of the tree and the serpent was not unique to Earth. It is a natural one, 'perhaps because life is said to originate in the waters of a planet, and the undulations of a serpent are a convenient symbol for the waves of a sea. The tree is almost always the symbol of life, beginning in the sea, rising to the atmosphere, and finally into space.' But there was another factor, he added, that perhaps was significant.

Your people, and some of mine, including myself, have, at least in part, a common ancestry. Tens of thousands of years ago, some of our ancestors lived upon this planet, Earth. There was, at that time, a small continent in a part of the now sea-covered area which you have named the Pacific Ocean. Some of your ancient legends refer to this sunken land mass as the 'Lost Continent of Lemuria, or Mu'.

Our ancestors had built a great empire and a mighty science upon this continent. At the same time, there was another rapidly developing race upon a land mass in the south-west portion of the present Atlantic Ocean. In your legends, this continent has been named Atlantis. There was rivalry between the two cultures, in their material and technological progress. It was friendly at first, but became bitter . . . In a few centuries their science had passed the point which your race has reached. Not content with releasing a few crumbs of the binding energy of the atom, as your science is now doing, they had learned to rotate entire masses upon the energy axis. Energies equal to 75

million of your kilowatt hours were released by the conversion of a bit of
matter about the mass of one of your copper pennies.

With the increasing bitterness between the two races . . . it was inevitable
that they would eventually destroy each other. The energies released in that
destruction were beyond all human imagination. They were sufficient to
cause major shifts in the surface configuration of the planet, and the resulting
nuclear radiation was so intense and so widespread that the entire surface
became virtually unfit for habitation, for a number of generations.

RETURN TO EARTH

Alan told Fry that further discussion had to await another meeting, as
time was up and they had landed back at White Sands. 'It is on the
ground now, and I will open the door,' he said. 'We will wait until you
are a little distance from the craft before we retrieve it . . . Take care of
yourself until we return.'

Dazed after such a fantastic experience, Fry stepped down and stum-
bled several paces through the sand before turning to look at the craft.

The door had closed behind me and, as I turned, a horizontal band of orange-
colored light appeared about the central, or widest part of the ship, and it
leaped upward as though it had been released from a catapult. The air rushing in to replace that which had been displaced upward impelled me a full
step forward and almost caused me to lose my balance. I managed to keep my
eyes on the craft while the band of light went through the colors of the spectrum from orange to violet. As the light went through the violet band, the
craft had risen several thousands of feet in the air, and I could follow it no
longer.

Fry then became thoroughly depressed. It was as if his life's work had
lost all significance. 'A few hours before, I had been a rather self-satisfied
engineer, setting up instruments for the testing of one of the largest
rocket motors in existence. While I realized that my part in the project
was a small one, I felt that, through my work, I was traveling in the fore-
front of progress. Now I knew that rocket motors . . . had been obsolete
for thousands of years. I felt like a small and insignificant cog in a clumsy
and backward science, which was moving toward its own destruction.'

Fry did not report or mention his experience to anyone, partly because
he had tacitly given his agreement to Alan not to do so, and also because
he was convinced no one would believe him. He continued rather unen-
thusiastically with his work, testing various types and sizes of rocket
motor. Alan had said he would return in a few months, and Fry grew
restless. After the first series of tests were completed, he went back home
to California, then returned to the base for a second series of tests.

One evening, after Fry had driven from his quarters at the 'H' building to the test site's instrument room, he saw an unusual glowing object,

about a foot in diameter. Walking towards it, he suddenly heard Alan's voice, as if beside him. 'Yes, Dan, it's ours. Since we are not using the sampling craft now, we thought it best to send down a small communications amplifier. We could get along without it, but it does [reduce] the chance of error in our communications almost to zero.' After allowing Fry to calm down following the shock, Alan went on to explain that he would eventually be able to successfully adapt his body to Earth's environment, but that he would need help.

'If you do not wish to assist us,' he continued, 'all memory of this meeting and the previous one will be erased from your mind . . . If, on the other hand, you do decide to assist us, you may find yourself in a situation that is not easy to endure . . . The only reward we can promise you is the inward satisfaction of having assisted in the survival of your race, and the acquisition of considerable knowledge and understanding that you would not otherwise be able to gain.' Fry gave his consent.

Alan first asked Fry to gather a number of textbooks dealing with the English language and mathematics. Regarding the latter, Alan explained that his system was based on multiples of 12, rather than 10, and that it would take him a while to master the 'new' system. These textbooks, he said, should be placed on a certain ledge at the test stand, where they would be 'collected' by a small sampling device, then analysed, copied, and returned 24 hours later. The arrangement worked well. On several occasions, Fry obtained books (including the Bible) for Alan from the base library, and on each occasion they were returned safely.

FURTHER COMMUNICATION

It was not until April 1954 that Alan once again re-established communications with Daniel Fry. At the time, Fry was relaxing in his retreat, deep in the woods of southern Oregon. As before, Alan was not present in person, but his voice was unmistakable. No device was used on this occasion; instead, communication was supposedly achieved directly via 'electronically boosted telepathy'.

Alan began by admonishing Fry for not having spoken publicly about his experience. Fry replied that, in the first place, he was afraid that any report by him might endanger Alan's proposed visit to Earth; secondly, he was unknown, apart from within the rocket industry. How therefore could he reach the people?

'Those who are not blind to the truth can recognize the value of a message regardless of the status of the messenger,' Alan responded. 'Our estimate of four years for adaptation to your environment was over-optimistic. The actual time will be closer to five. In the meantime, one of your major problems is becoming critical. Unless some small balancing force is applied in the right quarters, your entire civilization may wipe itself out . . . before we are in a position to be of assistance.'

Fry was asked to write down his experiences in a book, and to repeat through newspapers, radio and television what he had learned. 'You don't realize what you are asking,' said Fry. 'If I attempt to make public the information you have given me, it will only mean that I will be scorned and ridiculed . . . If I give a statement to our newspapers, they will either ignore it or print a comic distorted version.'

'Of course you will be ridiculed,' was the reply. 'Ridicule is the protective barrier which the timid or the ignorant erect between themselves and any possibility which frightens or disturbs them . . . It is the price exacted from every individual who is as much as one step in advance of his fellows.'

'Why don't you just set a small landing craft down on the White House lawn some morning, ask for worldwide communications facilities, and give your information and advice to the whole world at once?' asked Fry. 'Such a simple solution is only wishful thinking on your part,' replied Alan. 'If you think a little, you will see that there are many reasons . . . why such a course would not be successful.'

> . . . If we were to appear as members of a superior race, coming from 'above' to lead the people of your world, our arrival would seriously disrupt the ego balance of your society. Tens of millions of your people, in their desperate need to avoid being demoted to second place in the universe, would go to any lengths to disprove, or simply deny, our existence. If we took steps to force the acceptance of our reality upon their consciousness, about 30 per cent of the people would insist upon considering us as Gods, and would attempt to place upon us all responsibility for their own welfare. This is a responsibility we would not be permitted to assume, even if we were able to discharge it . . . Most of the remaining 70 per cent would adopt the belief that we were planning to enslave their world, and many would begin to seek means to destroy us. If any great and lasting good is to come from our efforts, they must be led by your own people, or at least by those who are accepted as such . . .

INFILTRATION

To travel about the Earth, Alan explained, he would of course require a passport (!) – unobtainable without a birth certificate. 'Since my origin was actually extraterrestrial, there is no legal way in which I can obtain either a birth certificate or a passport, yet I must have both,' he said. 'It was therefore necessary to find a County Registrar who could understand the need for my being here, and be willing to assist, even at some risk to himself . . . We will arrange for you to meet him, and you must become well acquainted since it will be up to you to conduct the negotiations.' He continued:

> We have made a careful analysis of the steps to be taken so that I may move

easily, and unnoticed, among your people . . . I must have a profession, or at least a gainful occupation, preferably one which is generally known to and accepted by the public, but which is normally conducted in private . . . The ideal occupation would be that of a purchasing agent in an international trading concern. Such a position would furnish a means of livelihood, a good background cover, and an excellent excuse to visit other countries whenever it might become necessary. It would also provide a non-political contact with most of the governments of your world, since every country, whether friendly or not, has things which it wishes either to buy or to sell.

Another problem would involve opening a bank account. 'It seems that most of your money systems are related to the value of gold,' Alan went on. 'I will therefore arrange to have a few pounds of the metal delivered to you here, so that you can exchange it for your currency and open an account in my name.'

'Not gold,' Fry objected. 'Gold has too many legal strings attached, and anyone who offers it for sale must be able to prove its source. If you happen to have some small ingots of platinum handy, they will do nicely. The demand for platinum somewhat exceeds the supply, so that it is not difficult to sell, and its present value is several times that of gold.'

'Very well,' agreed Alan, 'platinum it will be, although it seems strange to think of it as having so much value among your people . . . While platinum is an excellent substance for the plating of surfaces that will be exposed to corrosion, and most of our space ships are plated with it, it has few other uses in our technology.'

THE SURVIVORS

During this same lengthy communication in April 1954, Alan described what had happened to the survivors of the devastating conflict that had destroyed an entire civilization tens of thousands of years ago. On a high plateau in what is now Tibet, six of the surviving aerial craft were landed by their crews, to determine the fate of the survivors:

It was suggested that an attempt be made to reach another planet. The aerial craft then in use were capable of traveling in space, and had frequently been used to reach elevations of hundreds of miles, but no attempt had yet been made to leap the vast gulf between the planets, and the crew members were far from certain that such an attempt could succeed.

The planet which you now know as Mars was then in conjunction with the Earth, and preliminary estimates seemed to indicate that the crossing could be made. At that time, the surface conditions of temperature, atmosphere, water, etc. were somewhat better suited for human survival than the conditions your astronomers report to exist today. A vote was taken, and the members of the crews of four of the craft elected to take the huge gamble in the hope of preserving, thereby, at least a portion of the culture of the race. The remaining crews decided to remain on Earth. They believed that, because of the elevation of the plateau on which they were gathered, and the relatively low level of radiation at that point, they could continue to live in this area without suffering complete physical or mental degeneration in themselves or their descendants.

I can see the question forming in your mind, so I will explain that this race had achieved perfect equality of the sexes, and both were equally represented in this council, and in the crews of the ships. Of the four craft which essayed the great leap, three arrived safely at their destination. There is no record in our history as to the fate of the fourth. For many generations the grim struggle for survival demanded the entire time and energy of the people . . . As the battle for survival against the harsh environment was gradually won, the development of the material science resumed its normal pattern, and technology spurted ahead. With the lessons of the past constantly before the people, however, the material values were carefully maintained in their proper relationship to the social and spiritual values.

'The greatest need of your race, your civilization and your society today,' concluded Alan, 'is simple, basic understanding between man and his fellow man, between nation and nation, and between all men and that greater power and intelligence that pervades and controls all nature. Understanding is the key to survival for your race . . .'[1]

DISCUSSION

Was Dan Fry really taken for a ride – or was he taking us for a ride? I have dwelled at length on his story for two reasons: first, because I believe it to be fundamental to the understanding of certain extraterrestrial encounters; secondly, because I believe the story to be essentially true. That is *not* to say that I accept it unreservedly. There are a number of Fry's claims that require critical commentary.

I first met Fry in the early 1970s during a tour of the United States as a violinist with the London Symphony Orchestra. Then, in August 1976, he invited my friend and co-researcher Louise (Lou) Zinsstag and me to stay with him and his wife Florence at their home in Tonopah, Arizona. I recall that the temperature at the time went up as high as 113 degrees Fahrenheit, though, mercifully, the dry desert air made it feel less so. Dan and Florence were very kind hosts and gave us a lot of their time.

During the stay, which lasted nearly a week, I asked Dan many questions about his claims. In the first place, I wanted to know why he is listed, in most of his publications, as 'Dr Daniel Fry', holding a 'Ph.D. from St Andrews College of London, England'. There is no such college in London; furthermore, the ludicrous 'dignity degree' of 'Doctor of Philosophy (Cosmism)' was conferred on him in April 1960 by the 'St Andrews Ecumenical University Intercollegiate', as was evident when I saw the framed 'degree' in his home. Enquiries by Philip Klass, a vocal sceptic, revealed that the college was 'a sort of correspondence school', where anyone could apply for a Ph.D. by submitting a thesis and paying a modest fee.[2]

'I don't think that you could buy a degree there,' Dan replied. 'I certainly never paid anything for this one, neither was I ever asked to. It was given for the material in the first edition of my book, *Steps to the Stars*.'

'But Dan,' I argued, 'a Ph.D. is normally associated with a recognized university. You can surely sympathize with scientists when they become suspicious of a man who is trying to make his Ph.D. look like an accredited doctorate?'

'I've never attempted to make it look like that, it's been others who have done that,' he answered, somewhat defensively. 'It doesn't make any difference to me.'

It may not have made any difference to Fry, but by allowing the phoney Ph.D. to appear on the jackets of his books and publications, as well as on his stationery, it certainly made a huge difference to the degree of acceptance of his claims by the scientific community. Interestingly, though, *Steps to the Stars*[3] was taken seriously by at least one scientist. A letter to Fry from Parry Moon, Professor of Electrical Engineering at the Massachusetts Institute of Technology (a copy of which is in my possession) endorses the book thus:

> Your book, *Steps to the Stars*, was called to my attention by Alexander Mebane [who] had written a caustic criticism [and] I'm afraid I offended him by disagreeing completely with his analysis . . . *Steps to the Stars* seemed to me an admirable presentation for the layman. But, as in any popular exposition, some of the ideas were necessarily vague. A more rigorous treatment might be of great value to scientists . . .[4]

Another of my questions dealt with the fact that, in all editions of his first book, *The White Sands Incident* (first published in 1954),[5] the date of the incident is given as 4 July 1950. Yet 10 years after publication he admitted that the incident actually occurred on 4 July 1949.

'I had to change it,' he explained. 'In the last edition, I told the publisher to change it, but he decided this might not be good for publicity, and he kept it to 1950 after I'd said to change it, because there was now no need to hide the fact that it was 1949.'

'So there was a reason?' I asked.

'There was a reason,' Dan replied, 'because it turned out to be a year later than he had originally planned – that Alan could be here. He thought it could be done in four years; it actually took five years. Now, had the Pentagon, for example, taken this book seriously, at that time there was a pretty good chance they could have tracked him down. There had to be an escape mechanism. The fact is that on the evening of the Fourth of July 1950 I was not at White Sands – I didn't arrive there until later in July. And everyone in White Sands Proving Grounds and in Aerojet knows that.'

Fry held several security clearances during his time at White Sands, and when *The White Sands Incident* was published, he immediately took the first copy to the security section at Aerojet. 'I said, "Read this over, and if you find any violation of security, speak now." They read it over, and said there was no violation. And that copy was put into the Aerojet Technical Library – under non-fiction. It happened that the President of Aerojet served a term as Secretary of the Navy, and during that time was flying to Japan when he was supposedly buzzed by a UFO that circled his plane a number of times, in broad daylight and at close range. So this individual knew perfectly well that these things were not just swamp gas.'

Others at White Sands were equally convinced that UFOs were a reality. Commander Robert McLaughlin confirmed that on 21 April 1949 (the year of Fry's initial encounter) his team of US Navy scientists tracking a Skyhook balloon at the base observed an unusual silvery object. With the aid of theodolites and a stopwatch, the scientists were able to ascertain that the UFO was at an altitude of 56 miles, was 40 feet long, 100 feet wide, and that its speed when first seen was seven miles per second. 'I am convinced,' said McLaughlin, 'that it was a flying saucer, and further, that these discs are spaceships from another planet, operated by animate, intelligent beings.' On another occasion, he reported, two small discs, tracked from five observation posts at White Sands, were seen to pace an Army high-altitude rocket. After circling it briefly, the discs shot off at high speed.[6] McLaughlin is also quoted as saying: 'Many times I have seen flying discs following and overtaking missiles in flight at the experimental base at White Sands, New Mexico, where, as is known, the first American atom bomb was tried out.'[7]

Regarding the initial contact, Lou had always been incredulous about the exceptional courage shown by Fry. 'How could you just step into a dark contraption on a dark night,' she asked him, 'with nobody showing, and just a voice?'

He laughed. 'I did not have *all* that much courage; I mean, I tell you, my knees were knocking pretty well!'

'Well, they told him it was a sampling device,' added Florence, 'and he said after he got inside he wasn't sure that *he* might not be the sample!'

I have always been dubious about the authenticity of Fry's 16mm films of UFOs (copies of which are in my possession), particularly an object he said he saw in Oregon in May 1964, which to me looks like a couple of lampshades or similarly shaped devices fixed together and suspended with fine twine. He went into some detail as to the circumstances of the filming, and claimed that some frames show the limb of a cloud coming in front of the saucer. I remain unconvinced; the movement of the craft gives every indication of being a suspended fake. Perhaps I am wrong. But does this prove that Fry was lying about all his previous experiences? I think not. Most probably, he thought that a few fabricated movie films of 'saucers' would bolster his unprovable claims. I have come across a number of contactees who have done just that; Eduard (Billy) Meier being one.[8]

That aside, it seems to me that Fry did take at least one genuine photograph of a UFO (see plate section). The incident took place on the afternoon of 18 September 1954, as he was driving home from work on Garvey Boulevard, near Baldwin Park, California. He had no camera with him so, having just passed a drugstore, he made a U-turn and hastily purchased a Brownie box-camera and film. The UFO was photographed and within minutes the film given to the same store for processing.[9]

What is perhaps most striking about Dan Fry's account is that part which concerns the technology of flying saucer energy and propulsion. Regardless of his credentials, not only is Fry's description sophisticated because of his own technical career, but it is inherently more scientifically advanced and theoretically pristine than was anywhere available or studied in 1949–50. The kind of even now hypothetical physics and engineering that Fry says he was told about by Alan only began to be seriously examined in the late 1980s and in this decade by a few prescient academics and far-reaching experimental engineers (for example, Dr Hal Puthoff), writing in peer-reviewed and critically edited scientific journals that are widely read and highly respected repositories of leading scientific thought, literally on the edge of the twenty-first century.

Fry claimed to have had in-person meetings with Alan at about five-year intervals, but was reluctant to provide me with a description of the man. Concerning the success of his masquerade, Alan told Fry at their initial meeting, 'My first real test will come when I walk down the street of any of your larger cities, and if anyone – *anyone* – turns their head to look at me, then it would mean that I had failed somewhere.' Dan also remarked that the life expectancy of Alan's race was about two and a half times that of ours, and that two other individuals would be taking Alan's place eventually, in an attempt to keep a lid on the possibility of a nuclear holocaust.

Finally, Fry discussed the confusing variety of alien species with which we are confronted. 'Probably most of the confusion in the field,' he commented, 'has been caused by the need of many people to force all of the UFO phenomenon into one hole. Each individual case is an individual event that should be judged on its own merits. To think that all beings visiting this planet come from the same place, with the same purpose, and in the same type of craft, is like standing on the corner of 42nd Street and Broadway, watching all the people go by and assuming they all come from the same place, with the same purpose, and the same habits. We live in a galaxy that teems with life and intelligence in every direction – once we get our egos under control enough to admit the fact. We're just one of a series of evolutionary products – and by no means the leading one . . .'[10]

NOTES

1 Fry, Daniel W., 'The White Sands Incident' (revised edition), in *To Men of Earth*, Merlin Publishing Co., Merlin, Oregon, 1973.

2 Klass, Philip J., *UFOs Explained*, Random House, New York, 1974, pp. 248–9.

3 Fry, Daniel W., *Steps to the Stars*, CSA, Lakemont, Georgia, 1965.

4 Letter to Daniel W. Fry from Parry Moon, Professor of Electrical Engineering, Massachusetts Institute of Technology, 9 June 1958.

5 Fry, Daniel W., *The White Sands Incident*, New Age, Los Angeles, 1954.

6 McLaughlin, Robert B., 'How Scientists Tracked a Flying Saucer', *True*, March 1950.

7 *Flying Saucer Review*, vol. 5, no. 5, September–October 1959, p. 10.

8 See *Spaceships of the Pleiades: The Billy Meier Story* by Kal K. Korff (Prometheus Books, Amherst, New York, 1995) for an exposé of Meier's photographs and film. When I first spoke to Meier in 1965 (having heard about his earlier extraterrestrial encounters when I was in India the previous year), he seemed sincere, but on meeting him at his home in Switzerland in 1977 and examining the evidence, including film and photographs, I became deeply suspicious. For the record, I must also add that, in his interview with me published in *Light Years: An Investigation into the Extraterrestrial Experiences of Eduard Meier* (Atlantic Monthly Press, New York, 1987), author Gary Kinder excluded my negative conclusions about Meier's claims.

9 Norkin, Israel, *Saucer Diary*, Pageant Press, New York, 1957, p. 26.

10 Interviews with the author, Tonopah, Arizona, 24–27 August 1976.

Chapter 5

Proliferating Encounters

Bruno Facchini, a 40-year-old industrial worker, lived on the outskirts of Abbiate Guazzone, Italy, a short distance from the highway leading to Milan. On the evening of 24 April 1950, a violent storm struck the district of Varese in northern Italy. Just before 22.00 the rain stopped, and Facchini decided to take a breath of air outside his house. He was about to go inside when his attention was drawn to strange flashes in the sky a few hundred metres away. Because a high-voltage power line passed right over the spot and a pylon with electrical equipment stood in front of his house, Facchini became concerned that the storm might have damaged these structures.

Proceeding cautiously to avoid stepping on any 'live' cables, Facchini arrived at the spot. Seeing nothing amiss, he was on the point of returning to his house when he saw the strange flashing again. 'This time I could see that it was a little further away from where I stood,' he told investigator Antonio Giudici, 'so I decided to go closer.'

> When I did get closer, I caught sight of an enormous black shadow, almost round in shape (it looked like a ball with the top part flattened). In the middle of it I could see a little ladder, and from the top of the ladder was coming a greenish light. I was now able to have a close view of the source of the flashing; that is, I saw quite clearly an individual who, from the top of a pneumatic lift (of the type made with a base, an extensible shaft, and a platform on top), seemed to be standing and doing a welding job. I could see quite clearly that the individual . . . was wearing a diving suit and mask.
>
> My curiosity now aroused, I stepped closer, and now also saw two other individuals, likewise in diving suits and masks, moving about very slowly around the machine, which caused me to think that the suits they were wearing must be very heavy for them. The machine, which was of a dark colour, showed metallic reflections when lit up by the flashes coming from the welder.

Since Milan's La Malpensa airport, as well as the military airfields at Vergiate and Venegono, were not far away, Facchini presumed that an aircraft had made an emergency landing. 'I told the men that I lived close by and asked them if they needed any help,' he said. 'The only reply I got were some incomprehensible guttural sounds.'

The crew was wearing what looked like dark-grey, presumably heavy diving suits, with grey masks over their heads. A tube appeared to hang down from the level of the mouth, with an opening at the end, as if, Facchini reasoned, it could be joined to another tube or cylinder. During brief bursts of light from the welding, the skin of the crew looked to be light in colour. They were about 1.70 metres in height.

I tried to guess what their intentions were, and I got the impression that they wanted to invite me to go up into the machine. Then I heard a noise like the sound of a gigantic beehive, or perhaps it might be better to say like a big dynamo, and I saw, inside, another ladder going up, and all around – on the walls – tubes, cylinders and gauges. In that precise moment I realized that it couldn't be an aircraft, and I was seized by a sensation of panic and fled.

But after I had run a few paces I turned round, and saw one of the pilots grab a sort of camera that he was carrying round his neck and shoot a beam of light at me. I carried on running, and simultaneously I had the impression that I had been struck by a blunt instrument or, to put it better, by a powerful jet of compressed air, and I fell to the ground, landing, for further measure, right on top of one of the boundary stones marking the edges of the fields.

Suffering pain from his bruises, Facchini continued to watch the strange aircraft and its crew. 'It seemed as though they were no longer interested in me,' he said. 'I was sure that they did not desire to do me any harm.' Then it looked as if the crew was preparing to leave.

The individual who had been welding had now come down (the lift on which he was standing had in fact descended, its tubes re-entering) and the two others who had remained on the ground picked up the lift, now reduced in size, put it into a small box and stowed it inside the machine, the ladder was drawn in, and the door closed. Everything became dark. The noise like a beehive continued. Then, all of a sudden, it grew louder, and more powerful, and the machine rose at fantastic speed and vanished into the darkness.

All became silent. Facchini stood there, his eyes glued to the sky, but there was nothing to be seen. He returned home quietly and spent a sleepless night.

LANDING TRACES

The following day, Facchini returned to the landing site to look for his cigarette case which he had lost during the encounter. He discovered faint marks on the ground, consisting of four round impressions about one metre in diameter, set in a square, about six metres apart. He also noticed burnt grass and a few pieces of metal which he picked up, presuming them to be residue from the welding. Facchini reported the incident to police headquarters in Varese, and investigations were carried out at the landing site.

'I had an analysis made of the metal, which turned out to be an "anti-friction metal",' Facchini explained to Antonio Giudici. 'It was a shiny metal, with a granulous surface'[1] (see plate section). As is often the case with metallic samples found at UFO landing sites, nothing abnormal showed up in analysis. In its report, the Experimental Institute for Light Metals stated that:

the sample received consisted of three small metal fragments of a yellowish-white colour and with a total weight of 1.64 grams. The percential results of the chemical analyses made are as follows:

Copper	74.33%
Tin	19.38%
Lead	4.92%
Antimony	0.52%
Zinc	0.33%
Nickel	0.08%
Iron	0.02%

plus minimal traces of silver, aluminium, and magnesium. The fragments under consideration are thus of a 'lead bronze', with a high content of tin. The micrographical structure seems perfectly normal for a bronze of the type in question, in a cast state. The presence of no rare or abnormal element for an alloy of this type was detected. It is very probable that the fragments presented to us for examination come from the packing bed of a bearing that has had very heavy wear.[2]

It is unclear precisely what is expected when physical tests are taken of objects believed to have come from a craft of extraterrestrial origin. If they were not composed of material and elements that we otherwise know about from hundreds of years of scientific enquiry, then how would we be able to understand them? We have good reason to believe that the 10 metals listed above are found throughout the universe and are desirable construction materials for vehicles, regardless of where found and fabricated in the universe.

TRAUMA

A few days after the experience, Facchini began to have pains in the part of his back where the beam of light had struck him, and the area gradually turned black. The pain lasted for over a month. But it was the psychological trauma that persisted. 'What is more important is the fact that I never got over the shock that I suffered,' he said. 'Even today, years after, from time to time I feel hot flushes on the face, without any signs of fever.'[3]

In an interview in his home in 1981, Facchini – then 71 years old – discussed some further details of the 1950 incident and its aftermath. 'There seems to be no discrepancy in his fresh account as given to me,' reported investigator Ezio Bernardini. 'Talking of the appearance of the crew of the UFO, he described to me vividly his amazement when, on television years later, he saw the American astronauts walking about on the Moon: "They looked just the same!" . . . One thing on which he was adamant was that they were not "little men". They were of our size and build [and were] just like us, and they could pass anywhere here as men of our Earth.'

Over the past 30 years, said Facchini, many people, including engineers and technicians, had visited him to learn more about his experience. Some military and civilian experts told him that attempts had been made to construct similar craft here, but that for various reasons all had been failures. 'You must realize that that machine, that UFO, had not only all those tubes everywhere inside and outside, but also two big holes,' Facchini expounded, 'and the technical experts have explained to me that, by expelling and compressing the air in the tubes, the UFO could move laterally and, sending the air out through the two big holes, the machine could go up or down.'[4]

Here we have a technical account not as aerodynamically sound nor as advanced as others reviewed in this book. However, just as several propulsive technologies co-exist in various states of sophistication in conventional use by mankind variously, so might unconventional technologies co-exist on a broader planetary basis than just our own.

Among Facchini's visitors was a high-ranking Italian Navy officer who, after hearing his story, said: 'You are a lucky man! How much I'd have given to have been in your place and to have observed that marvel of technology!'

'What, *me* a lucky man?' said Facchini, bitterly. 'I'd have liked to pass on to *him* all that I've had to put up with!' He went on to describe the tremendous upheaval in his life caused by the 1950 experience; the long-drawn-out mental trauma inflicted upon him by the authorities with their endless interrogations, by the hordes of curiosity-seekers, by the endless arguments with journalists, and by all the ridicule he had encountered. 'If I'd have known all that was coming, I'd have kept jolly quiet about it . . .'[5]

DEVELOPMENTS IN ARGENTINA

Leonard Stringfield, the former United States Air Force intelligence officer who later specialized, as a civilian, in cases involving crashed unidentified aerial craft, was given details of an extraordinary incident that took place in a remote area of Bahía Blanca, Argentina, on 10 May

1950. The witness, Dr Enrique Caretenuto Botta, Italian by birth, was an ex-Second World War pilot, an aeronautical engineer, and, at the time of the incident, an architectural engineer with a Venezuelan real-estate company engaged in a construction project on the pampas. Dr Botta first related his story to the Venezuelan researcher, Horacio Gonzales who, in turn, put Stringfield in contact with Botta. In the following account, I have incorporated some of the information supplied to the late UFO researcher Coral Lorenzen, of the former Aerial Phenomena Research Organization (APRO), by Gonzales, who was APRO's representative in Caracas.

Dr Botta was driving along the highway about 75 miles from his hotel (at a point he gave as 64° Longitude West Greenwich; 37° 45′ South Latitude, 800 feet above sea level) when his attention was drawn to a metallic disc-shaped object resting on the ground off the highway, with a flashing light on top. He stopped to investigate. After a few moments waiting to see if anything would happen, he approached the object.

'The object was resting on the ground in an inclined position,' Dr Botta wrote to Stringfield in October 1955. 'The disc was 32 feet in diameter, the surface was slippery and brilliant. The height was about 13 feet, the tower [dome?] with windows was six feet high . . .' Dr Botta decided to enter the vehicle through a small hatch in the side.

> Three little men were seated in soft armchairs. They were dead. One of the three, the pilot (I believe), was seated in the center of the tower. In front of him was a large panel with bright instruments. His hands were resting on two levers. They were about four feet in height. In appearance they were human, equal to ourselves with eyes, nose and mouth. The color of their hair was gray-chestnut, cut short. Their skin was bronze, their faces were dark [or charred]. They were dressed in overalls of a gray-lead color . . .[6]

According to the information supplied by Dr Botta to Gonzales, the pilot sat on what appeared to be a 'control chair' while the other two 'were lying on lounges along the curved wall of the ship . . . All three were dressed in brown, tight-fitting overalls that exposed only the hands and the face, and their feet were encased in some kind of boots . . . their skin [was] a tobacco-brown, their eyes light colored.'[7]

Dr Botta's report to Stringfield continues:

> I touched the bodies which were rigid. In the tower there was a smell [like] ozone and garlic. In the roof of the cabin there was an intermittent small light of orange-whitish color . . .
>
> There were no cables, no pipes, only the panel of the controls. Above the panel there was a small [glass-like] sphere with a circle. To the right of the pilot there was an apparatus similar to a TV screen. I remained five minutes in the tower but the absence of the [presumed] fourth person impressed me

so much that I went out of the machine very stunned.

Dr Botta headed straight back to his hotel and there related his adventure to two colleagues. Arming themselves with revolvers and taking a camera, the engineers at first considered going to the site immediately, but as night was approaching they decided to wait until morning. On reaching the spot then, all that could be seen was a pile of ashes. Dr Botta took a photograph of the ash and one of the group scooped up some of it. His hand turned purple, the colour remaining indelible for several days.

As the engineers wandered around the site looking for more evidence, one of them glanced up and saw three objects: one cigar-shaped, high up, and the others smaller and disc-shaped. One of the discs, about 10 metres in diameter, was hovering above the group at an estimated altitude of 600 metres. In haste, Dr Botta managed to take five photographs, only two of which showed the objects with sufficient clarity. The two discs then shot upwards, apparently merged with the cigar, which flew for a short distance, turned 'blood red', made an 80-degree turn, and disappeared in seconds.

Dr Botta said that although the craft had a metallic appearance, it felt (very untypically) 'resilient, like rubber'. Also, there were holes or vents in the floor. 'For weeks he suffered from a fever which no doctor in the area could diagnose, and his skin was covered with blisters,' reported Gonzales. 'He had entered the disc with dark green eyeglasses (used by pilots) and the outline of the glasses was marked around both eyes. A doctor tested him for radiation but could not find any traces . . . Because of the character of the man and his professional standing, it is difficult to believe the story is a hoax.'[8]

Dr Botta's observation that the faces of the aliens were 'dark or charred' is interesting. During a conversation with Dr Rolf Alexander at Mexico City airport in 1951, General George Marshall, US Army Chief of Staff in the Second World War and later US Secretary of State, allegedly revealed that American authorities had recovered alien craft and bodies. On three (unspecified) occasions, said Marshall, there had been landings which had proved disastrous for the occupants: breathing the heavily oxygenated atmosphere of Earth supposedly had incinerated the visitors from within.[9] (If General Marshall did say this, his remark is technically correct: what we call 'burning' or incineration is a process of rapid, often destructive oxidation.)

UNIDENTIFIED SUBMERGIBLE OBJECTS

One night in late June 1950, Romero Ernesto Suarez was walking along the coastal road between Rio Grande and San Sebastian in Tierra del Fuego, Argentina, when he heard an unusual noise of turbulence in the sea. Because there was no wind or storm that could have given rise to such activity, Suarez became alarmed. Peering into the darkness in the direction of the noise, he saw suddenly a huge, luminous, oval-shaped object emerge from the sea about 500 metres from the shore. The craft ascended vertically to a certain altitude, made an abrupt 90-degree turn, then disappeared rapidly towards the northwest. Fifteen days later, again at night, Suarez witnessed a similar occurrence, this time between Rio Gallegos and Santa Cruz. Four small, luminous domed discs emerged vertically from the sea in perfect formation, levelled, then flew up the coastline before turning in the direction of the Cordillera de los Andes.[10]

Over the years, many sightings of what I call unidentified submergible objects (USOs) have been reported from the coast of Argentina and elsewhere, examples of which appear in subsequent chapters. Lieutenant Colonel Philip Corso, who served on President Eisenhower's National Security Council staff and who once headed the US Army's Office of Research and Development's Foreign Technology desk at the Pentagon, confirms that USOs were of considerable concern to military authorities:

> Unidentified Submerged Objects [were] a worry in naval circles, particularly as war planners advanced strategies for protracted submarine warfare in the event of a first strike. Whatever was flying circles around our jets since the 1950s, evading radar at our top-secret missile bases . . . could plunge right into the ocean . . . and surface halfway around the world without so much as leaving an underwater signature we could pick up. Were these USOs building bases at the bottom of the oceanic basins beyond the dive capacity of our best submarines . . .?[11]

A BRIEF REPAIR STOP IN FRANCE

Another 'repair' case from 1950 is alleged to have taken place on the night of 23 (or 24) July, in Guyancourt, 20 kilometres from Paris, France. At about 23.00, Claude Blondeau, the proprietor of a small café, was taking some air before retiring to bed when he heard a slight noise 'like the wind'. Turning, he noticed two greyish discs hovering just above the ground about 100 metres away.

Each disc had a row of rectangular 'portholes' that encircled the circumference. As Blondeau watched, a thick oval hatch opened in the bottom part of each disc. From each hatch emerged a 'man' of about 1.7 metres in height, dressed in dark-blue or brown 'flying suits'. They wore no headgear and – as in the Facchini case – appeared quite like humans. Joining each other at one of the two discs, the men began to replace one of a number of 'plates' located on the underside of each disc. (A remarkably similar operation was also observed in the USA in 1964 – see Chapter 13). This replacement of a presumably defective unit was carried out with bare hands and without the aid of any tools.

Although apprehensive, Blondeau approached the two pilots and asked if they had had to make a forced landing. 'Yes, but not for long,' one of them replied, in rather halting French. Through the open hatch, Blondeau had a brief view of the inside of the craft. A 'formidable' light filled the interior of the circular cabin, in the centre of which could be seen a chair, similar to a dentist's chair, covered in red material. In front of the chair was something that looked like a radio transmitter with a number of buttons, and on a pillar was what appeared to be a large oval 'steering wheel' or control column with projecting handles at opposite sides. Other apparatus was visible on 'blocks' or 'panels' around the 'control chair'.

Blondeau asked some questions about the purpose of the buttons on the control panels. 'Energy!' came the curt response. Without further comment, the men re-entered their respective discs, took their places and c'osed the hatches. The portholes became luminous and within seconds the two craft tilted up on end and accelerated upwards at very high speed, making the same wind-like noise as when they had first appeared. The entire incident lasted no more than two minutes.[12]

NORTH ATLANTIC ENCOUNTER

In his book *Flying Saucers: Top Secret*, Major Donald Keyhoe described an important sighting made 'in 1956' by the crew of a US Navy aircraft en route from Iceland to Newfoundland. Keyhoe's source for the report was Captain James Taylor, a retired US Navy officer, whose name had been given to him by Admiral Delmer S. Fahrney, a friend of the Major. Taylor learned of the incident from a Navy pilot who was one of the primary witnesses, a lieutenant at the time, identified in the book by the pseudonym, George Benton.[13] Years later, members of Keyhoe's National Investigations Committee on Aerial Phenomena (NICAP) conducted an interview with Commander 'Benton' to glean further details. It was the first time he had confided his experience to anyone other than military intelligence people and to his friends.

The incident actually occurred in the early morning of 8 February 1951 (Keyhoe gave the date in his book as 1956 perhaps to protect identities, perhaps by mistake). Lieutenant Benton was piloting an R5D four-engine transport aircraft—a Navy version of the famous Douglas DC-4 airliner—on a flight from England to the United States, via Iceland and Newfoundland. The weather was excellent, the visibility good. Seated next to Benton in the co-pilot's seat was the plane's commander, Lieutenant Commander F.K. (identified as Peter Mooney in Keyhoe's book). Back in the cabin, asleep, were two extra crews, one a relief crew for Benton's men and the other on board as passengers. The aircraft was flying at over 200 knots ground speed at an altitude of 10,000 feet. About three and a half hours out of Iceland, the plane, on automatic pilot, passed over a weather ship off the coast of Greenland. The ship reported conditions normal.[14] Then it happened. Benton's report follows:

Lt. Cmdr. F.K. and myself were on constant watch for other aircraft. I observed a yellow glow in the distance about 30 to 35 miles away, at about the 1 o'clock position and below the horizon. My impression was that there was a small city ahead, because it was the same glow you get from a group of lights on the surface before you get close enough to pick them out individually.

Knowing that we had passed the tip of Greenland, my first thought was that we were behind schedule and had drifted north, but remembering that we had passed over the weather ship, I knew this was not the case. I called F.K.'s attention to the glow and asked him what he thought it was. He said that it looked like we were approaching land. I asked our navigator to check his navigation. He did, and replied that we were on flight plan and course.

The lights were farther away than we thought because it took us from 8 to 10 minutes to get close enough to where the lights had a pattern . . . about 15 or 18 miles away. At that time, due to the circular pattern of lights, I got the impression that possibly two ships were tied up together and that lights were strung between them for either transferring cargo from one to the other, or that one was in some kind of trouble. I asked the navigator to check his ship plot. He replied that there were no ships plotted in this area and that we were not close to the shipping lanes anyway. The radioman also went on the air to the weather ship, which verified that there were no ships in the area.

Since it was time for Lt. J.'s crew to relieve us, I had the plane captain awaken them. When Lt. J. and Lt. M. came up forward, I pointed the lights out to them. Their only comment was that it had to be a ship because it was on the water and we were overtaking it fast. At this time, we were 5 to 7 miles away; it was about 30 degrees to our right, and we had to look down at about a 45-degree angle. The lights had a definite circular pattern and were bright white.

Collision Course

'Suddenly, the lights went out,' continued Benton. 'There appeared a yellow halo on the water. It turned to an orange, to a fiery red, and then started movement toward us at a fantastic speed, turning to a blueish red around the perimeter.'

Due to its high speed, its direction of travel, and its size, it looked as though we were going to be engulfed. I quickly disengaged the automatic-pilot and stood by to push the nose of the plane over, in hopes that we could pass under it due to the angle it was ascending. The relief crew was standing behind us; everyone began ducking, and a few heads were hit on objects.

It stopped its movement toward us and began moving along with us about 45 degrees off the bow to the right, about 100 feet or so below us and about

200 to 300 feet in front of us. It was not in a level position; it was tilted about 25 degrees. It stayed in this position for a minute or so. It appeared to be from 200 to 300 feet in diameter, translucent or metallic, shaped like a saucer, a purple-red fiery ring around the perimeter and a frosted white glow around the entire object The purple-red glow around the perimeter was the same type of glow you get around the commutator of an auto-generator when you observe it at night.

When the object moved away from us, it made no turns, as though it was backing up about 170 degrees from the direction it approached us, and was still tilted. It was only a few seconds before it was out of sight. [Speed estimated in excess of 1,500 m.p.h.] All of our cameras were within reach, but no one was calm enough to think about taking a picture. Most of us were wondering what it was. Our impression was that it was a controlled craft. It was either hovering over the water or sitting on it, then it detected us and came up to investigate.

MILITARY INTELLIGENCE PREDILECTION

'After Lt. J.'s crew had taken over,' said Benton, 'I proceeded aft and learned that most of the passengers had observed the same thing. Since I was unable to identify the object, I asked Dr M., Commander, US Navy, if he had observed the object. He replied that he had and that he did not look because it was a flying saucer and he did not believe in such things.'

> I immediately returned to the cockpit and informed the crew to keep quiet about what we had observed because it might have been our first sighting of a saucer – during those years, when you mentioned you had such a sighting, you were believed to be crazy. Lt. J. informed me that it was too late because he had called Gander airfield in Newfoundland to see if the object could be tracked by radar.
>
> When we landed at Argentia (Newfoundland), we were met by intelligence officers. The types of questions they asked us were like Henry Ford asking about the Model T. You got the feeling that they were putting words in your mouth. It was obvious that there had been many sightings in the same area, and most of the observers did not let the cat out of the bag openly. When we arrived in the United States, we had to make a full report to Naval Intelligence. I found out a few months later that Gander radar did track the object in excess of 1,800 mph. . .[15]

During a two-hour interrogation by US Air Force intelligence officers, the pilots tried to obtain some information about the flying discs. 'What's behind all this?' asked Lt. Cmdr. F.K. 'Up to now, I believed the Air Force. You people say there aren't any flying saucers. After a scare like that, we've got a right to know what's going on.'

'I'm sorry,' replied the captain in charge of the interrogation, 'I can't answer any questions.'

The completed intelligence reports were flashed to four commanders, with an information copy to the Director of Naval Intelligence (DNI). After the aircraft reached its final destination at Patuxent Naval Air Station, Maryland, the crews were interviewed again, at the request of the Office of Naval Intelligence (ONI).

Five days later, Benton received a phone call from a scientist in a government agency, asking for a meeting to discuss the sighting in more detail. Benton agreed. The following day the scientist turned up, showed his credentials, and listened intently to the pilot's report. He then unlocked a dispatch case and took out some photographs. 'Was it like any of these?' he asked. At the third photo, Benton recognized the same type of craft he and the other crew members had seen. 'That's it,' he said. 'Somebody must know the answers, if you've got photographs of the things.' Like the Air Force captain, the scientist apologized for being unable to discuss the matter, and left.[16]

This is a lucid, compelling account by a senior career military aviator. The operational and technical intelligence collection methods described are squarely in line with ordinary practices of that time and later. Interestingly, the account describes how another aviator became alarmed because he was unable to accept the novelty of flying saucers – as many contemporary aviators still do not.

SECRET US NAVY INVESTIGATIONS

According to information supplied by a navigator who formerly flew many secret strategic reconnaissance missions for the US Navy during the late 1940s and early 1950s, UFOs were the subject of frequent investigations at that time. 'Back in those days,' the source told researchers William Jones and Dr Irena Scott, 'UFOs were considered to be a very important matter. When the areas of the Soviet Union which were assigned to us were clouded over, we were frequently assigned to fly to various locations around the world to help indigenous personnel conduct UFO investigations. Many of these missions were to South America. There was a lot going on there, apparently.'

These assignments were for specific investigations. According to Jones and Scott: 'They would fly into the country, pick up a team of investigators and then fly them into the region where the report or reports were to be taken. He said he didn't learn much about the reports since he spoke neither Spanish nor Portuguese.'

In effect, it seemed that his team acted as 'truck drivers' for groups of indigenous military personnel who did not have access to their own aircraft. The final reports were flown back with them to Turkey [Adana] and then sent back to the Pentagon (apparently not to Project Blue Book) in Washington, DC, through Oslo, Norway. When asked why they were sent via this [unusual] route, he replied, 'For diplomatic purposes.' They were classified

'Secret' before being forwarded. His group did no analysis of the reports they collected and no information about what they were used for ever came back to them.[17]

AERIAL ENCOUNTER NEAR MOUNT KILIMANJARO

Another aerial encounter took place in February 1951, when the crew and passengers aboard a Lockheed Lodestar of East African Airways observed a 'bullet-shaped' flying object near Mount Kilimanjaro, Tanganyika. The object was first seen by the radio operator, Dennis Merrifield, who pointed it out to Captain Jack Bicknell. As Captain Bicknell reported:

> The morning was clear and cloudless, visibility was good and the weather perfect. I timed the Saucer for 17 minutes while the Lodestar kept its course. Twice it rose vertically to a final height of 40,000 feet, then it moved east towards the coast at a terrific speed. There was a large fin-like object attached to the rear, although it wasn't clearly defined. There was no apparent propelling power [and] definitely no vapour trails.

On landing, Bicknell prepared an affidavit and signed it, together with Merrifield and the nine passengers. One passenger, Captain H. B. Fussell, said that through his powerful binoculars the object appeared to be bullet-shaped. 'Its colour was whitish-silver with three vertical black bands down the side,' he reported. 'For 10 minutes it remained stationary, then it suddenly rose vertically by 5,000 feet. Again it became stationary, and then a minute later it rose again and moved laterally away at a great speed . . . the object was definitely metallic.' Another passenger, Charles Vernon, said that the object must have been immense, 'two or three times the size of the largest passenger plane.'[18]

Investigator Waveney Girvan learned that a ciné film of the object had been taken by an American passenger, a Mr Overstreet, who had taken his film to the 'authorities' in the United States. Efforts to locate Overstreet or a copy of his film proved fruitless.[19]

A CONTACT NEAR CAPE TOWN

Bizarre encounters with unknown flying machines and their occupants continued to occur in the early 1950s, though few were reported: it was not until 1953 that stories by 'contactees' began to spread into the public domain. The following case, for example, did not come to light until 1977, when the Spanish journalist Juan José Benítez published his interview with a man who claimed to have established contact with extraterrestrials in South Africa in 1951.

The witness, identified only as 'H.M.' by Benítez, was a British engineer who, in 1977, had been working for twenty years as an engineer specializing in instrumentation, such as development and construction of automatic pilots for aircraft. At the time of the interview, H.M. was employed by one of Spain's leading firms dealing with advanced technology. It was only after a great deal of effort that Benítez was able to persuade the witness to talk about his experience, and even then, on condition that his identity was not revealed.

At the time of the incident – in the early spring of 1951 – H.M. was working for Contactor, a subsidiary of the British Rheostat Company, and he was living with his wife in the small town of Paarl, some 32 kilometres outside Cape Town. One day, after their car had been out of use for some weeks, they found that the battery had run down, so H.M. decided to 'jump-start' the car down a slope near their house and charge the battery by taking a drive around the neighbourhood. He got into the little car and set off in the direction of the Drakensteen mountain, about 10 to 12 kilometres away. 'My idea was a simple one,' said H.M. 'It was just to go as far as a level area up near the top of this mountain and then come back. The run would be more than enough to top up the battery. And so, in fact, at about 11.15 p.m. I arrived on this mountain. The traffic along the road at that time of night was virtually nil. The area where I now was lay at an altitude of about 900 feet above sea-level, and forms a sort of small table-land running right up to the foot of one of the great cliff faces of the mountain.

'There was a moon that night, and I remember how the vast shadow of the mountain fell across a large part of the table-land, so that this area was plunged into what, by contrast, seemed to be an accentuated darkness.' Then began his encounter with the unknown:

I was just about to start back for home when I saw a man waving his arm to me. He indicated that I should pull up . . . I pulled up and asked him what was the matter. He came up to my window and said: 'Have you any water?' I replied that I hadn't, except for what was in the radiator. The man looked upset when I said this, and went on: 'You see, we need water!' I could see how keen he was to get this water, so I suggested that I take him to a stream that crosses beneath the road a little further down the hill. Then the man asked: 'Is it far to this stream?' I replied that it wasn't, that it was, maybe, half a kilometre or so, and that [it] was a mountain stream. I told him that it was very good water too, because it came straight from the mountain above us. At that the man seemed satisfied.

'In what language did the man speak to you?' asked Benítez. 'In English,' replied H.M. 'But he had a rather strange accent . . . In South Africa, as you know, there are all sorts of people apart from the English-and Afrikaans-speaking folk; there are Americans, Germans, Dutch, Indians, French, Malays, Chinese, and so on. And pretty well everybody speaks English, each of them with a different sort of accent according to his nationality. But this man's accent was strange. Anyway, I invited him to get in the car. Which he did. And we set off for the stream . . .

'I asked him if he had any sort of receptacle to hold the water. And he said he had not. "All right," I said. "I've got an oil-can with me which maybe will do." My companion's manner was brief. "That will be all right," he said. So we arrived at the stream, and the two of us set about washing out the can and filling it with water. When the operation was completed, we returned to the car and set off back to the place where I had met him.'

THE CRAFT AND COSMONAUTS

'I pulled up at a certain distance from the foot of the mountain,' continued H.M., 'and the man pointed into the dark area formed by its shadow. "There, please, there!" he said, meaning that I should drive nearer to the rock face. And, as we moved into the shadow and my eyes got used to the darkness, I perceived that there was a strange object there . . . it was about 100 metres or so from the road, and in the zone of shadow cast by the mountain.'

It was quite big. Its diameter may have been between 10 and 15 metres or so. It wasn't very high. Maybe, from the feet up to the top it could have measured say four metres or so. In the under part I could see an opening which was lit up and some steps which, as I was able to ascertain shortly afterwards, led to the interior of the machine [see Fig. 4]. I stood there, dumbfounded.

The man invited me to enter . . . I won't deny that I felt afraid. So I just said nothing, like someone who feels distrust. But the man insisted, and with a friendly gesture invited me to go with him into the machine. And I stepped inside ahead of him. Inside the object, which was completely circular, there were other men. A total of four more, in fact. One of them was lying stretched out. Apparently, so my companion explained to me, they had had a slight

Fig. 4. A sketch of the craft by the British engineer 'H.M.'.

accident, and the recumbent man had got burnt. Then I replied that I would like to get a bit closer so as to be able to see the wounded man. But my companion said no, that I must not move from the spot where I was. So there I stayed, just by the entrance . . . It was a circular room. There were square windows all around it, and under these windows a sort of circular couch going all the way round.

The men were all shorter than I am . . . maybe 1.50 metres or 1.60 metres. In the area where the windows were, the ceiling was somewhat curved. In the centre of the room there were some levers, like the ones used in railway signalling boxes. These levers were set in a small rectangular area and were about one metre in height. The top of these levers ended in a sort of 'fork', like those on the hand-brakes in the older types of cars . . . Maybe there were eight, set in two rows. What I can say is that each lever emerged from the inside of the machine. I could see the rectangular slot quite clearly. And over on the other side of the room I saw a sort of table. But it wasn't a table . . . I thought it might possibly be an instrument-panel of some sort, but I could see no instruments on it. This is my own line of work, so I can assure you that I took a good look at it.

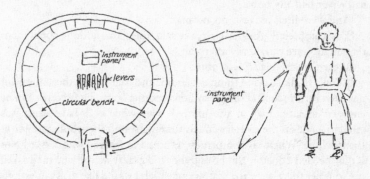

Fig. 5. Sketches made by 'H.M.'.

Another detail which surprised the witness was the lighting – described as 'very white' – the source of which could not be discerned. 'I couldn't see any lights anywhere,' he explained to Benítez. 'It was just as though the light was coming from the walls or the ceiling or from everywhere, all at once.' This interesting feature correlates with descriptions given by a number of contactees over the years.

'And what were the crew doing?' asked Benítez. 'Well,' replied H.M., 'the man who had accompanied me set down the can of water near where the other four were, and then came back to where I was standing.'

The men were all dressed in a sort of beige-coloured laboratory overall which fell to below the knees, fastened with a belt. The rest of the clothing was not 'abnormal' in any way – though perhaps abnormal in its normality! 'They were wearing trousers and shoes,' said H.M. 'I imagine that if these had been different from ours I would have noticed it.'

'What might have been the age of the man who talked to you?' asked Benítez.

'He was a bit older than the others,' replied H.M., 'maybe he might have been 40-plus.'

'And how were their faces?'

'I noticed nothing strange about them. Maybe their foreheads were a bit more pronounced . . .'

'And their hair?'

'Not very long. And the same on all of them. It wasn't black hair – maybe chestnut coloured. As I said, it wasn't strange in any way.'

'Were they muscular in their build?'

'No. Rather on the slim side. Their hands reminded me somewhat of the hands of women.'

'Did you notice if they had beards?'

'No. They had no beards at all. It is curious; it seemed as though they had never had any beard.'

'And were their movements normal?'

'Yes, completely. As I have already said, there was nothing about them that might have caught my attention.'

'Were they talking to each other?'

'I don't think so. One thing is certain, and that is that those four did not even turn to look at me when I entered the craft. They seemed to be engaged in attending to the injured man who, as I have said, was stretched out on the circular couch running round the whole interior of the machine. When my companion returned to where I was, I asked him if they needed a doctor. But he said they did not. It was he who then asked me whether there were any matters on which I would like to ask questions and be given information. And I said yes, naturally.'

PROPULSION AND ORIGIN
'I said that, being an instrumentation engineer, I was puzzled to see no panels or navigation instruments,' continued H.M. 'I also asked him how the machine worked.'

I asked: 'Where are the engines?' To which he replied: 'We don't have any engines.' So I asked: 'Then how do you navigate?' and at this he pointed to the levers and said: 'We have a different system. We nullify gravity. That is how we rise.' I asked him how they overcame gravity, and he replied that they used a very heavy fluid, which circulated in a tube. And with this system they created a magnet . . . That is to say, somewhat as we do with electromagnets, except that they, instead of using electricity, were using this fluid . . . So then I thought of mercury. Meanwhile, the man continued his explanation to me.

Apparently this fluid was subjected to a velocity similar to the velocity of light; that is to say, the velocity of electricity. But, I answered: 'That is impossible inside a tube.'

'No,' he replied, 'it is simple. When the fluid is leaving the tube, it is already entering at the other end. Thus, its relative speed is infinite . . .' So it seems that, on the basis of this system plus a few 'magnets' of a kind which clearly do not exist on our planet, these beings had achieved enormous velocities and were able to conquer gravity . . .

I asked where they came from. He pointed towards the stars and said: 'From there.' I even insisted on wanting to know from which cardinal point in the sky they came, but he simply kept repeating: 'From there.' And then he rapidly changed the subject. It was obvious that he did not want to say any more about that. So, after we had chatted for about 15 or 20 minutes, he pointed in a friendly manner towards the door and invited me to leave – and I did. I went down out of the machine and departed . . .

A DREAM?

The witness estimated that the entire episode had lasted for about 45 minutes. 'And I can assure you,' he told Benítez, 'that they were the strangest minutes of my whole life.' Apparently, he simply left the craft standing there in the dark: there is no mention of his having observed it take off. 'Next day,' he said, 'thinking it had all been a strange dream, I went back to the spot. And there were some very strange marks there. And, on top of that, there was the matter of my can, which we had to carry the water in, and which now was missing . . . Had it been a dream, I would have forgotten about it straight away. But this was something very different from a dream. And I remember it all still with absolute perfection, and in all its details . . .'[20]

ABSURDITIES AND CORRELATIONS

This fascinating story may yield some important clues as to the elusive nature of the UFO phenomenon. There is no reason to believe that the witness fabricated the story. The details are so striking as to be guaranteed to inspire typical disbelief or even ridicule. What hoaxer would have described, for example, aliens dressed in very terrestrial-type coats and trousers, and a flying saucer with two rows of levers 'like those on the hand-brakes in the older types of cars'? Even by 1951 standards, this seems primitive. Surely an alien race sufficiently advanced to travel from another planet would have the wherewithal to locate and obtain such a basic requirement as water, without flagging down a passing motorist? If deception was involved, it is more likely that the aliens were responsible. On the other hand, we usually believe that aliens are omnipotent and omniscient. Why should that be axiomatic?

Despite apparent absurdities, there are interesting correlations with other cases. H.M. said, for example, that the hands of the men reminded him 'of the hands of women'; that their foreheads were 'a bit more pronounced'; and that 'it seemed as though they had never had any beard'. These descriptions – and there are others – closely correlate with those given by other contactees. The corollary could be, of course, that H.M. was familiar with the literature on the subject and had simply incorporated some of its features into his yarn. I reject this hypothesis because it is unlikely that any hoaxer would incorporate so many unbelievable terrestrial elements into the story.

Why should it be incredible that something so fundamental to survival as water would be needed by the craft's occupants? I am reminded here of the celebrated Eagle River, Wisconsin, contactee case of 18 April 1961, investigated by Dr J. Allen Hynek and Major Robert Friend of Project Blue Book. The witness, Joe Simonton, claimed that a flying saucer (see Fig. 6) hovered just above the ground on his property and one of two human-type occupants handed him a sort of jug and indicated they required water. Shortly after Simonton had obliged, he was handed some pancakes which the aliens had been cooking on a flameless grill! Absurd, to be sure, yet there is no evidence to suggest that Simonton made up the story. Moreover, certain features of his description of the aliens correlate well with some other contact stories; for example, he reported that: 'They were about five feet tall and about 120 lbs and looked . . . of Italian descent. Very nice-looking fellows about 25 to 30 years old. Each one was very well built in proportion to their size. [They] had a complexion much finer than any woman I ever saw.'[21]

Regarding the instruments described by H.M., there are several

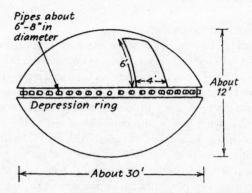

Fig. 6. A sketch of the object encountered by Joe Simonton.
(*FSR Publications*)

accounts of exotic instrumentation, including putatively official accounts, but these are usually associated with operator-occupants whose morphology is unlike that of humans. This raises an interesting point that the contactee experiences tend to be with extraterrestrials whose morphology, dress and mannerisms are closely similar to those of ordinary humans, while the accounts of abductees *usually* involve aliens who, though humanoid, are still rather 'other-worldly' in physique and behaviour.

THE THREE GIANTS

The witness in the following case, identified as 'Rose C.', was 24 years old at the time of her initial encounter, which occurred on 11 April 1952, near the French town of Nîmes (Gard). Living with her father at the time in one of his small outhouses, Rose was awoken in the middle of the night by the growling of their dogs. Concerned, she went outside.

Rose found herself in the presence of four strangers: three very tall men of about 2.30 to 2.40 metres in height and one normal-sized man. The normal man, speaking in perfect French and acting as translator between Rose and the three 'giants', began by explaining that they had come from a faraway world. Apart from their extraordinary height, the giants could have passed for Hindus, claimed Rose. Two looked very athletic while the third, who acted as 'leader', seemed to be rather older. This man had what looked like a 'black half-marble' in the centre of his forehead, and around his neck he wore a strap holding a kind of box with buttons. Rose was shown their craft, which was hovering about a metre off the ground. It was enormous, slate-grey and shaped like a straw hat. She declined an invitation to go with them, explaining that she had an elderly father and a young daughter to look after.

The normal man, a former teacher, said that he himself had been contacted by these beings 20 years earlier (1932), at the age of 25, and, having no ties, had accepted an invitation to live with them. When Rose commented on his youthful appearance, he replied that 'up there time passes much less quickly'. After explaining to Rose that she had nothing to fear, he went on to ask if they could take some books away with them! Obligingly, Rose took the visitors to her father's other outhouse and gave them a copy of Alexandre Dumas' *The Count of Monte Cristo* and some old magazines.

While in the outhouse, the leader gave Rose a demonstration of their ability to levitate and teleport objects. Using the buttons on the small box, he dematerialized a rock, which then reappeared outside without the door having been opened. Furthermore, large rocks, lying at some distance away from the outhouse, were made to float in the air like balloons. It was not just the technological tricks that impressed Rose. The leader also evinced telepathic ability, connected perhaps, she wondered, with the device on his forehead. He seemed also to emanate great wisdom.

The teacher explained to Rose that the extraterrestrials had 'established' Earth for the use of its human inhabitants: this had been rather like a 'penal colony', consisting of banished individuals, from whom humans are descended. He went on to say that 11,357 years ago (9405 BC), because of Man, there had been a cataclysm on Earth. (Many other contactees have been told that Earthmen destroyed their civilization thousands of years ago.)

As to the purpose of their current visits, it was explained that the extraterrestrials were here to take vegetation and soil samples to evaluate the consequences of our atomic explosions. The destructive, senseless behaviour of humans, and disregard not only for our contemporaries but also for future generations on this beautiful planet, was commented on. Asked by Rose why the extraterrestrials did not intervene, the teacher replied that no good had ever come out of such attempts in the past.

The visitors then told Rose that they had to go, owing to the fact that time spent on Earth exhausted the giants. Having warned Rose to stay clear of the craft and to hold the dogs, they boarded their craft and it took off, making a droning sound and creating a strong draught of warm wind.[22] [23]

There is a great deal more to this fascinating story, which was investigated by Charles Gourain and published in France. Gourain was impressed by Rose's sincerity, as was Guy Tarade, who wrote the preface to the book. 'I have known Rose for several years, and I don't hesitate to say that hers is a unique case,' he states, adding that, in his opinion, extraterrestrial encounters never seem to be accidental but are always predetermined. 'It is,' he says, 'a genetic inheritance which motivates the encounters between the extraterrestrials and us . . .'[24]

NOTES

1 Giudici, Antonio, 'The Case of Bruno Facchini', translated by Gordon Creighton, *Flying Saucer Review*, vol. 20, no. 6, November–December 1974, pp. 30–1.
2 'Examination of Some Metallic Fragments Attributed to a Flying Saucer', Report No. I.S.M.L. N.530954/4157, Istituto Sperimentale dei Metalli Leggeri, Novara, 30 September 1953.
3 Giudici, op. cit., p. 31.
4 Bernardini, Ezio, 'A Classic Case from 1950', translated by Gordon

Creighton, *Flying Saucer Review*, vol. 32, no. 4, published June 1987, pp. 12–13.

5 Ibid.

6 Stringfield, Leonard H., *Situation Red, The UFO Siege!*, Doubleday, Garden City, New York, 1977, pp. 80–3.

7 Lorenzen, Coral E., *The Great Flying Saucer Hoax: The UFO Facts and Their Interpretation*, William-Frederick, New York, 1962, pp. 54–6. As Lorenzen reports, Dr Botta's story first appeared in the Caracas daily newspaper *El Universal* on 7 May 1955, which led Horacio Gonzales to interview the witness in depth.

8 Stringfield, op. cit., pp. 83–4.

9 Good, Timothy, *Beyond Top Secret: The Worldwide UFO Security Threat*, Sidgwick & Jackson, London, 1996, p. 511.

10 Sanchez-Ocejo, Dr Virgilio, and Stevens, Lt-Col. Wendelle C., *UFO Contact from Undersea*, UFO Photo Archives, 3224 S. Winona Circle, Tucson, Arizona 85730, 1982, pp. 165–6.

11 Corso, Col. Philip J., with Birnes, William J., *The Day After Roswell*, Pocket Books, New York and London, 1997, p. 54.

12 Guieu, Jimmy, *Les Soucoupes Volantes viennent d'un autre Monde*, Fleuve Noir, Paris, 1954, pp. 230–3. Translated by Lex Mebane and included in an article by Ted Bloecher, *Flying Saucer Review*, vol. 20, no.3, December 1974, pp. 26–7.

13 Keyhoe, Maj. Donald E., *Flying Saucers: Top Secret*, G.P. Putnam's Sons, New York, 1960, pp.15–20.

14 *UFO Investigator*, National Investigations Committee on Aerial Phenomena, September 1970, p. 3.

15 Ibid., October 1970, p. 3.

16 Keyhoe, op. cit., pp. 19–20.

17 Jones, William E., and Scott, Dr Irena, 'US Navy Support of UFO Research', Mid-Ohio Research Associates (MORA), 5837 Karric Square Drive, Box 162, Dublin, Ohio 43017.

18 *Sunday Dispatch*, London, 25 February 1951.

19 Girvan, Waveney, *Flying Saucers and Common Sense*, The Citadel Press, New York, 1956, pp. 80–1.

20 Benítez, Juan José, 'The Ufonaut's Plea for Water', translated from the Spanish by Gordon Creighton, *Flying Saucer Review*, vol. 24, no.2, March–April 1977, pp. 3–6.

21 Clark, Jerome C., *The UFO Encyclopedia*, vol. 3, Omnigraphics, Penebscot Building, Detroit, Michigan 48226, 1996, pp. 168–77.

22 Mesnard, Joël, 'The French Abduction File', translated by Claudia Yapp, *MUFON UFO Journal*, no. 309, January 1994, pp. 5–7.

23 C., Rose, *Rencontre avec les Extraterrestres*, Editions du Rocher, 1979.

24 Ibid.

Chapter 6

The Space People

Nothing is more certain to provoke derision during a conversation about UFOs than the name of George Adamski, whose initial story of his contact with an extraterrestrial in 1952 was the first contactee case to become known to the general public. This fact alone is important. As we have seen, there have been a number of stories by others claiming contacts prior to 1952 but, as far as I am aware, none of these was published widely until the mid-1950s.

Adamski's 'Aryan' space people have never gone down well. Indeed, in some circles they inspired rumours that Hitler was alive and well and living in Patagonia, together with teams of Nazis who were building fleets of flying saucers in preparation for world domination. Adamski's space people were not all fair-skinned, however. As he pointed out during a private meeting in Denmark in 1963: 'People on other planets are coloured too, just like we find here. Orthon himself is dark: not as dark as a Negro, but darker than you and me.'[1] This brings to mind the 'Hindu-looking' spacemen reported by Rose C. and other contactees.

BACKGROUND

George Adamski was born in Poland (of Polish and Egyptian parents) in 1891, and when he was two his parents emigrated to Dunkirk, New York State. In 1913 he joined the 13th US Cavalry Regiment, stationed at Columbus, New Mexico. According to his FBI file, he worked in 1916 as a caretaker and painter at the Yellowstone National Park, then in 1918 entered the National Guard and was stationed at Portland, Oregon. Following discharge from the Army, his various jobs included working at a flour mill and in the concrete business. In 1921 he began lecturing on philosophy in California.[2] Adamski founded the monastery of the 'Royal Order of Tibet' at Laguna Beach in 1934, where he taught 'Universal Laws' and 'Universal Progressive Christianity'. (Though some of Adamski's philosophy is eclectic, much seems to be original.) A number of his talks were broadcast on radio stations in Los Angeles and Long Beach. At this time, Adamski's students conferred on him the title of 'Professor'. In 1940 he moved with his wife and some devotees to Palomar Gardens, on the southern slopes of Mount Palomar, California.

One of his students gave him a telescope – a six-inch Tinsley reflector – through which he took many remarkable photographs of spacecraft.[3]

Adamski also possessed a 15-inch reflector telescope, which was housed under a dome at Palomar Gardens. This gave rise to considerable confusion at times, since the famous Palomar Observatory at Mount Palomar – housing the 200-inch Hale telescope – was situated 11 miles from Adamski's home. It is said by some detractors that, to give credence to his claims, 'Professor' Adamski used to boast that he was employed by the Observatory. It is true that during a meeting with researchers Jim and Coral Lorenzen in the spring of 1951, he claimed full access to the Observatory's telescopes. 'A letter I wrote to the Director of the Palomar Observatory elicited an answer which was most revealing,' wrote Coral Lorenzen. 'Adamski's claims of free access to the Palomar telescopes and his inferred relationship with the Observatory personnel had caused them considerable embarrassment and the added burden of answering correspondence to deny his claims.'[4]

Adamski was the subject of a number of investigations by the FBI, partly as a consequence of allegedly unpatriotic and pro-Communist statements that were reported by various people and also because Adamski himself instigated meetings with the FBI to keep them informed about his experiences; on one occasion, ironically, to inform on a couple of his associates whom he considered might be unpatriotic! Regarding Adamski's alleged claims of having been employed at the Observatory, one FBI document (see Fig. 7) is damning. It states in part:

> [name deleted] at the Palomar Observatory, advised that he had been acquainted with Adamski since 1943, at which time Adamski had called himself 'the Reverend Adamski' and had held Easter services in the valley . . . [name deleted] further advised that Adamski claimed to have worked at the Observatory at Mount Palomar, but stated that Adamski had never been employed at the Observatory.[5]

In his first book, published in 1953, Adamski (presumably aware of the official denial) did state: 'To correct a widespread error let me say here, I am not and have never been associated with the staff of the Observatory. I am friendly with some of the staff members, but I do not work at the Observatory.'[6]

Adamski has been referred to by his numerous detractors as a mere 'hamburger seller'. This is true, to the extent that he did indeed help serve at the Palomar Gardens Café, which was owned by one of his devotees, Alice Wells, though an FBI file mistakenly names him as owner and operator of the café.[7]

ENCOUNTER AT DESERT CENTER

In Chapter 1, I cited Adamski's detailed description of 'Orthon', who he claims stepped out of a flying saucer in the Californian desert, 10 miles from Desert Center, at 12.30 on 20 November 1952. Although the basic story is well known, most readers will be unfamiliar with the details, as published in *Flying Saucers Have Landed*, the best-selling book by Desmond Leslie and Adamski, which has long been out of print.

The contact was partly observed from a distance by six friends of Adamski, all of whom signed affidavits testifying to this event. Adamski, who had already succeeded in taking some remarkable photographs through his telescope of what he called 'scoutcraft' and 'mother ships', had made a number of unsuccessful trips to the Californian desert in 1952, in hope of making contact with the extraterrestrials. On 20 November, he arranged for Alice Wells and Lucy McGinnis to drive him to a certain destination, which turned out to be the contact site near Desert Center. Accompanying them, in a separate car, were Alfred and Betty Bailey and George and Betty Hunt Williamson. They arrived at their destination shortly after 08.00, but it was not until after 12 noon that the adventure began. At this time, a twin-engined aircraft passed low overhead then disappeared in the distance. 'Suddenly and simultaneously we all turned as one,' related Adamski, 'looking again toward the closest mountain ridge where just a few minutes before the first plane had crossed.' And there it was:

> Riding high, and without sound, there was a gigantic cigar-shaped silvery ship, without wings or appendages of any kind. Slowly, almost as if it was drifting, it came in our direction, then seemed to stop, hovering motionless . . . At first glance it looked like a fuselage of a very large ship with the sun's rays reflecting brightly from its unpainted side, at an altitude and angle where wings might not be noticeable.

Excitedly, binoculars were passed round and attempts were made to photograph the craft. Adamski considered unpacking his telescope (to which was attached a camera) from the car, but thought better of the idea, since his hunch was that this was not the place where contact would take place.[8]

On the twenty-seventh anniversary of this event, in 1979, I interviewed Lucy McGinnis, Adamski's former secretary, whom I found to be honest and objective. 'Here came this great big ship that looked like a dirigible,' she confirmed. 'And George said, "Quick, get me up there! I want to go and set the telescope up." So I drove him [and Al Bailey] up to where he said we should go.'

> I kept looking out of the car. And that ship turned and just followed us. And he said, 'Here. Stop!' So I stopped, and he got out, and that dirigible stopped

– quite a ways away. I couldn't very well judge how far away it was. And he set up the telescope. And after he got everything set up, he said, 'Now you go back.'[9]

Adamski remained by himself at this new site (which I have visited), observed by the others from an estimated distance of between half a mile and a mile away. As the car left, the large cigar-shaped object turned its nose and was lost from sight, 'but not before a number of our planes roared overhead in an apparent effort to circle this gigantic stranger,' reported Adamski. Five minutes later, another craft appeared:

. . . my attention was attracted by a flash in the sky and almost instantly a beautiful small craft appeared to be drifting through a saddle between two of the mountain peaks and settling silently into one of the coves about a half a mile from me. It did not lower itself entirely below the crest of the mountain. Only the lower portion settled below the crest, while the upper, or dome section, remained above the crest and in full sight of the rest of my party who were back there watching. Yet it was in such a position that I could see the entire ship.[10]

Meanwhile, the others strained to see the craft. 'It seemed to me like it was a kind of light,' Lucy told me, 'but it was so dim and the sun so bright that I couldn't be sure. Anyway, I saw it come down . . . I looked through the binoculars, but they weren't adjusted to my eyes and I couldn't tell for sure. I could see better without them, because I have very good eyes, and it wasn't such a long distance away. Out there in the desert you can see a long way.'[11]

Without taking adequate time to focus through the ground glass on the back of the old German plate-camera which was attached to his telescope, Adamski began to take several photographic plates. Regrettably, these did not come out well, though some additional shots he took with a hand-held Brownie box-camera came out better, if rather indistinct owing to the distance of the object. As the small craft moved away and disappeared, two military aircraft roared overhead.

THE SPACEMAN

A short while later, Adamski noticed a man standing at the entrance to a ravine between two hills, about a quarter of a mile away. 'He was motioning to me to come to him,' he said, 'and I wondered who he was and where he had come from. I was sure he had not been there before.'

As I approached him a strange feeling came over me and I became cautious. At the same time I looked round to reassure myself that we were both in full sight of my companions. Outwardly there was no reason for this feeling, for the man looked like any other man, and I could see he was somewhat smaller than I and considerably younger. There were only two outstanding differences that I noticed as I neared him: 1. His trousers were not like mine. They were in style much like ski-trousers and with a passing thought I wondered why he wore such out here on the desert. 2. His hair was long, reaching to his shoulders . . . But this was not too strange for I have seen a number of men who wore their hair almost that long.

Suddenly, the feeling of caution left Adamski as the truth began to dawn on him. 'By this time we were quite close,' he continued. 'He took four steps toward me, bringing us within arm's length of each other. Now, for the first time, I fully realized that I was in the presence of a man from space – *a human being from another world!* I had not seen his ship as I was walking toward him, nor did I look around for it now . . . I was so stunned by this sudden realization that I was speechless.'

> The beauty of his form surpassed anything I had ever seen. And the pleasantness of his face freed me of all thought of my personal self. I felt like a little child in the presence of one with great wisdom and much love . . . He extended his hand in a gesture toward shaking hands. I responded in our customary manner. But he rejected this with a smile and a slight shake of his head. Instead of grasping hands as we on Earth do, he placed the palm of his against the palm of my hand, just touching it but not too firmly.

A full description of the spaceman appears in Chapter 1, but a repetition here of salient details will not go amiss. Orthon was about five feet six inches in height, appearing to be about 28 years old, with an extremely high forehead, grey-green eyes slightly aslant at the corners, slightly high cheek bones, with a rather tanned complexion and no signs of any facial stubble. He wore a chocolate-brown seamless one-piece garment of a very fine woven material which had a sheen to it, with no signs of fasteners.

Orthon's shoes were described by Adamski as high, 'like a man's Oxford', fitting closely around the foot, with the opening on the outer side about halfway back on the heel, between the arch and the back of the heel. They seemed to be fastened with two narrow straps, without buckles or fasteners. The heels were slightly lower than normal, and the toe of the shoe was blunt.[12] It was the soles of the shoes which were of great significance, for these left hieroglyph-type impressions in the desert sand, plaster casts of which were later made.

COMMUNICATIONS

After Adamski's various attempts at communication in English, sign language and telepathy, it was learned that the visitor supposedly came from the planet Venus. Orthon even pronounced this word (and occasionally others) in English, with a slightly high-pitched voice. (At a later meeting, he explained that he understood the language well, but was probing

Adamski's telepathic abilities.)

In response to Adamski's question about the purpose of the extrater-restrials' visit, Orthon indicated by means of gestures and expressions that they were concerned about radiation going out into space from our nuclear bombs.

Questioned about his craft, Orthon indicated that it had been brought into the Earth's atmosphere by the large carrier ship seen earlier by all the witnesses. Communication followed at some length about these craft and other matters. Orthon made Adamski understand that there were people coming to Earth from other planets in our solar system, as well as from planets of other systems. 'I remembered reports of men being found dead in some saucers that have been found on Earth,' said Adamski, 'saucers that had apparently crashed. So I asked if any of their men had ever died coming to Earth. He nodded his head in the affirmative, and made me understand that things had on occasion gone wrong within their ships . . .'

Picking up his Brownie camera, Adamski asked if he could take a photograph of Orthon. 'I am sure that he understood my desire, since he was so good at reading my mind. Also I am positive that he knew I would do him no harm because he showed no signs of fear when I picked up the Kodak. Nevertheless, he did object to having his picture taken.' He added:

> I have heard many times that men from other worlds are walking the streets of Earth. And if this be true, I could easily understand his desire not to be photographed, because there were a few distinguishing points about his facial features. Normally these would not be noticed. But in a photograph they would be conspicuous and serve as points of identification for his brothers who have come to Earth.[13]

Eventually, however, Adamski was able to obtain a photograph of Orthon, presumably on another occasion. During a conversation with Adamski in 1959, his co-worker Lou Zinsstag asked him about a painting of Orthon which depicted the spaceman as rather effeminate and undis-tinguished. 'You've got a good hunch there,' replied Adamski. 'Orthon did not look like that at all. He had a very manly, highly intellectual face, but as his features were so distinct and characteristic it would have been dangerous for him to have had them published. He was in Los Angeles several times . . .'

'And to my great surprise,' reported Lou, 'George drew a little wallet from his pocket, and for a few moments I was allowed to gaze at a photo-graph of Orthon's face in profile. It was indeed very different from the painted version.'[14] Lou told me that Orthon's most striking feature was his pronounced chin, a feature also reported, for instance, by the Spanish abductee Julio Fernández in 1978.[15]

Adamski continued communicating with Orthon. One of his many questions dealt with death.

> I wanted to know if they die, as Earth men die . . . He pointed to his body and nodded in the affirmative – that bodies do die. But pointing to his head, which I assumed to mean his mind, or intelligence, he shook his head in negation; this does not die. And with a motion of his hand, he gave me the impression that this – the intelligence – goes on evolving. Then pointing to himself, he indicated that once he lived here on Earth, then pointing up into space, [that] now he is living out there . . .

Finally, Orthon kept pointing to his feet and talking in a language that was incomprehensible. 'It sounded like a mixture of Chinese with a tongue that I felt could have sounded like one of the ancient languages spoken here on Earth,' commented Adamski. 'I have no way of knowing this as fact. It was only my reaction as I listened, and his voice was indeed musical to listen to.'

> From his talk and his pointing to his feet, I felt there must be something very important there for me. And as he stepped to one side from the spot where he had been standing, I noticed strange markings from the print of his shoe left in the earth. He looked intently at me to see that I was understanding what he wanted me to do. And as I indicated that I did, and would comply, he stepped carefully on to another and another spot. Thus he made three sets of deep and distinct footprints. I believe his shoes must have been especially made for this trip and the markings heavily embossed on the soles to leave such deep imprints.[16]

THE SCOUTCRAFT

Orthon then motioned for Adamski to follow him, and they both walked towards the waiting craft. 'It was a beautiful small craft,' said Adamski, 'shaped more like a heavy glass bell than a saucer. Yet I could not see through it [though] it was translucent and of exquisite colour.'

> However, that no mistake may be given here, let me say that I definitely do not believe this ship was made of glass such as we know it. It was a specially processed metal . . . I believe they know how to bring their primary elements from the opaque stage to a translucent stage, yet practically indestructible in hardness, as is the diamond. And it was of such a material that this space craft was made . . . Also it is this translucent quality, along with the power they use, that makes them appear as different coloured lights without definite form.
>
> The ship was hovering above the ground, about a foot or two at the far side from me, and very near to the bank of the hill. But the slope of the hill was such that the front, or that part of it closest to me, was a good six feet above the earth. The three-ball landing gear was half lowered below the edge of the flange that covered them, and I had a feeling this was a precautionary act just in case they had definitely to land. Some of the gusts of wind were pretty

strong and caused the ship to wobble at times. When this took place, the sun reflecting on the surface of the ship caused beautiful prismatic rays of light to reflect out from it, as from a smoky diamond . . .

Nearing the ship, I noticed a round ball at the very top that looked like a heavy lens of some kind. And it glowed . . . The top of the craft was dome shaped, with a ring of gears or heavy coil built into and encircling the side wall at the base of this domed top. This, too, glowed as though power was going through it. There were round portholes in the side wall, but not all the way round, because immediately above one of the balls of landing gear I noticed that the wall was solid. Whether this was true over the other two balls I cannot say because I did not walk around the ship. The covered portholes must have been made of a different quality or thickness of material for they were clear and transparent. And once, for a fleeting second, I saw a beautiful face appear and look out. I felt that whoever was inside was looking for the one who was still out with me, but no word was spoken . . .

The lower outside portion of the saucer was made like a flange, very shiny yet not smooth as a single piece of metal would appear. It seemed to have layers of a fashion, but they couldn't be used as steps because they were in reverse to what steps should be . . .

My space-man companion warned me not to get too close to [the craft] and he himself stopped a good foot away from it. But I must have stepped just a little closer than he, for as I turned to speak to him, my right shoulder came slightly under the outer edge of the flange and instantly my arm was jerked up, and almost at the same instant thrown back down against my body. The force was so strong that, although I could still move the arm, I had no feeling in it as I stepped clear of the ship.

My companion was quite distressed about this accident . . . However, he did assure me that in time it would be all right. Three months later, his words have been proved true for feeling has returned and only an occasional shooting pain as of a deeply-bruised bone returns to remind me of the incident.

At the time, Adamski was more concerned about damage to the exposed negative plates still in his jacket pocket, on the side where he had been closest to the craft. As Adamski moved these to his other pocket, Orthon indicated that he would like one of the plates. 'This he placed in the front of his blouse,' said Adamski, 'but I still didn't see any opening or pocket of any kind. As he did this, he made me understand that he would return the holder to me, but I did not understand how, when, or where.'

A request by Adamski to take a ride in the ship was politely turned down. Then Orthon entered the craft.

With a few graceful steps he reached the bank at the back of the ship and stepped up on to the flange. At least that is the way it looked to me. Where the entrance was, or how he went into the ship, I do not know for sure, but as it silently rose and moved away, it turned a little and I saw a small opening about the centre of the flange being closed by what looked like a sliding door.

Also I heard the two occupants talking together, and their voices were as music, but their words I could not understand.

As the ship started moving, I noticed two rings under the flange and a third around the centre disk. This inner ring and the outer one appeared to be revolving clockwise, while the ring between these two moved in a counter clockwise motion.[17]

THE SEQUEL

After the craft had disappeared, Adamski returned to the footprints. 'As I was walking back to them,' he reported, 'I noticed that both his footprints and mine were visible as we had walked toward the hovering ship. But his were deeper in every instance than mine.' Adamski waved his hat to the others in a prearranged signal, and they came to meet him at a nearby roadside, leaving the cars there because of the rough terrain leading to the contact site. The entire encounter had lasted for one hour. And all the while, military aircraft circled overhead – also for some time afterwards.

George Hunt Williamson set about making plaster casts of the footprints. There was much excitement, and everyone wanted to ask questions at once. Adamski was beside himself – literally. 'I felt as though I was only moving bodily here on Earth,' he said, 'and my answers to the questions were given in a daze.'[18]

Lucy McGinnis's affidavit, identical to those of the others, reads as follows:

> I, the undersigned, do solemnly state that I have read the account herein of the personal contact between George Adamski and a man from another world, brought here in his Flying Saucer – 'Scout' ship. And that I was party to, and witness to the event as herein recounted.[19]

I asked Lucy how much she had been able to see. 'You couldn't see very much detail that far away,' she explained, 'they were far away enough to look like fenceposts! But they stood talking to each other, and we saw them turn and go back up to the ship. Now, I didn't see [Orthon] get into the ship.'

> And when it left, it was just like a bubble or kind of like a bright light that lifted up. Then George went out on to the highway and he motioned for us to come out. He told us that he had got too close and his arm had got caught in the radiation from the craft. And he suffered from that for quite a while . . . You could see where the two of them had walked on the ground. There's no question about that at all.[20]

Objections have been voiced concerning the fact that Adamski seemed rather too well prepared for the Desert Center contact (e.g., the plaster of Paris). A clue has been provided by Sergeant Jerrold Baker, a former Air

Force instructor who had been staying at Palomar Gardens at that time. Several days before the party left for the desert, Baker claims he inadvertently heard a tape recording being replayed of a 'psychic communication' received through Adamski. 'From this brief behind-the-scenes listening,' he said, 'I was able to determine that the desert contact was not a mere stab in the dark or a picnic in the desert, but a planned operation.' Furthermore, Baker refuted Adamski's claim to have spent untold hours watching the skies and waiting for a chance to photograph the space ships. 'This is not true,' he continued. 'I know that he knows exactly when a ship is coming, and is there at the precise instant to snap the picture.'[21]

Adamski was convinced that the symbols on the footprints (see plates) held important clues for mankind. He proffered that ancient civilizations once lived on Earth, whose development and understanding of the universe were far superior to those of contemporary mankind. The symbols, he said, were probably of a universal nature and might be understood perhaps as 'guideposts in space, presently being used by men of other worlds in interplanetary travel. And thus a helpful hand is extended to Earthmen as they turn their thoughts and efforts outwardly toward space travelling.'[22] Similar symbols were reproduced and discussed by Professor Marcel Homet in his book *Sons of the Sun*, first published in 1958. Homet held that the symbols dated back at least 10,000 and probably 20,000 years.[23]

THE RETURN VISIT

Because of Orthon's promise to return the photographic plate, Adamski kept himself in a state of constant alertness. With the plate camera attached, he set up his telescope at a spot on the Palomar Gardens property giving an unobstructed view of the far distance, including a long expanse of the Pacific Ocean shoreline.

On the morning of 13 December 1952 he was alerted by the roar of jets overhead. In the distance he saw a flash, and remarked to others with him at that time that it might be the spacecraft. Then, at 09.00, he saw another flash in the sky, and tried to aim his telescope at it. 'Sure enough,' he reported, 'I was able to observe it gliding noiselessly in my direction – an iridescent glass-like craft flashing its brilliant colours in the morning sun . . .'

As it came over the nearby valley, it seemed to stop and hover motionlessly. With utmost will power I restrained my excitement in an effort to get a really good picture this time. Quickly I took two shots. Then realizing that the ship being that near was too large to get the whole thing in the picture with the camera in that position, I turned the camera on the eyepiece and took another shot while it was still hovering. I shot the fourth picture just as the ship was beginning to move again . . . the first three of these pictures proved to show good detail, while the fourth – taken in motion – turned out fuzzy, but is still good.

While changing the position of the camera on the eyepiece, Adamski made some quick calculations of the craft's dimensions, comparing it with known distances. Instead of being 20 feet in diameter, as he had guessed three weeks earlier, he estimated that it was probably about 35 to 36 feet in diameter and 15 to 20 feet in height.

> As it approached probably within 100 feet of me, and to one side, one of the portholes was opened slightly, a hand was extended and the selfsame holder which my space-man friend had carried away with him on November 20 was dropped to the ground. As the holder was released, the hand appeared to wave slightly just before the craft passed beyond me. I watched the holder drop and strike a rock as it hit the ground.

Adamski picked up the holder, dented from its impact with the rock, and wrapped it in his handkerchief so as not to damage any potentially important evidence, such as fingerprints. Meanwhile, the craft crossed a small ravine on the Palomar Gardens property as it moved towards the base of the mountains to the north. Then dropping below the level of the treetops, it flew close to a cabin where it was allegedly seen by some other witnesses and photographed by one.[24]

The photograph, supposedly taken with a Brownie box-camera by Jerrold Baker, who had been staying on the property since the end of October, was (along with Adamski's photos) the subject of much controversy. The saucer certainly looks blurred (see plates), but given the slow shutter speed of the Brownie, even slight movement of the craft would have led to this result. Baker signed a statement testifying to the event, which, in part, states:

> Suddenly in the corner of my eyes, I saw a circular object skim over the treetops from the general area where the Professor was located . . . I waited momentarily, mostly because of shock I guess, as it continued coming closer. It then hung in the air not over 12 feet high at the most, and about 25 feet from where I was standing. It seemed as if it did this knowing I was there waiting to photograph it. I quickly snapped a picture and as I did it tilted slightly and zoomed upwards over the tree faster than anyone can almost imagine . . .
> These things I know for certain: 1. The saucer made no sound. 2. It was guided by a superior intelligence. 3. There was a slight odor present as the saucer sped upwards. 4. It had portholes and three huge ball-bearings, presumably landing gears.[25]

According to a later affidavit, Baker denied having taken the picture. 'I did not take the alleged photograph accredited to me,' he wrote. 'The alleged photograph was taken with the Brownie camera along with three or four similar photos by Mr George Adamski on the morning of Dec. 12th 1952, and not on Dec. 13th, 1952.'[26]

And in a letter to investigator James Moseley, Baker elaborated:

> It was my suggestion that [Adamski] be located at one spot with his telescope and camera while I or any other individual be located at another spot on the property with a different type of camera . . . Much to my amazement, within a week after this suggestion, George Adamski early one morning disclosed the fact that he had taken pictures with the Brownie camera, adjacent to his cabin.

Baker went on to give the names of two other people who he claimed could verify these matters.[27] While it is known that Baker later turned against Adamski, due in part to the latter's displeasure at Baker's behaviour while staying at Palomar Gardens (including plans for operating a gadget to 'draw down flying saucers and airplanes', which matter was reported to the FBI),[28] it is possible that Adamski may have twisted the evidence. Perhaps he took the Brownie photo and then asked Baker to take the credit for it, to bolster the evidence. However, Lucy McGinnis, who was there at the time, disagreed vehemently. 'That whole thing was Baker's doing,' she told me. He was raring to do anything to prove the point. He didn't deny it before me, that I remember.'[29]

On the same day the photographs were taken, Adamski took the photographic plates, with the exception of the one returned by Orthon, for developing and printing by a Mr D.J. Detwiler in Carlsbad, 40 miles away. All of the pictures turned out perfectly. Uncertain for a while as to who should develop the plate returned by Orthon, Adamski finally decided to take it to his local photographer. 'When the finishing was done, and with witnesses present, and a print was made,' wrote Adamski, 'there were indications of the original photo – which I had taken before the space visitor took the holder – being washed off; and this was replaced by a strange photograph and a symbolic message, which to this day has not been fully deciphered. Several scientists are working on it [and are] still working on deciphering the markings of the footprints.' He decided against having any fingerprints taken.

At Adamski's request, representatives of two government agencies (the FBI and the Air Force's Office of Special Investigations, OSI) came to visit him at Palomar Gardens. 'These men listened intently to my detailed description of all that had taken place,' he reported, 'but they registered no surprise. Nor did they express any doubt regarding the truthfulness of my statements. They did not even question me . . . They did take a couple of my photographs of the craft, as well as a print from the dropped negative, which I gave to them.'[30] The FBI memorandum pertaining to this meeting, sent to the Director, J. Edgar Hoover (Fig. 7), confirms that Adamski did indeed give the investigators a detailed description of the Desert Center encounter (and much else as well). Some extracts follow:

> Adamski stated that he took a picture of the space ship and the space man, but the space man could evidently read his thoughts inasmuch as he motioned to him not to take the picture and when the space man left he took the 'plate' with the negative on it with him . . .

This differs from Adamski's probably more accurate version of events, as published in *Flying Saucers Have Landed*, wherein he stated that he did not actually take a photo of Orthon. The FBI memorandum continues:

> Adamski advised that on December 13, 1952, the space ship returned to the Palomar Gardens and came low enough to drop the plate which the space man had taken from him . . . and had then gone over the hill . . . Adamski stated that as the space ship was leaving, [Jerrold Baker] also took a picture of the ship . . . Adamski furnished the writer with copies of the space writing and photographs of the space ship.

As to Adamski's assertion that the agents expressed no doubts concerning the truthfulness of his statements, there is no such confirmation by either the FBI or the OSI. On the contrary. 'No further investigation is being conducted,' concludes the memo. 'This case is being [*sic*] considered closed.'[31] Another FBI memo states that: 'OSI of the Air Force has done considerable investigation . . . and lends no credence to the truthfulness of Adamski's statements.'[32]

AN ADMONISHMENT FROM THE FBI

During a speech before the Corona, California, Lions Club, on 12 March 1953, Adamski boasted that, according to a local newspaper report, his material had been 'cleared' by the FBI and OSI, apparently supporting this extravagant claim on other occasions with a doctored copy of a paper signed by FBI and Air Force agents. For this misrepresentation, Adamski was visited by FBI agents from San Diego and soundly admonished. As FBI Director J. Edgar Hoover wrote:

> At the time of the interview, Mr Adamski denied making the statement. In the Agents' presence he wrote a letter to the Editor of the Riverside 'Enterprise' pointing out the incorrectness of the article with respect to FBI and Air Force Intelligence clearance of his material . . . At the same time, Adamski, in a statement to the interviewing agents, advised he had not made and did not intend to make statements to the effect [that] the United States Air Force or Federal Bureau of Investigation have approved material used in his speeches.[33]

ADAMSKI further advised that on November 20, 1952, on the California Desert, at a point ten and two-tenths miles from Desert Center on the road to Parker and Needles, Arizona, that he had made contact with a space craft and had talked to a space man, ADAMSKI stated that he, ███████ ████████████████ and his wife MARY, had been out in the desert and that he and the persons with him had seen the craft come down to the earth. ADAMSKI stated that a small stairway in the bottom of the craft, which appeared to be a round disc, opened and a space man came down the steps. ADAMSKI stated he believed there were other space men in the ship because the ship appeared translucent and could see the

SD 100-8382

shadows of the space men. ADAMSKI described the space man as being over 5' in height, having long hair like a woman's and garbed in a suit similar to the space suits or web suits worn by the U. S. Air Force Men. ADAMSKI stated that he and the space man conversed by signs and that there appeared to be a certain area around the space ship which consisted of magnetic or electric lines of force, inasmuch as when he got too close, some of the lines went through his arm and momentarily paralyzed his arm. ADAMSKI stated that he took a picture of the space ship and the space man, but the space man could evidently read his thoughts inasmuch as he motioned to him not to take the picture and when the space man left he took the "plate" with the negative on it with him.

ADAMSKI further advised that he had obtained plaster casts of the footprints of the space man and stated that the casts indicated the footprints had designs on them similar to the signs of the Zodiac,.

(b)(7)(c)

On January 12, 1953, ADAMSKI advised that on December 13, 1952, the space ship returned to the Palomar Gardens and came low enough to drop the plate which the space man had taken from him, ADAMSKI, and had then gone off over the hill. ADAMSKI stated that he saw the space ship and that as the space ship was leaving, ███████████████ also took a picture of the ship.

ADAMSKI stated that when he had the negatives developed at a photo shop in Escondido, California, that the negative that the space man had taken from him contained writing which he believed to be the writing of the space men. ADAMSKI furnished the writer with copies of the space writing and photographs of the space ship.

(b)(7)(c):(

███████████████████████ at the Palomar Observatory, advised that he had been acquainted with ADAMSKI since 1943, at which time ADAMSKI had called himself "the Reverend ADAMSKI" and had held Easter Services in the Valley. ███████████ stated that in talking to ADAMSKI, ADAMSKI had told him, ███████ that he had had a "cult" or colony at Laguna Beach, California, previous to 1943, and that he had also been interested in metaphysics and astrology.

(b)(7)(c)

███████████ further advised that ADAMSKI claimed

Fig. 7. Excerpts from an FBI office memorandum to J. Edgar Hoover regarding some of George Adamski's claims, dated 28 January 1953. (*FBI*)

SD 100-8382

(b)(7)6:(

to have worked at the Observatory at Mount Palomar, but stated that ADAMSKI had never been employed at the Observatory.
████████ stated that ADAMSKI also claims to have been associated with ████████████, formerly with the Observatory at Mount Palomar, but now located in Pasadena, California. ████████ stated that ████ had known ADAMSKI
for quite some time and that ████ address was:

████████████████████████
Pasadena, California

Telephone: Sycamore ████

(b)(7)(d)

████████ stated he considered ADAMSKI to be
more qualified in astrology than astronomy. He continued
that he had never viewed any space ship and believed ADAMSKI
to be an opportunist.

(b)(7)(c)

Copies of the letters of ████████ and prints
of the space writing and flying saucers, are being enclosed
to the Bureau, for informational purposes only.

Copies are also being enclosed to the Cleveland
Office for any action that they may desire to take.

This information is not being furnished to (b)(7)(c)
the U. S. Air Force inasmuch as ████████ OSI, was along
at the time of the interview, and is cognizant of all facts
contained herein.

ADAMSKI furnished the following information
concerning himself:

Born:	4-17-1891
Place:	Eland
Father:	JOSEPH ADAMSKI (deceased)
Mother:	FRANCES ADAMSKI (deceased)
Sister:	████████
	(phonetic)
Address:	Lackawana, New York
Sister:	Mrs ████ ( ),
Address:	Dunkirk, New York
Sister:	Mrs. ████████
Address:	Dunkirk, New York

THE SCOUTCRAFT PHOTOGRAPHS

Most researchers have denounced Adamski's famous 'fifties' scoutcraft
photos as fakes. Descriptions of the 'model' used cover a wide range of
utensils: 'lampshade'; 'operating theatre lamp'; 'saucepan lid with ping-
pong balls'; 'tobacco humidor'; 'chicken feeder'; 'the top of a 1937
canister-type vacuum cleaner'; and 'a bottle-cooler made in Wigan,
Lancashire'. The problem is that no one has yet produced examples of
any of the above items which resemble proportionately the pictured craft.
Adamski, incidentally, offered $2,000 to anyone who could prove his
photos were fakes. There were no takers.

Desmond Leslie, a former Second World War fighter pilot who wrote the first (and longer) part of *Flying Saucers Have Landed*, made a strong case for the authenticity of the photographs:

> Anyone with experience of tele-photography who has obtained an original print made from the original Adamski negatives will at once notice a factor that has to be taken into account by all film directors using models to represent the full-size objects, a factor known sometimes as 'atmospheric softening'. This phenomenon is due to moisture and dust in the atmosphere, so that it is impossible to match up a foreground model with a distant background (however sharp your depth of focus) unless certain partial gauzings and screenings are used. The effect through a tele-photo lens is to produce a certain greying and flattening which is practically impossible to reproduce artificially. Tele-photography also slightly alters the perspective, hence the flattening and greying effects clearly noticeable in Adamski's pictures.

Leslie also points out, based on his experiments using Adamski's actual telescope at the site where the photographs were taken, that it is evident that atmospheric distortion was, in one of the photos, responsible for making one of the three balls under the craft appear to be larger than the others. This would not have happened if a model had been used. Leslie further adduces the fact that Adamski's old German Hagee-Dresden Graflex plate-camera could only be used in conjunction with the six-inch telescopic lens: no other lenses were supplied with it and to fake a photo using a model, a much lower focal length would have been used.

Leslie gave the Adamski enlargements to one of Hollywood's most revered directors, John Ford, who stated his opinion that the saucer was a large object shot through a telephoto lens of about six inches focal length. These findings were confirmed by Joseph Mansour, whose job it was to make photos of model aircraft appear to be the real thing.[34] And Pev Marley, Cecil B. de Mille's leading special-effects cameraman, who had served as a photographer with Enemy Intercept Command in the Second World War, reportedly testified at a meeting of US Air Force Reserve officers in 1953 that Adamski's pictures, if faked, were the cleverest he had ever seen.[35] Later, he denied having made such a statement.[36]

In *Beyond Top Secret*, I discussed the extraordinary 8mm colour movie film of a scoutcraft, identical to the one photographed at Palomar Gardens, taken by Adamski in the presence of Madeleine Rodeffer and three other witnesses at Silver Spring, Maryland, on 26 February 1965 (a frame from which is reproduced in the colour plates). According to William Sherwood, an optical physicist and former senior project development engineer at Eastman Kodak, the film is authentic.[37] My own investigations into the film began in 1966: I can state unequivocally, based on those investigations and a friendship with Madeleine which spans three decades, that the film is genuine. (Interestingly, as *might* have been the case with Jerrold Baker, Adamski asked Madeleine to take credit for the film.)

Also in *Beyond Top Secret*, I cited the photographs taken by Stephen Darbishire, in the presence of his cousin Adrian Myers, at Coniston, Lancashire, in February 1954. The best of these two photographs shows a 'glass-like' craft which is identical in proportion to Adamski's scout-craft.[38] I am not hesitant to say that this photograph also is authentic. Darbishire insisted that the craft had what appeared to be a series of port-holes arranged in sets of four, whereas the published photos of Adamski showed what appeared to be a set of only three. Adamski's slightly blurred fourth photo of the scoutcraft, unpublished in 1954, shows a fourth porthole. Darbishire was unaware of this until Desmond Leslie showed him the photo. Another photograph of an identical craft, taken in 1973 in Peru, shows a fourth porthole (see plates).

INSIDE THE SPACE SHIPS

Following the Desert Center encounter and the return of the scoutcraft at Palomar Gardens, Adamski's experiences with the space people pro-liferated. It was then that *some* of his claims became increasingly absurd. In his second book, *Inside the Space Ships*, also long out of print, he begins by describing how, on 18 February 1953, two unknown men approached him at a hotel in Los Angeles.

> I noted that both men were well proportioned. One was slightly over six feet and looked to be in his early thirties. His complexion was ruddy, his eyes dark brown, with the kind of sparkle that suggests great enjoyment of life. His gaze was extraordinarily penetrating. His black hair waved and was cut according to our style. He wore a dark brown business suit . . .
>
> The shorter man looked younger and I judged his height to be about five feet nine inches. He had a round boyish face, a fair complexion and eyes of greyish blue. His hair, also wavy and worn in our style, was sandy in colour. He was dressed in a grey suit and was also hatless. He smiled as he addressed me by name. As I acknowledged the greeting, the speaker extended his hand and when it touched mine a great joy filled me. The signal was the same as had been given by the man I had met on the desert . . . Consequently, I knew that these men were not inhabitants of Earth.

After Adamski had accepted an invitation to accompany them, he was taken in a black Pontiac sedan to a destination some way out of Los Angeles. During the journey, the strangers introduced themselves as coming from 'Mars' and 'Saturn'. Adamski asked himself how they man-aged to speak English so well. To this unspoken thought, one of the men responded:

(LEFT) Ensign Rolan D. Powell, beside his Grumman F6F Hellcat on board the aircraft carrier USS *Yorktown* during the Battle of Leyte Gulf, Second World War. In the summer of 1945, while based at the US Naval Air Station, Pasco, Washington, Powell and five other Hellcat pilots were scrambled to intercept an unknown flying craft 'the size of three aircraft carriers side by side' hovering at high altitude above the top-secret plutonium production facility for the world's first nuclear bombs at Hanford. *(US Navy)*

(BELOW) A Grumman Hellcat. During the interception above Hanford, the pilots were ordered to force their aircraft to 42,000 feet—well above their rated ceiling—but were unable to reach the unknown craft. *(© Timothy Good)*

(LEFT) Gösta Carlsson at the site in Angelholm, Sweden, where in May 1946 he claimed to have encountered a landed craft and extraterrestrial men and women. *(Göteborg Information Centre on UFOs)*

(MIDDLE LEFT) Daniel Fry, the rocket engineer who claimed to have been taken for a ride in an uninhabited spacecraft which landed at the White Sands Proving Grounds, New Mexico, in 1949, and who later claimed periodic contacts with an extraterrestrial.

(MIDDLE RIGHT) Enlargement of a probably genuine photo taken by Fry near Baldwin Park, California, in September 1954.

(BELOW LEFT) Bruno Facchini, showing the jacket and boots he wore during his encounter with a craft grounded for repairs near Milan, Italy, in April 1950. *(Domenica del Corriere)*

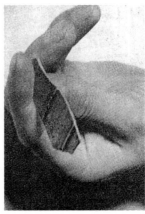

(LEFT) One of the pieces of metal found by Facchini at the landing site, believed to be residue from the repair operation he witnessed. *(Domenica del Corriere)*

Two of Adamski's celebrated series of photos of the 'Venusian scoutcraft', photographed through his 6-inch reflector teloscope at Palomar Gardens, California, on 13 December 1952. Adamski stressed that the photographs (RIGHT AND BELOW) should be printed showing both the field of view of the telescope as well as the unusual light forms. (© G. A. F. International)

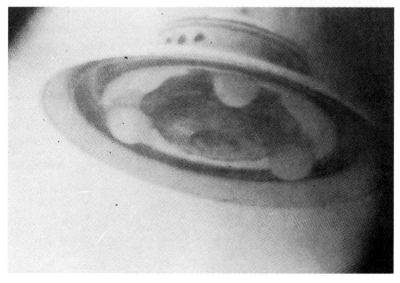

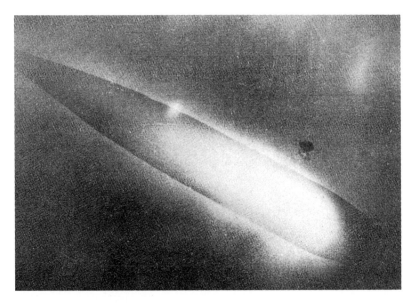

According to Adamski, this giant carrier-craft functioned as a submarine, as well as a spacecraft. Photographed through Adamski's telescope on 9 March 1951. *(© G. A. F. International)*

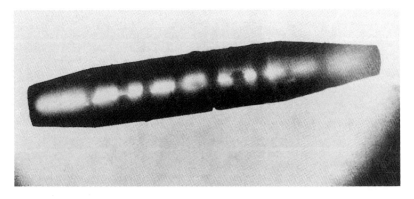

A giant carrier-craft photographed by Adamski through his telescope on 1 May 1952. The craft was hovering over a mountain some 30 miles away, enabling its length to be estimated as between 1,200 to 1,500 feet. Adamski asserted that the small notch (amidships bottom) is the exit port for smaller 'scout craft' which, on return, enter by a hatch on the upper deck. *(© G. A. F. International)*

The controversial photograph reportedly taken by Sgt.
Jerrold Baker with a Kodak Brownie camera, shortly after
Adamski's famous telescopic pictures of the 'scout-craft'
were taken at Palomar Gardens on 13 December 1952.
Although Baker signed a statement testifying that he had
taken the picture, he later recanted.
(© G. A. F. International)

The famous photographic plate
with substituted hieroglyphic-
type symbols, given to Adamski
by 'Orthon' in December 1952.
(© G. A. F. International)

One of the plaster casts of Orthon's footprints, in the possession of
Desmond Leslie. *(©Timothy Good)*

George Adamski (right) with Lou Zinsstag. In the early 60s, disillusioned by Adamski's increasingly extreme claims, Zinsstag resigned from her position as his Swiss co-worker. To the end of her life, however, she supported his original claims.

Lucy McGinnis, one of the six witnesses to Adamski's encounter with an extraterrestrial near Deser Canter, California, in November 1952. McGinnis worked as Adamski's voluntary secretary for fourteen years, but resigned in the early 60s. Like Lou Zinsstag, she supported Adamski's original claims until her death. (© *Timothy Good*)

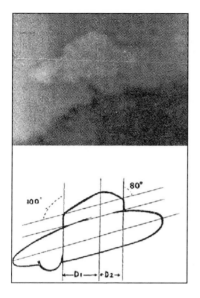

(ABOVE RIGHT) An Adamski-type craft photographed by Stephen Darbishire at Coniston, Lancashire, in February 1954, and (ABOVE LEFT) his seldom seen second photograph, showing a distortion effect similar to that visible in the Silver Spring film. To this day, Darbishire, now a successful painter, stands by the authenticity of his photos. (© *Stephen Darbishire*)

Stephen Darbishire (left) and his cousin, Adrian Myers, who also witnessed the encounter.

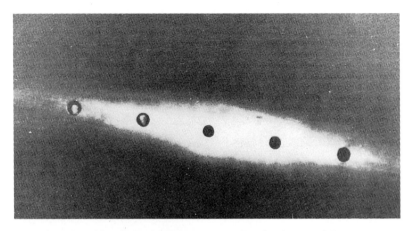

One of a series of four Polaroid photos purportedly taken by one of the extraterrestrials from a 'scout-craft', showing Adamski (second from left) looking through a porthole in a carrier-craft during a trip into space in April 1955.
(© *G. A. F. International*)

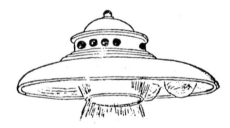

(BELOW) An enlargement of a Polaroid photo showing an Adamski-type craft, taken by architect Hugo Vega 31 miles from Lima, Peru, in October 1973. The sketch is also by Vega. (*Hugo Luyo Vega/UPI*)

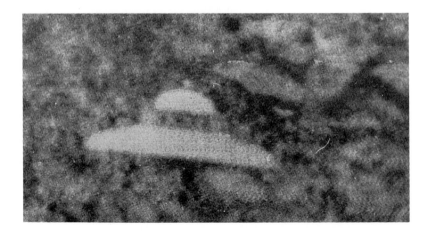

Air Marshal Sir Peter Horsley, former Deputy Commander-in-Chief of Strike Command and Equerry to HM The Queen and Prince Philip for seven years, who in 1954 had a two-hour meeting with an apparently extraterrestrial man in London. Sir Peter is shown here in the cockpit of a Hawker Hunter T7 at RAF Valley in May 1975, with his nephew, Flt.-Lt. Gjertsen, in the co-pilot's seat.
(© Crown Copyright)

Howard Menger, a charismatic contactee who rose to prominence in the late 1950s, shown here during an interview with the author in 1978.
(© Timothy Good)

One of Menger's Polaroid photos of a landed craft, taken in 1956. (© *Howard Menger*)

An alleged spaceman, also photographed with a Polaroid camera by Menger in 1956. Note the 'aura' surrounding the craft, explained by the spacemen as due to an 'electro-magnetic flux' emanating from the craft, which also gave rise to various optical distortions (see, for example, the dome). (© *Howard Menger*)

A supposed spacewoman photographed with a Polaroid camera by Howard Menger in 1956. The light is not a flashlight but a gadget attached to the woman's belt, which, when touched, rendered her invisible. (© Howard Menger)

(RIGHT) Frames from a 16mm film taken by Howard Menger in the Blue Mountains, Pennsylvania, in the late 1950s. (© Howard Menger)

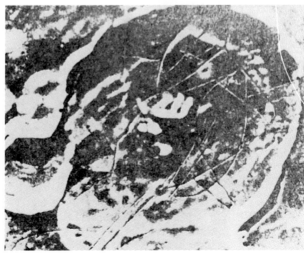

A light-enhanced photo of the Gassendi crater on the Moon, taken in the 1930s through a 100-inch telescope, showing some interesting features. Howard Menger, as well as George Adamski and other contactees, claimed that the extraterrestrials had bases on the Moon. (Mount Wilson Observatory)

One of the several unusual flying craft photographed by the science journalist, Bruno Ghibaudi, on the shores of the Adriatic coast at Pescara, Italy, in April 1961. A few months later, Ghibaudi claimed to have had meetings with space people.

This photograph was submitted anonymously to the Italian magazine *Domenica del Corriere* on 23 June 1963. The craft supposedly had landed in woodland on a hill in Genoa. 'I saw it together with a garage worker not many days ago,' claimed the photographer. 'For personal security reasons, I cannot give you my name.' *(Domenica del Corriere)*

(ABOVE) The modified Pre-Columbian ring given to Pallman by his extraterrestrial friend, 'Satu Ra'. Just below the centre can be seen the metal inset, which, when it glowed, alerted Pallman to the nearby presence of his strange friends.

(ABOVE) One of the natives (a tribal chief) whose habits were studied by the extraterrestrials. Among other tasks, the Indian head-hunters were employed at the plantation as very effective security guards! They regarded their alien masters simply as eccentric foreigners. *(Ludwig Pallmann)*

(RIGHT) Ludwig F. Pallmann (centre), the German businessman who claimed to have met alien anthropologists in India in 1964, and to have stayed at their base—a plantation in the Amazon jungle of Peru—in 1967. In this photo, Pallmann is shown in Benares, India.

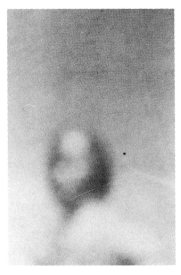

Carroll Wayne Watts and his wife Rosemary, following his disturbing encounters with alien beings near Wellington, Texas, in 1967. After publication of his story and some of his photos in February 1968, Watts received several threats to his life. *(Associated Press)*

An enlargement of a photograph showing what Watts claimed was the hairless head and shoulders of one of the alien beings. 'The man appears to have hair, but it was a shadow on the back of his head,' Watts explained.
(Carroll Wayne Watts)

A Polaroid photograph of one of the craft Watts claims to have encountered, taken on 11 June 1967. The craft was about 100 feet in length. *(Carroll Wayne Watts)*

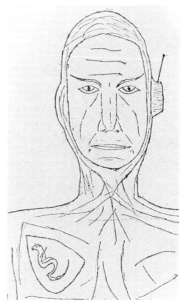

Herbert Schirmer, the police patrolman who claims to have been taken aboard an alien craft in Ashland, Nebraska, in December 1967, and (right) his drawing of the crew leader. *(© Warren Smith & Herbert Schirmer)*

Jan Siedlecki, who claimed to have boarded an alien craft in Cross Gates, Leeds, in 1976. *(© Timothy Good)*

Walter Rizzi, who claims to have encountered a landed alien craft and its occupants in the Dolomite Mountains of Italy in July 1968. This picture was taken in 1997. *(© Timothy Good)*

(LEFT) The two 'bio-robots' seen in February 1991 by former police officere Luis Torres and his wife, together with two police colleagues and their wives, in the El Yunque National Forest of Puerto Rico. This sketch by investigator Jorge Martín was approved by the witnesses. *(Jorge Martín)*

(ABOVE) Carlos Mercado, who claims to have been abducted by alien beings in June 1988 and taken to their base in the Sierra Bermeja, beside the Laguna Cartegena, Puerto Rico. *(© Timothy Good)*

Harold Westendorff, a champion aerobatics pilot, beside his Embraer 712 in which he circled a huge, unidentified aerial craft at Pelotas, Rio Grande do Sul, Brazil, in October 1996. *(GPCU, Pelotas)*

We are what you on Earth might call 'contact men'. We live and work here ... We have lived on your planet now for several years. At first we did have a slight accent. But that has been overcome and, as you can see, we are unrecognized as other than Earth men.

Adamski began to wonder to himself why he had been singled out by the space people, to which unspoken thought came the reply:

You are neither the first nor the only man on this world with whom we have talked. There are many others living in different parts of the Earth to whom we have come. Some who have dared to speak of their experiences have been persecuted ... Consequently, many have kept silent. But when the book on which you are now working reaches the public, the story of your first contact out in the desert ... will encourage others from many countries to write you of their experiences.

The Pontiac turned off the highway and went along a rough track in the desert, at which point Adamski caught sight of a craft, similar but larger than the one three months earlier, glowing soft white in the distance. 'As we came to a stop, I noticed that a man was standing beside the glowing craft,' he said, '[who] appeared to be working on something connected with it. The three of us walked towards him and, to my great joy, I recognized my friend of the first contact ... He was dressed in the same ski-type flying suit that he had worn on the first occasion, but this suit was light brown in colour with orange stripes at top and bottom of the waistband.' His hair, on this occasion, was cut short.

After greetings were exchanged, Orthon explained that he was repairing a small part of the scoutcraft. As he did so, he emptied the contents of a small 'crucible' on the sand. 'I stooped and cautiously touched what appeared to be a very small amount of molten metal which he had thrown out,' Adamski reported. 'Although still quite warm, it was not too hot to be handled, and I carefully wrapped it in my handkerchief ...'

Asked by an amused Orthon why he wanted the material, Adamski explained that it might furnish proof of the reality of their visit; that people usually demanded concrete evidence. 'Yes,' replied Orthon, 'and you *are* a race of souvenir hunters, aren't you! However, you will find that this alloy contains the same metals found on your Earth, since they are much the same on all planets.'[39] Desmond Leslie gave this piece of metal to George Ward, Britain's Air Minister at the time, who arranged for an analysis. The sample proved to be composed predominantly of a very high-grade aluminium, combined with trace elements probably collected when in the molten state.[40]

No names were given to Adamski for any of the space people he met, the reason being, he claimed, that they had 'an entirely different concept of names as we use them'. In *Inside the Space Ships*, pseudonyms were used for identification purposes, and these were thought up by Adamski together with Charlotte Blodget, who was the book's ghost-writer. They included 'Firkon' for the 'Martian' and 'Ramu' for the supposed Saturnian (though Adamski implies that this was the latter's real name).

The repair completed, Adamski was invited to enter the craft with the others. He found himself in a one-room cabin.

I was aware of a very slight hum that seemed to come equally from beneath the floor and from a heavy coil that appeared to be built into the top of the circular wall. The moment the hum started, this coil began to glow bright red but emitted no heat . . .

I marvelled anew at the unbelievable way in which they were able to fit parts together so that the joinings were imperceptible . . . there was no sign of the door that had closed behind us . . . I estimated the inside diameter of the inside cabin to be approximately eighteen feet. A pillar about two feet thick extended downward from the very top of the dome to the centre of the floor. Later I was told this was the magnetic pole of the ship, by means of which they drew on Nature's forces for propulsion purposes . . .

I noticed that a good six feet of the central floor was occupied by a clear, round lens through which the magnetic pole was centred. On opposite sides of this huge lens, close to the edge, were two small but comfortable benches curved to follow the circumference. I was invited to sit on one of these and Firkon sat beside me to explain what was going on . . . while Orthon went to the control panels. These were located against the outer wall between the two benches . . . a small flexible bar fell into place across our middles . . .

'Sometimes,' explained Firkon, 'when a ship is thoroughly grounded, a sharp jerk is experienced when breaking contact with Ear 'though this does not happen very often, we are always prepared.'

Adamski turned his attention to what appeared to be graphs and charts that covered the walls for about three feet on either side of the invisible door. 'They were fascinating,' he said, 'entirely different from anything I had seen on Earth.'

There were no needles or dials, but flashes of changing colours and intensities. Some of these were like coloured lines moving across the face of a particular chart . . . others took the forms of different geometric figures . . . The wall for a distance of about ten feet directly behind the benches on which we sat appeared to be solid and blank, while on those beyond . . . were other charts somewhat similar, yet differing in certain ways from those I have described.

The pilot's instrument board was unlike anything I could have imagined . . . it looked rather like an organ. But instead of keys and stops there were rows of buttons. Small lights shone directly on these, so placed that each illuminated five buttons at a time. As far as I can remember, there were six rows of these buttons, each row about six feet long.

In front of the board was a pilot's seat, similar in shape to the other benches. Close beside it was a peculiar instrument connected to the central pole, which apparently functioned as a sort of periscope.

Adamski found it impossible to discern where the light came from. 'It seemed to permeate every cavity and corner with a soft pleasing glow,' he said. 'There is no way of describing that light exactly . . . it seemed to consist of a mellow blend of all colours, though at times I fancied one or another seemed to predominate.'

Only a very slight sensation was felt as the craft left the ground. Adamski's attention was drawn to the large lens at his feet. 'We appeared to be skimming over the rooftops of a small town; I could identify objects as though we were no more than a hundred feet above the ground. It was explained to me that actually we were a good two miles up and still rising, but this optical device had such magnifying power that single persons could be picked out and studied, if so desired . . .' It was further explained that the central pillar not only served as a powerful telescope, with one end pointing at the ground and the other – in the dome – pointing at the sky, but also provided most of the power for the craft.

PROPULSION

Four cables appeared to run through, or immediately below, the floor lens, joining the central pole in the form of a cross. Firkon explained their purpose:

> Three of those cables carry power from the magnetic pole to the three balls under the ship which, as you have seen, are sometimes used as landing-gear. These balls are hollow and, although they can be lowered for emergency landing and retracted when in flight, their most important purpose is as condensers for the static electricity sent to them from the magnetic pole. This power is present everywhere in the Universe. One of its natural but concentrated manifestations is seen displayed as lightning.
>
> The fourth cable extends from the pole to the two periscope-like instruments, the one beside the pilot's seat and the other directly behind his seat but close to the edge of the centre lens . . . They can be switched on and off, or adjusted at will, so that both members of the usual crew can have full use of the telescope without interfering with each other.[41]

In his last book, *Flying Saucers Farewell*, Adamski expounded on the principles of what he called the 'three-point electrostatic propulsion control system'. 'As we use retro-rockets to steer a rocket vehicle,' he explained, 'the saucers use their variable three-point system to manoeuver by regulating the charge. In horizontal flight within a planet's ionosphere, saucers travel along the planet's geo-magnetic lines of force. They turn abruptly by shifting the ball-charge.'[42]

In conversation with Captain Edward Ruppelt, the scientific and technical intelligence officer who had headed the Air Force's Project Blue Book investigation into UFOs, Adamski claimed that the saucers only use 10 per cent of the power they harness from nature, the excess being dissipated from the skin of the ship. 'Particles that would hit the ship are repelled by the negative radiation from the skin of the ship,' he explained, 'so they never touch anything, not even a meteorite.'

'But why the saucer shape?' asked Ruppelt.

'You see, they don't have to make a turn as we do,' Adamski replied. 'To us it looks like they make a right-angled turn, but they don't. They can cut one [ball-gear] off or the other. Whichever one they cut off, that's the way the ship is going to go.'[43]

Ann Grevler, who claimed to have been taken aboard a scoutcraft by a similar group of extraterrestrials in the then Eastern Transvaal, South Africa, in 1957, provides further details of this particular type of propulsion system:

> The general idea of their propulsion is that cosmic power (electricity?) is drawn out of the surrounding air, through the top of the central column . . . This power is then irradiated via a pump at the bottom of the central pillar, over powdered quartz of a kind, which is spread over the largest possible field within the ship. The result is ionized air [which] is pumped through the three hollow rings around the outside base of the cabin structure as well as circulating through the three balls underneath – these latter being used for motive power and direction and not used primarily as landing gear.[44]

The machinery for motive power, Adamski claimed, was housed mostly beneath the floor of the cabin and under the lower exterior of the craft (see Fig. 8). 'I did not actually see any of it,' he explained, 'but I was shown into a very small room which served both as an entrance to the compartment which contained the machinery, and as a workshop for emergency repairs.'[45]

Earlier, Adamski had found it almost impossible to believe that the visitors could have accidents, or needed to repair their craft from time to time. 'I had to remind myself that, after all, they too were *human beings*,' he wrote, 'and, no matter how far advanced beyond us, must still be subject to error and vicissitude.'[46]

THE MOTHER SHIP

Alerted by Orthon that the scout was about to land on the mother ship, Adamski watched in astonishment as, in the solid wall behind the bench, a round hole began to open out, similar to the iris of a camera, until it was about 18 inches wide. This turned out to be one of the portholes, arranged in series around the cabin. Adamski reasoned there must be four on each side of the craft, making a total of eight (though a total of 16

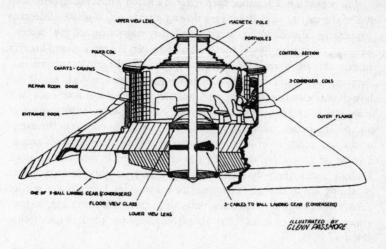

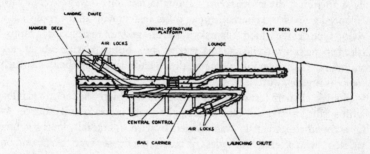

Fig. 8. Diagrams of the scout- and carrier-craft based on Adamski's descriptions.
(*G.A.F. International*)

is indicated in the photographs taken in December 1952). Through the portholes, Adamski could see the giant carrier craft, 40,000 feet above the Earth.

As we came nearer, its huge bulk seemed to stretch away almost out of sight, and I could see its vast sides curving outward and downward. Slowly, very slowly, we drew nearer until we were almost on top of the great carrier. I was not astonished when my companion told me she was about one hundred and fifty feet in diameter and close to two thousand feet in length. The spectacle of that giant cigar-shaped carrier ship hanging there motionless in the stratosphere will never dim in my memory.

The scoutcraft descended then entered a hatch which had appeared on top of the carrier ship and glided down on two rails, the rate of descent apparently controlled by friction and the magnetism of the saucer's flange. The craft no longer under its own power, Adamski found himself subjected to the normal forces of gravity, and once nearly lost his balance. Arriving on the platform of a 'huge hangar or storage deck' inside the bowels of the carrier ship, the craft was met by a crew member – of a dark complexion and wearing a beret-type cap – holding something that looked like a clamp attached to a cable. This was connected to the flange of the scoutcraft in order to 'recharge' it. 'These smaller craft are incapable of generating their own power to any great extent,' explained Firkon, 'and make only relatively short trips from their carriers before returning for recharge. They are used for a kind of shuttle service between the large ships and any point of contact or observation, and are always dependent on full recharging from the power plant of the mother ship.'

Adamski and his companions proceeded to a large control room, rectangular in shape but with rounded corners. With the exception of two door openings, the entire room – about 45 feet long – was covered with coloured graphs and charts. Among other instruments in the room was a robot-type device which Adamski was cautioned not to describe, a miniature version of which he had observed in the scoutcraft. After a brief look round, he was shown into a lounge, the splendour yet simplicity of which took his breath away.

Not less than 40 feet square and about 15 feet high, the room was filled with a 'soft, mysterious blue-white light', with no trace of its source. As he stepped in, Adamski was greeted by two 'incredibly lovely young women', who offered him a glass of water, of a denser type than found on Earth. His description of these women reads like something out of a fairy tale:

> The one who had brought me the water was about five feet three inches in height. Her skin was very fair and her golden hair hung in waves to just below her shoulders in a beautiful symmetry. Her eyes, too, were more golden than any other colour . . . Her almost transparent skin was without blemish of any kind, exquisitely delicate, though firm and possessed of a warm radiance. Her features were finely chiselled, the ears small, the white teeth beautifully even. She looked very young . . . Her hands were slender, with long, tapering fingers. I noticed that neither she nor her companion wore make-up of any kind . . .

Adamski called this 'Venusian' woman 'Kalna' and the other – a 'Martian' – 'Ilmuth'. The latter, a brunette, was taller than Kalna, with large black 'luminous' eyes. Both women wore robes of a veil–like material which fell to their ankles, bound to their waists by a girdle of a contrasting colour, and tiny sandals.

The room was furnished with a long table surrounded with chairs, divans and settees of different designs and sizes, but lower and more comfortable than those on Earth. 'They were covered in a material of a deep soft nap with a brocade effect,' Adamski claimed. 'The colours varied . . . rich, warm and subdued.' Beside the chairs were low glass- or crystal-topped tables with decorative centre-pieces. The entire floor was covered by a single luxurious rug which reached to the walls, of a plain medium-brown colour.

Various pictures were placed around the walls, evidently depicting scenes from another planet, with completely different architecture. One picture showed a large mother ship. As the thought passed through Adamski's mind that this was the craft they were on, Kalna corrected it 'No, our ship is really very small in comparison,' she said. 'That one is more like a travelling city than a ship, since its length is several miles, while ours is only two thousand feet.' Adamski struggled to comprehend such a fantastic size. 'Many such ships have been built,' Kalna continued. 'However, they are not intended for the exclusive use of any particular planet, but for the purpose of contributing to the education and pleasure of all citizens in the whole brotherhood of the Universe.'

Adamski was taken up one level to the flight deck, where Firkon explained that the mother ship carried many pilots, working in shifts of four – two men and two women – as well as twelve scoutcraft and many mechanical devices, including pressurizing equipment installed between the walls. This particular ship, said Ilmuth (who was on the point of taking the controls), had four such walls or skins.

Suddenly, openings like portholes appeared in the walls, and both pilots took their places in small seats on opposite sides of the flight deck. 'I felt a slight movement and the ship seemed to nose upward,' reported Adamski. 'We are now about 50,000 miles from your Earth,' said Ilmuth.[47]

Firkon then invited Adamski to look out of one of the portholes. What he said he saw in 1953 is corroborated by views described by some astronauts years later. With other corroborations, it persuades me that not all of George Adamski's encounters with the 'space people' were fantasies – nor were they all the product of deception.

NOTES

1 Petersen, Hans C., *Report from Europe*, Scandinavian UFO Information (SUFOI), Jylland, Denmark, 1964, p. 91.

2 FBI Office Memorandum to the Director, FBI, from the Special Agent in Charge, San Diego (100-8382), 28 January 1953.

3 Zinsstag, Lou, and Good, Timothy, *George Adamski: The Untold Story*, Ceti Publications, Beckenham, Kent, UK, 1983 (out of print).

4 Lorenzen, Coral E., 'Looking Back – at Adamski', *The APRO Bulletin*, Aerial Phenomena Research Organization, Tucson, Arizona, vol. 31, no. 6, 1984, pp. 1–3.

5 FBI Office Memorandum, 28 January 1953.

6 Leslie, Desmond, and Adamski, George, *Flying Saucers Have Landed*, Werner Laurie, London, 1953, p. 171.

7 FBI Office Memorandum to the Director, FBI, from the Special Agent in Charge, San Diego (100-8325), 28 May 1952.

8 Leslie and Adamski, op. cit., pp. 187–9.

9 Interview with the author, Escondido, California, 20 November 1979.

10 Leslie and Adamski, op. cit., pp. 192–3.

11 Interview with the author, 20 November 1979.

12 Leslie and Adamski, op. cit., pp. 193–7.

13 Ibid., pp. 197–203.

14 Zinsstag and Good, op. cit., pp. 6–7.

15 Ribera, Antonio, 'The Soria Abduction', Parts I, II and III, *Flying Saucer Review*, vol. 30, nos. 3, 4 and 5, 1985.

16 Leslie and Adamski, op. cit., pp. 204–5.

17 Ibid., pp. 205–10.

18 Ibid., pp. 210–13.

19 Ibid., opposite p. 192.

20 Interview with the author, 20 November 1979.

21 Letter from Jerrold Baker to Frank Scully, 31 January 1954, published in *Nexus*, edited by James Moseley, January 1955, pp. 15–16.

22 Leslie and Adamski, op. cit., pp. 212–13.

23 Homet, Marcel F., *Sons of the Sun*, Neville Spearman, London, 1963, p. 185.

24 Leslie and Adamski, op. cit., pp. 217–18.

25 'Adamski's Answer to Baker', *Mystic Magazine*, June 1955, pp. 96–7.

26 Affidavit from Jerrold Baker, 29 June 1954, published in *Nexus*, vol. 2, no. 1, January 1955, p. 14.

27 Letter from Jerrold Baker to James Moseley, 18 November 1954, published in *Nexus*, vol. 2, no. 1, January 1955, pp. 14–15.

28 FBI Office Memorandum to the Director, FBI, from the Special Agent in Charge, San Diego (100-8382), 28 January 1953.

29 Interview with the author, 20 November 1979.

30 Leslie and Adamski, op. cit., pp. 219–20.

31 FBI Office Memorandum, 28 January 1953.

32 FBI Office Memorandum to the Director, FBI, from the Special Agent in Charge, San Diego (100-8382), 15 December 1953.

33 Letter to an enquirer (name deleted) from John Edgar Hoover, Director, FBI, 17 December 1953.

34 'Desmond Leslie Answers David Wightman', *Flying Saucer Review*, vol. 6, no. 5, 1960, pp. 3–5.

35 Meeting at Veterans' Administration Building, Quontset T-26 (city not named), 1 June 1953, as published in Leslie and Adamski, op. cit., p. 229.

36 *Nexus*, vol. 2, no. 1, January 1955, p.13.

37 Good, Timothy, *Beyond Top Secret: The Worldwide UFO Security Threat*, Sidgwick & Jackson, London, 1996, pp. 441–4.

38 Ibid., pp. 444–5.

39 Adamski, George, *Inside the Space Ships*, Arco Spearman, London, 1956, pp. 33–40.

40 Leslie, Desmond, 'Commentary on George Adamski', in *Flying Saucers Have Landed* by Desmond Leslie and George Adamski, revised edition, Neville Spearman, London, 1970, p. 245.

41 Adamski, op. cit., pp. 41–8.

42 Adamski, George, *Flying Saucers Farewell*, Abelard-Schuman, London, p. 39.

43 Recorded discussion between George Adamski and Captain Edward Ruppelt, Palomar Terraces, 9 April 1955.

44 'Anchor', *Transvaal Episode*, The Essene Press, Corpus Christi, Texas, 1958, p. 37.

45 Adamski, *Inside the Space Ships*, p. 48.

46 Ibid., p. 44.

47 Ibid., pp. 48–67.

Chapter 7

Claims, Contradictions and Corroborations

'I was amazed to see that the background of space is totally dark,' reported George Adamski, as he gazed in wonderment through one of the giant space carrier's portholes. 'Yet there were manifestations taking place all around us, as though billions upon billions of fireflies were flickering everywhere, moving in all directions, as fireflies do. However, these were of many colours, a gigantic celestial fireworks display that was beautiful to the point of being awesome.'[1]

On 20 February 1962, nine years after Adamski's first alleged flight in space, astronaut John Glenn, orbiting the Earth in the Mercury VI space capsule, described a similar scene:

> At the first light of sunrise – the first sunrise I came to – I was still faced back towards the direction in which I had come from with normal orbit attitude and just as the first rays of sun came up on to the capsule I had glanced back down inside to check some instruments and do something and when I glanced back my reaction was that I was looking out into a complete star field – that the capsule had probably gone up while I wasn't looking out the window and that I was looking into nothing but a new star field. But this wasn't the case, because a lot of the little things that I thought were stars were actually a bright yellowish green about the size and intensity as looking at a firefly on a real dark night . . . there were literally thousands of them.[2]

Corroboration was provided by Soviet cosmonauts Vladimir Komarov, Konstantin Feoktistov and Boris Yegorov in Voshkod 1 on 12 October 1964:

> The luminous particles were visible only against a black sky with the sun shining from the side . . . Their movement is strange. Sometimes we saw two particles moving towards each other. The general feeling was that these tiny particles came from our ship; apparently, these are simply dust particles that are found everywhere, even in the cosmos.[3]

There remain dissenting opinions regarding the nature of the 'firefly effect'. Lieutenant General Thomas Stafford, US Air Force (Retired), who flew two Gemini missions in 1965/6 and was Commander of Apollo X, which orbited the Moon in May 1969, told me in 1996 that he believes the effect is caused by sunlight shining on sublimated particles ejected from the thrusters used for positioning the spacecraft, which (on Gemini) were fuelled by hydrogen-peroxide gas. He also pointed out that water and urine dumped from our manned spacecraft can give rise to similar effects.[4] This is true, though micrometeoroids ('space dust') remains the better explanation in other cases.

A REMARKABLE DEVICE

During this first flight on board one of the giant carrier craft, Adamski claimed he was shown an instrument, no larger than an ordinary (fifties) cabinet radio and with a screen similar to that of a television. 'With this,' explained Firkon, 'we can picture and register anything taking place on the Earth, or on any planet over which we either pass or hover. Not only do we hear the spoken words, but pictures are picked up and shown on the screen. An internal mechanism breaks these down into sound vibrations, which are simultaneously translated into words of our own language, all of which are recorded in a manner similar to your own tape-recordings.'

Firkon went on to explain that all words are made up of vibrations or scales similar to a musical octave, just as all melodies are composed of certain notes, thus unknown languages can be learned in a relatively short time. When strange vibrations become apparent, these are transposed into picture form, showing exactly what the strange words or their vibrations mean.[5]

HARMFUL RADIATION BELTS

Adamski was introduced by his hosts to an older man, described as 'a greatly evolved being', unbelievably aged 'close to one thousand years', who warned of the harmful effects of Earth's nuclear detonations:

Even though the power and radiation from the test explosions have not yet gone out beyond your Earth's sphere of influence, these radiations are endangering the life of men on Earth. A decomposition will set in that, in time, will fill your atmosphere with the deadly elements which your scientists and your military men have confined into what you term 'bombs'.

The radiations released from those bombs are now only going out so far, since they are lighter than your own atmosphere and heavier than space itself. If, however, mankind on Earth should release such power against one another in full warfare, a large part of Earth's population could be annihilated, your soil rendered sterile, your waters poisoned and barren to life for many years to come. It is possible that the body of your planet itself could be mutilated to an extent that would destroy her balance in our galaxy . . . For us, travelling through space could be made difficult and dangerous for a long time to come, since the energies released in such multiple explosions would then penetrate through your atmosphere into outer space.[6]

SECOND FLIGHT INTO SPACE

During his second alleged flight into outer space, on 21 April 1953, Adamski was taken on board a different ('Saturnian'!) scoutcraft, four times the diameter of the 'Venusian' one, prior to a flight in a larger carrier ship. Shown into what he was told was a kitchen, Adamski could see no fittings or cupboards of any type. The appearance proved deceptive. 'Zuhl', the host, explained that the walls were lined from top to bottom with cupboards and compartments which, like all doors on the craft, were invisible until opened. In these cupboards, food and everything necessary for its preparation was stored. A small glass-like door set into one of the walls proved to be an oven. No burners of any description could be discerned. Zuhl explained:

> We do not cook our food in the same way as you. Ours is done quickly by means of rays or high frequencies, a method with which you are now experimenting on Earth. However, we prefer most of our food in the state in which it is grown, and live chiefly on the delicious fruits and vegetables which abound on our planets. To all intents and purposes we are what you call 'vegetarians', but in emergencies, if no other food is available, we do eat meat.

Ushered into a nearby lounge, Adamski noted that the floor-covering was yellow-grey in colour, with a texture similar to thick sponge rubber. As before, couches and individual seats were scattered about. 'I saw no books, papers or reading matter of any kind,' he said, 'nor did I see any shelves or cases in which something of this kind might be kept.'[7]

On board the carrier ship were people similar to those he had encountered on the previous trip. Although none of the women looked to be older than their early twenties, Adamski learnt later that their ages varied from thirty to two hundred years. On this occasion they were wearing 'beautiful, sheer gowns' with wide belts decorated with gems that 'sparkled with a softness and vitality such as I had never seen on Earth'. The men wore 'gleaming white blouses' with long full sleeves drawn in tight at the wrists. The trousers were also loose, 'very similar to our own styles', but made of an unknown material.

> The men's height varied from about five to six feet, and all were splendidly formed, with weight in proportion. Like the women, they varied in colouring, but I noticed that the skin of one was definitely what we would call copper-coloured. All had neatly trimmed hair, although it differed in length and cut to some degree, as here on Earth. None wore long hair . . . The men's features, though uniformly handsome, were not greatly different from those

of Earth men, and I am positive that any of them could come amongst us and never be recognized as not belonging here.[8]

REMOTELY CONTROLLED DISCS

On this carrier ship, one 'laboratory room' housed twelve small, unmanned discs (similar to the 'foo-fighters' discussed in Chapter 1). 'I guessed immediately that these were the registering discs or small, remotely controlled devices sent out by the mother ships for close observation,' said Adamski.

> They were about three feet in diameter, of shiny, smooth material, and shaped rather like two shallow plates, or hub-caps, turned upside down and joined at the rims so that the central part was a few inches thick. I learned, however, that such discs varied in size from about ten inches to twelve feet in diameter, depending on the amount of equipment carried . . . they contained highly sensitive apparatus which not only guided each little saucer perfectly in its desired path of flight, but also transmitted back to the mother ship full information on every kind of vibration taking place in the area under observation . . . sound, radio, light – and even thought waves . . .

At a flight deck, six women worked quickly and nimbly as they fed instructions and flight data to the waiting discs. 'I remember noting the resemblance to six women playing in a pantomime, a silent concerto,' commented Adamski. 'It was fascinating to see how, when a disc had received full "instructions", one of the trap doors would open and the disc would slide smoothly into the orifice, passing through air-locks before hurtling away into outer space on its mission.'

Meanwhile, back at the laboratory, screens registered what the small discs were recording. 'I noticed on one of the screens varying lines shaping, disappearing and reappearing in new formations,' said Adamski. 'The lines would then be replaced by round dots and long dashes, which would quickly form into various geometrical figures.' The purpose of these was explained by Adamski's extraterrestrial companions:

> The discs are now hovering above a certain spot on Earth and registering the sounds emanating from that spot. This is what you are seeing on the screen as shown by the lines, dots and dashes. The other machines are assembling this information and interpreting it by producing pictures of the meanings of the signals, together with the original sounds . . .
>
> Everything in the Universe has its own particular pattern. For example, if someone speaks the word 'house', the mental image of a dwelling of one kind or another is in his mind. Many things, including human emotions, are registered in the same way. By the use of these machines, we know even what your people are thinking, and whether or not they are hostile toward us . . .
>
> On each side of the mother ship, just below the disc-launching ports, is a magnetic ray projector. When a disc goes out of control, a ray is projected to

disintegrate it. This accounts for some of the mysterious explosions that take place in your skies . . . On the other hand, if a disc goes out of control near the surface of a planet where an explosion might cause damage, it is allowed to descend to the ground where a milder charge is sent into it [which] causes the metal to disintegrate in slow stages. First it softens, then turns into a kind of jelly, then a liquid, and finally it enters a free state as gases, leaving not a wrack behind.[9]

It was alleged that, by means of these discs, the space people were alerted to an abnormal condition building up on the fringe of our atmosphere – 'a condition constantly increasing with every atomic or hydrogen bomb that is exploded on Earth'.[10]

THE MOON

It was on this second purported trip into space that Adamski was shown – on a viewing screen – the surface of our Moon. His descriptions have given rise to ridicule, yet some are intriguing:

> I was amazed to see how completely wrong we are in our ideas about this, our nearest neighbour. Many of the craters are actually large valleys surrounded with rugged mountains, created by some past terrific upheaval within the body of the Moon . . . True, some of the craters had been formed by meteorites hitting the Moon's surface, but in every such case, these craters showed definite funnel bottoms. And as I studied the magnified surface of the Moon upon the screen before us, I noticed deep ruts through the ground and in some of the imbedded rock, which could have been made in no other way than by a heavy run-off of water in times past. In some of these places there was still a very small growth of vegetation perceptible. Part of the surface looked fine and powdery, while other portions appeared to consist of larger particles similar to coarse sand or fine gravel . . .[11]

With the obvious exception of the 'very small growth of vegetation', all these descriptions were confirmed years later. Prior to unmanned landings on the Moon, astronomers argued about its surface structure. Dr Thomas Gold of the Greenwich Royal Observatory, for example, stated in 1955 that the lunar *maria* ('seas') were covered with a layer of dust so thick that anyone attempting to land in one might be swallowed up.[12] Neil Armstrong, speaking from first-hand (or first-foot) experience, dispelled those speculations as he stepped down from Apollo XI's lunar excursion module on that memorable 20 July 1969. 'The surface is fine and powdery,' he reported.

Adamski's most ardent detractors have to be impressed by the identical descriptions in each case; however, his description of 'a small animal . . . four-legged and furry'[13] that he saw on the lunar surface through the viewing screen has not been confirmed!

Adamski's perhaps most outlandish claim was made relative to his third claimed flight into space, on 23 August 1954, when he was again shown the Moon on a viewing screen. On this occasion he was shown views of the other side, supposedly depicting a temperate section around the equator, with snow-capped mountains, forests, lakes, rivers and even a 'fair-sized city' where, he was informed, human beings could live comfortably, given sufficient depressurization.[14]

In defence of this claim, Desmond Leslie pointed out that several photographs taken by the Apollo VIII crew show pronounced greenish hues on the lunar surface, giving the impression that one is looking at high-altitude forests. One picture shows what looks like a beautiful blue lake. If the best cameras on Earth recorded the Moon in this way, Leslie argued, then Adamski can be forgiven for falling victim to an 'optical illusion'. He also points out that Adamski had poor sight (he suffered from a cataract), but hated to admit it and never carried his glasses around with him.[15] The poor eyesight may explain some inconsistencies in his descriptions, but it hardly explains the fact that he claimed to have been *told* by his hosts about the forests, lakes and rivers, and so on.

Adamski reported that during a trip to Holland in May 1959, he was listening in his hotel room to a BBC Radio programme – *The News of Europe* – when he was surprised to hear a report by a Russian scientist stating that the Moon was not composed of volcanic dust, but rather of granite formations similar to Earth. Furthermore, many green spots that looked like vegetation had been observed on the other side of the Moon.[16]

The greenish hues in some of the Apollo VIII photographs are remarkable. I have a superb exhibition print of the Schmidt crater. There is no denying the impression of possible 'moss-type' vegetation conveyed by the dark-green colour surrounding the crater, while the crater itself is a mixture of white, fawn and pinkish-coloured areas (see colour plates). Even astronomer Patrick Moore credited the possibility of vegetation on the lunar surface. 'On the whole moon there is no living thing,' he stated, 'apart perhaps from a few scattered patches of lichens or moss-type vegetation on the floors of some of the craters.'[17]

John McLeaish of NASA informed me that the greenish tint is due to a slight underexposure of the Ektachrome (SO–368) film:

> Color films tend to produce a color when viewing a slightly underexposed neutral subject . . . In general the far side of the moon is topographically higher than the side we view from earth. A major exception is the crater Tsiolkovsky and its mare. It is the darkest location, densitometrically on the moon (greenish in some photos) and it is believed to be the lowest topographic point on the moon by many scientists.[18]

While it is correct that improperly exposed film can produce false colour, the extent depicted in some Apollo VIII photos seems extreme, and my photograph of the Schmidt crater, for example, has been perfectly exposed.

A few weeks after hearing from McLeaish, I received an unsolicited letter from Paul D. Lowman Jr. of NASA's Planetology Branch:

> Your interesting letter on the apparent green color of the lunar surface has been supplied to me . . . I doubt if the green color is authentic. I base this opinion primarily on the eye-witness accounts given by the flight crews. For example, Bill Anders (Apollo 8) was emphatic that there was essentially no color on the lunar surface. Other crews have modified this, saying that it appears a brownish color. However, none have reported any green . . .

I asked General Thomas Stafford what colours, if any, he had seen while orbiting the Moon. He replied that in the early morning the mountain peaks showed a reddish glow as the sun came up; later, the surface assumed a light tan colour, changing to a dazzling white at noon. He reports never seeing green colours, and believes the Apollo photos showing such colours are caused by photographic processing effects.[19]

Fred Steckling, a staunch supporter of Adamski who claimed to have had several meetings with the 'space people',[20] argued that the areas on the Moon which Adamski described as being inhabited were in fact protected by giant invisible domes created by 'magnetic rays' which effectively maintained the air pressure at 7.5 pounds per square inch, thus shielding the occupants from the vicissitudes of the lunar environment (temperatures range from 230 degrees Fahrenheit at noon on the equator to -290 degrees Fahrenheit at night; the atmosphere is practically non-existent, and the gravity is one-sixth that of Earth's).

The French researcher René Fouéré earlier proposed a similar hypothesis to account for Adamski's description of rivers, lakes and forests. If the extraterrestrial colonists on the Moon were technologically superior to us, he argued, might they have been able to produce and contain an artificial atmosphere – 'a giant atmospheric bubble, within which lakes could be created, rivers made to flow, and snow made to fall . . .'?[21] Fouéré proposes another interesting hypothesis to account for Adamski's description; a hypothesis to which I shall return later in this chapter. Pointing out that the images of the Moon were projected on to a screen, Fouéré goes on to speculate that these might have been faked.

> If Adamski really did meet extraterrestrials, one might have thought the latter had deliberately shown him a false picture so that our men of science, reading Adamski's books later, might be convinced of the author's intellectual folly and dishonesty and, at the same time, of the non-existence of extraterrestrial craft. After all, the extraterrestrials, if they exist, may perhaps not be so keen that we should believe in their existence . . . It is possible that

they are out not to draw attention to themselves and that, if they had indeed had dealings with Adamski, they might have been able to condition him psychically in such a manner that, once out of their hands, he would go off and spread incredible fables around the world.[22]

LUNAR ANOMALIES

Despite absurdities and contradictions in Adamski's claims about the Moon, some mysteries remain. There are, for example, areas which suggest artificially constructed grooves or 'rilles'. These are usually explained geologically in terms of ancient river beds, bygone seismic activity, or collapsed 'lava tubes'; nonetheless, the rilles on the Gassendi crater (see plates) look artificial.

Also of interest are so-called 'Transient Lunar Phenomena' (TLP). Astronomer Patrick Moore, who coined the term, speculatively attributed the TLPs to 'gaseous emissions caused by moonquakes'. Even more interesting is the fact that water vapour has been detected, 'erupting like geysers through cracks on the lunar surface', according to instruments put there by Apollo XI and XII.[23] A water vapour cloud of more than 10 square miles was apparently discovered, and while Apollo VIII was in lunar orbit, astronaut Frank Borman was reported as having said: 'It looks like clouds down there.'[24] This bolsters Adamski's assertion that tenuous clouds occasionally form.

In late 1996 the Pentagon announced that radar soundings taken in mid-1995 by Clementine, an unmanned spacecraft, detected what seemed to be an enormous lake of frozen water at the bottom of the solar system's largest crater, known as the Aitkin basin, on the Moon's south pole. The crater itself is more than seven miles deep and the apparent ice lake is estimated to be tens of feet deep, covering an area of 30 to 50 square miles. Scientists at the Pentagon's Ballistic Missile Defense Organization, which sponsored Clementine, stated that the radar soundings of the Moon's polar areas had convinced them that the substance in the crater is indeed water ice, though further data were needed before this can be confirmed. An immense and permanent shadow in the crater prevents the ice from evaporating. Nothing like it was found in the Moon's north pole nor in another southern area exposed directly to the Sun. The presence of water on the otherwise apparently dry Moon is probably due to a collision with a comet (composed largely of ice) perhaps 3.6 billion years ago, the scientists speculated. This important discovery will facilitate living on the Moon, when eventually we establish colonies there.[25]

Stranger still are what appear to be artificial constructions occasionally observed on the Moon. Frank Halstead, Curator of the Darling Observatory in Duluth, Minnesota, described an object he and his assistant observed through a telescope on 6 July 1954. It looked like a 'straight black line' on the floor of the crater Piccolomini, where no such line had been noted before. It has not been seen since. The object was also confirmed by the Tulane Observatory.[26] Halstead also reported seeing a cigar-shaped craft about 800 feet long, later joined by a disc estimated to be 100 feet in diameter, while crossing the Mojave Desert in California by train with his wife, on 1 November 1955.[27]

Further evidence for constructions on the Moon came from the British astronomer Dr H. P. Wilkins (whose UFO sighting is described in Chapter 8), when he reported what looked like a curved 'arch or bridge', two miles long, on the Mare Crisium, following its discovery by the American science writer John O'Neill, in 1953. Using a less powerful telescope than Wilkins's, O'Neill mistakenly assumed it had a span of 12 miles. 'It looks artificial,' stated Wilkins on BBC Radio in December 1953. 'It's almost incredible that such a thing could have been formed in the first instance (by nature), or if it was formed, could have lasted during the ages in which the moon has been in existence.'

The lunar 'bridge' was explained as an illusion caused by the Sun's rays shining obliquely through a gap between two rocky promontories. Adamski had another, equally unlikely explanation: the bridge, he claimed, was a mile-long mother ship undergoing maintenance. When something goes wrong with one of these gigantic craft, he explained, it is brought down between two mountain tops. If some of the larger machinery needs to be moved, another mother ship straddles the two mountains and is used as a crane to lift the heavy machinery. 'This will give the appearance of a bridge,' he said, 'but as soon as they have repaired the other ship, they both leave, so your so-called bridge has gone.'[28]

On balance, some evidence supports Adamski's claim that alien bases – if not rivers, lakes, forests and so on – exist on the Moon. Like others, it is a claim that we would be unwise to dismiss out of hand, ridiculous though it seems. Lieutenant Colonel Philip Corso, US Army (Retired), who served in the Army staff's Research and Development directorate at the Pentagon in the early 1960s, claims that, in 1961, NASA 'agreed to co-operate with military planners to work a "second-tier" space program within and covered up by the civilian scientific missions [and] to open up a confidential "back-channel" communications link to military intelligence regarding any hostile activities conducted by the EBEs [extraterrestrial biological entities] against our spacecraft even if those included only shadowing or surveillance'. Corso claims that the US Army and Air Force possessed at least 122 photos taken by astronauts on the Moon 'that showed some evidence of an alien presence'.[29] He adds:

An extraterrestrial presence on the moon, whether it was true or not in the 1950s, was an issue of such military importance that it was about to become a subject for National Security Council debate before Admiral Hillenkoetter and Generals Twining and Vandenberg pulled it back under their [UFO] working group's security classification. The issue never formally reached the National Security Council, although Army R&D [Research and Development] under the new command of General Trudeau in 1958 quickly developed preliminary plans for Horizon, a moon base construction project designed to provide the United States with a military observation presence on the lunar surface . . . Horizon was supposed to establish defensive fortifications on the moon against a Soviet attempt to use it as a military base, an early-warning surveillance system against a Soviet missile attack, and, most importantly, a surveillance and defense against UFOs.

'Years later,' continues Corso, 'there was even some speculation among Army Intelligence analysts who had been out of the NASA strategy loop that the Apollo moon-landing program was ultimately abandoned because there was no way to protect the astronauts from possible alien threats.'[30]

MEETINGS WITH VIPs

Following the success of his books, Adamski became a celebrity. He was invited all over the world to lecture as well as to meet many prominent people. In 1959, he was invited by Queen Juliana of the Netherlands to Soestdijk Palace at The Hague. The meeting, which took place on 18 May, was attended by dignitaries and experts, including Prince Bernhard; Lieutenant-General H. Schaper, Chief of the Royal Netherlands Air Staff; and Professor Jongbloed, an expert in aviation medicine.

After a question from the Queen about one of his trips around the Moon, Adamski was asked some facetious questions by General Schaper and by an astronomer. 'I have known of no major officials of our Air Force, and few astronomers, who have told what they actually know about the visitors from space,' he responded. 'It is a known fact that the secret files and confidential reports of the Air Force have never been released to the public, or even to high officers in the government. I am inclined to believe this applies to all governments.'

Although many questions were asked about the space people, most questions dealt with mankind's future in space. Going well over its 45-minute allotted time, the meeting lasted two hours. Later, on arrival at the hall where he was to give a lecture, Adamski was besieged by the press and members of the public determined to know what had been discussed at the meeting. 'This I could not do,' he said, 'for the meeting had been on a level of dignity that denied me the privilege of speaking until the

Queen spoke first.'[31] The press was furious, and most of the newspapers heaped ridicule on the meeting – without knowing what had been discussed. 'New Scandal in the Court of Holland' was the banner headline in France, Belgium and Switzerland.

Carol Honey, an aerospace engineer who was Adamski's right-hand man for seven years, told me that although he did not accept all of the contactee's claims, he was present on several occasions at Palomar Terraces (Palomar Gardens having been sold in the mid-1950s) when Adamski was visited by various government and military officials.[32] Adamski also claimed that he was consulted by the Air Force regarding an incident in which one of their jets was three hours overdue. Eventually it came in for an unannounced landing, minus pilot and co-pilot! According to Adamski, the incident took place in Washington on 6 July 1956, and the information came to him in the form of a letter from a base in the Panama Canal Zone. The Air Force wanted to know what had happened. Adamski replied that he would check with 'the boys' (as he later called his space contacts), and two weeks later informed the Air Force that the pilots had been 'picked up' by a spacecraft and given the choice of returning to Earth or going with its crew. Apparently, they chose the latter option, because a number of pilots who had been similarly 'abducted' in the past and then returned to tell their stories had been mistreated and ridiculed. To show good faith, the 'boys' returned the aircraft back to its base under remote control. The Air Force purportedly confirmed it had arrived at a similar conclusion.[33]

In March 1960, Adamski claimed to have received a telegram summoning him to official meetings in New York and Washington. One of these meetings, he related to his co-workers, was with an aide to Dag Hammarskjöld, then Secretary-General of the United Nations, as well as others. 'In New York,' he wrote, 'I had the great honor of dining and visiting for one and a half hours with Dag Hammarskjöld's right-hand man in the UN. Original arrangements had been made for me to meet Mr H. but the African conflict took him away at just that time. But I learned a great deal of interest to us all . . .'

During the same period, Adamski had a 15-minute meeting with Senator Margaret Chase-Smith (Maine), Chairman of the Senate Finance Committee for Space Research at the time. 'I gave her as much information as possible,' he reported, 'for which she thanked me. This was an accomplishment that I little expected . . .'[34]

POPE JOHN XXIII

On 31 May 1963, Adamski claimed to have had a private meeting at the Vatican with the ill Pope John XXIII, to deliver an important package which he said had been given to him by one of the space people in

Copenhagen. The claim has been roundly denounced, of course, yet circumstantial evidence suggests that such a meeting did take place.

Expected to be at a certain entrance at 11.00, Adamski was accompanied as far as the steps of St Peter's by his co-workers May Morlet from Belgium and Louise (Lou) Zinsstag from Switzerland. Lou, whom I knew as a close and trusted friend for nearly 20 years, described the occasion as follows:

> May and I got him there in good time. We walked slowly up the broad central stairway, looking around us. Within a few minutes, George cried out: 'There he is – I can see the man! Please wait for me here in about an hour's time.' He descended the steps swiftly, turning to the left. I had looked to the right because I expected him to be admitted through the well-known gate where the Swiss Guards were posted. Yet, without any hesitation, he walked to the left of the Dome where I now noticed a high wooden entrance-gate behind the open doorway, with a small built-in door. This door was partly opened and a man was standing inside it, gesturing discreetly to George. He wore a black suit but not a priest's robe. On his chest I noticed some kind of coloured material in white, green and red . . . I made a mental note that Adamski's being received at a gate other than the usual one where the Swiss Guards checked on every visitor meant that he would not be registered on the daily visitors' list, and that his visit would probably not be recorded officially by the Vatican. This, I realized, was very interesting in itself but would not be helpful if we had to look for proof.

> May and I returned an hour later. There was George already, grinning like a monkey. I never saw his face as happy as that; his eyes shining like beautiful topazes, something I shall never forget. 'We have done it,' he said. 'I was received by the Pope. He gave me his blessing and I gave him the message.'

> When later in the day we lunched with George he told us that the Pope was not lying in the room above St Peter's Square, as the people had been told, but that his bedroom faced the most beautiful part of the Vatican garden. And he added confidentially: 'If you ask me, the Pope is hardly a dying man. I have seen several people dying of cancer but the Pope's skin has still got a fine texture like a child's. They haven't yet tried to operate on him but I'm sure that's what they will do soon. He is not too old for that.' George added that the Pope even had rosy cheeks, and had said that he did not feel so bad.

> George had been helped on with a kind of cassock over his suit before he entered the bedroom. The Pope gave him a nice smile and said: 'I have been expecting you.' When George handed him the sealed message from Copenhagen, he said – also in English: 'That is what I have been waiting for.' He then spoke to his visitor in a very low and soft voice for a few minutes. Adamski had to bend his head down close to the Pope's, whose last words were: 'My son, don't worry, we will make it.' After receiving the papal blessing, Adamski was ushered out.

During lunch, Adamski took out of his breast pocket a small plastic wallet, lilac in colour. 'It bore the most singular inscription I have ever seen, protected by a transparent cover,' said Lou. 'The written characters were of a very unusual kind; certainly neither Roman nor Gothic, nor were they Russian, Chinese, Japanese, Arabic or Hebrew. But beneath the text was the date of the interview – 31 May 1963 – written in Roman letters. We sat spellbound when George opened the little wallet.'

Embedded in white cotton was a most beautiful golden coin with the Pope's head in profile on it. As I discovered later, it was a new ecumenical coin, ready for sale but not yet on the market, the impending Ecumenical Council having been postponed due to the Pope's illness. Weighing the coin in my hand I felt sure it was of at least 18- if not 22-carat gold (my father was a goldsmith). Two weeks later the coin was on sale in European banks and I went to have a look at it. Its price was between 300 and 400 Swiss francs. At the time of our visit to Rome, Adamski could not have bought it even if he had had the money, which he certainly did not, and, as in Basle, he had never left our hotel without us and he had never entered a shop.[35]

Two days later the Pope died. Not surprisingly, there has been no official confirmation of Adamski's meeting. Following an ambiguous response from the Vatican to Ronald Caswell, one of Adamski's British co-workers, stating that they were unable to provide the required information (see Fig. 9), I followed up with another request. 'With regard to the alleged private audience granted by Pope John XXIII on 31 May 1963,' I was informed by an officer of the papal court, 'I would assure you that no such private audience ever took place.'[36] I received no response to my enquiry as to how Adamski managed to obtain the coin.

Despite Lou Zinsstag's misgivings about a number of Adamski's later claims, leading to her resignation as one of his co-workers, she remained convinced for the rest of her life that Adamski had been granted a papal audience. 'I knew for certain that he was expected and received by Pope John XXIII,' she wrote, 'which also made me inclined to believe him when he told us of other secret meetings in other important buildings, such as the White House.'[37]

PRESIDENT JOHN F. KENNEDY
According to Madeleine Rodeffer, with whom he stayed for the last few months of his life in 1965, Adamski had at least one clandestine meeting with President John Kennedy. She related to me how in May 1963 Kennedy allegedly visited Adamski late one night at the Willard Hotel, close to the White House. There is no substantiation for this claim, though it is a fact that Adamski possessed a US Government ordnance card which gave him access to all US military bases and to certain restricted areas, and this might lend support to his claim of having once visited the White House itself. Lou Zinsstag was also given the story:

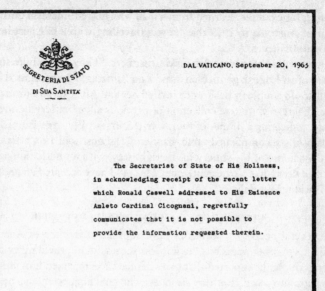

DAL VATICANO, September 20, 1963

The Secretariat of State of His Holiness,
in acknowledging receipt of the recent letter
which Ronald Caswell addressed to His Eminence
Amleto Cardinal Cicognani, regretfully
communicates that it is not possible to
provide the information requested therein.

Fig. 9. A letter from the Vatican in response to a request by Ronald Caswell for details of George Adamski's private audience with Pope John XXVII in May 1963. (*Ronald Caswell*)

He told me that he had been entrusted with a written invitation for President Kennedy to visit one of the space people's huge motherships at a secret airbase in Desert Hot Springs, California, for a few days. In order to keep this visit absolutely secret, Adamski was to take the invitation direct to the White House through a side door ... where a man he knew was ready to let him in. Adamski later learned that Kennedy had spent several hours at the airbase after having cancelled an important trip to New York, and that he had had a long talk with the ship's crew, but that he had not been invited for a flight.[38]

A WELL-CONNECTED GENTLEMAN

That Adamski was well connected is supported by others. Dr Jacques

Vallée, for example, learned from a man who hosted Adamski during his tour of Australia in 1959 that he was travelling with a passport bearing special privileges.[39]

Also remarkable were Adamski's manners. 'They were, quite simply, those of an English gentleman,' said Lou Zinsstag, who recounted how a well-to-do couple in Basle once invited her and Adamski to a formal dinner. There was quite a collection of precious silver cutlery beside each plate, indicating a dinner of four to five courses. 'We were astonished at Adamski's accomplished table manners,' the host told Lou afterwards. 'He made use of his cutlery in the right way, without hesitation, and he ate and drank like a true gentleman. He could have accepted an invitation to Buckingham Palace.'[40]

FURTHER ATTEMPTS TO SUBSTANTIATE THE CLAIMS

On several occasions, witnesses have substantiated some of Adamski's claims. On other occasions, he was less successful in providing evidence. For example, he reported that two scientists accompanied him on one of his trips into space. 'Both are scientists who hold high positions,' he wrote to Charlotte Blodget. 'However, the way things are nowadays with everything classified as security, for the time being they must remain in the shadow. When they believe that they can release the substantiation they have without jeopardizing either the national defense or themselves, they have said that they will do so through the press.'[41] Regrettably, the two scientists have not come forward, and it is doubtful if they are still living. Researcher Richard Ogden alleged that his friend Dr David Turner was one of the scientists.[42] If so, he was backward in coming forward.

In the postscript to *Inside the Space Ships*, Adamski wrote that on 24 April 1955 he was taken for a ride into space, specifically to fulfil his request to take some photographs inside the space ships. 'We can guarantee nothing for reasons which will be clear to you later,' one of his hosts told him, 'but we shall try to get a picture of our ship with you in it. This would be simple enough if we could use our own method of photography, but that would not serve your purpose. Our cameras and film are entirely magnetic and you have no equipment on Earth that could reproduce such pictures. So we must use yours and see what we can get.'

Adamski had brought along a new Polaroid camera, and explained its workings to his hosts. Orthon and Adamski allegedly stood in the carrier ship, looking through the very thick glass portholes – an estimated six-foot gap lay between the outer and inner windows – while a spaceman took photographs through a porthole of a nearby scoutcraft. 'From her ball top,' Adamski reported, 'she was throwing a beam of bright light upon the larger craft. Sometimes this beam was very intense, and again not so intense.'

ER-7-4372 A

OCT 4 1955

Honorable Gordon H. Scherer
House of Representatives
Washington 25, D. C.

Dear Mr. Scherer:

Thank you for your letter of 15 September 1955 with which you enclosed a letter to you from Thomas H. Eickhoff.

The questions which Mr. Eickhoff has raised in his letter to you
are largely outside of the jurisdiction of this Agency. Section 102(d) of
the National Security Act of 1947 provides that CIA shall have no police,
subpena, law-enforcement powers, or internal-security functions. Insofar as Mr. Eickhoff appears interested in pursuing the problem of
mail fraud in connection with George Adamski's book entitled "Inside
The Space Ships", it would appear to be a problem of law-enforcement,
from which we are specifically barred by statute.

CIA, as a matter of policy, does not comment on the truth or falsity
of material contained in books or other published statements, and therefore it is not in a position to comment on Mr. Adamski's book or the
authenticity of the pictures which it contains.

The subject matter of Mr. Adamski's book would appear to be more
in the jurisdiction of the Department of Defense and the National Science
Foundation, and it may be that you would wish to refer some of these
questions to them for consideration.

Mr. Eickhoff's letter is returned herewith for your files.

With kindest regards.

OGC:WLP/blc (20 Sept. 55)   Sincerely,
Orig. - 1 Add.
2 - Signer
2 - Legislative Counsel w/basic

Fig. 10. A letter from Allen W. Dulles, Director of the CIA (1953–61), to the
Honorable Gordon H. Scherer, House of Representatives, 4 October 1955,
regarding an allegation by Thomas Eickhoff of 'mail fraud' perpetrated by George
Adamski relating to his statement in *Inside the Space Ships* of having on one occasion
travelled aboard extraterrestrial spacecraft with two scientists. Nonetheless, as
revealed in *Beyond Top Secret*, Dulles reportedly stated that he would prevent
anyone from testifying in court concerning Adamski's book 'because maximum
security exists concerning the subject of UFOs'. (*CIA*)

As the photographs show, they were experimenting with the amount of light
necessary to show the mother ship and at the same time penetrate through the
portholes to catch Orthon and myself behind them. While this was going on,

radiation from both the mother ship and the Scout had been cut to a minimum. I learned later that the men had been obliged to put some sort of filter over the camera and lens in order to protect the film from the magnetic influences of the craft.

When the scoutcraft returned, Adamski studied the pictures. Although pleased, he cursed himself for not bringing more film along, and his hosts were less than satisfied with the results. While the four pictures (one of which is reproduced in the plate section) purportedly show the hull of the carrier craft (and one the edge of the scoutcraft's porthole) the faces peering through the portholes are not clear. Nonetheless, I see no evidence of photographic trickery.

The spacemen decided to attempt shots of the interior of a small flight deck with two pilots sitting at the controls. 'But these two attempts failed, due to the greater magnetic power in the carrier in comparison to that in the Scout,' Adamski explained. 'Without some as yet undeveloped filter system for our film, it is impossible to get clear photographs within the space ships. When I asked if a better camera with a finer lens might be more successful, I was told that any appreciable improvement was unlikely because of the type of film used.'[43]

In addition to the metal 'slag' given him by Orthon, Adamski claimed to possess another sample of alien alloy, which he had had analysed chemically:

> When I first telephoned to ask the result, this man sounded very excited. But when I saw him later in his laboratory, he . . . tried to brush the whole thing off lightly. When he said it was nothing that could not be picked up in any old scrapyard, naturally I persisted in demanding an explicit statement of his findings. He then admitted that there were 'slight differences' in composition from any usual alloy, but said that could have happened by a variation in heating or some 'slight accident' which had gone unnoticed at the time, thereby making duplication of the alloy improbable.[44]

In December 1958 Adamski claimed that a spaceman picked him up by car from his delayed train near Kansas City, Missouri, and drove him a short distance to a grove of trees above which a craft was hovering. The method of boarding was dramatically different from anything hitherto, and is worth recording here because of the similarity to descriptions provided in much later years by some abductees and contactees:

> I had the experience to be lifted up into the space craft while the ship was hovering. It feels as if something is surrounding you like a transparent or plastic curtain, yet you can't touch it and you don't see it, and like a magnetic force it lifts you just like an elevator into the ship. And they can do this from a thousand miles away if they want, but usually it is only two to three hundred feet. You can take baggage and everything with you, as if you are stand-

ing on a platform, even though you can't see it. This only works in the open and the person being lifted is visible the whole time.[45]

Once aboard, Adamski asked his friends if they would lower the craft and land him during daylight so that hundreds of people could witness the event, but it was explained that although they themselves could escape harm from our military, he would be arrested and held incommunicado. Even if hundreds of people did see him land and walk out of the ship they would quickly be silenced. The landing in Davenport, Iowa, where Adamski's train was heading, was therefore reportedly delayed until nightfall.[46]

MORE CLAIMS, MOUNTING DOUBTS

In 1960, according to Adamski, the US military exploded a nuclear device in the Earth's upper atmosphere. The Russians had a similar idea, he told Lou Zinsstag, but after the first American test the project was cancelled on both sides, supposedly due to warnings from the space people. (In actuality, the US Navy's Project Prime Argus exploded three such nuclear devices, in 1958.) Adamski also claimed that the space people were involved in neutralizing harmful radiation from our nuclear tests.

> From one of their space laboratories in a mothership, green balls are being sent out (they have been observed all over the world). This is being done in order to counteract or neutralize or even absorb concentrations of radiation created through our bomb experiments.[47]

I do not know about the frequency of sightings of 'green balls' in 1959, when Adamski made this statement, but I do know that from 1948 to 1950, at least, there was a plethora of reports of 'green fireballs' – particularly in the vicinity of nuclear test sites and installations – and that these were the subject of secret studies by military and scientific intelligence personnel, as I have recounted in *Beyond Top Secret* (see also Figs. 11, 12). As an Army Intelligence memo from the period reveals:

> Agencies in New Mexico are greatly concerned over these phenomena. They are of the opinion that some foreign power is making 'sensing shots' with some super-stratosphere device designed to be self-disintegrating.[48]

Adamski further claimed in 1959 that secret studies were being conducted into various methods of cancelling gravity. 'They already have models for anti-gravity propelled flying objects in disc form,' he wrote, 'but none of these methods for application of a free kind of energy must be revealed to the public, because such a society in possession of these advanced methods would soon escape from economic control.'

From 1960 onwards Adamski began warning his co-workers about a new and 'foreign' group of space visitors. Though admitting that it was

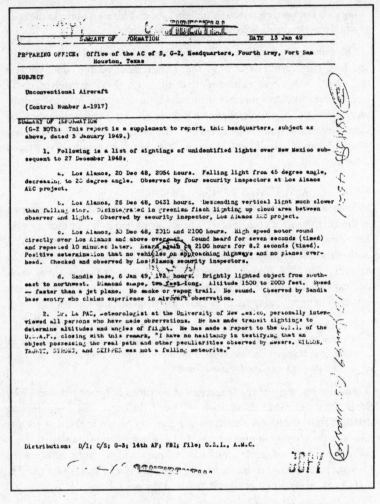

Fig. 11. A US Army Intelligence report, dated 13 January 1949, relating to sightings above Los Alamos. (*US Army*)

not easy for him to guess their purpose nor to estimate their numbers, he nonetheless speculated:

Who is to say that defense will not be needed? Since we are in the process of evolving, people of other planets throughout the Cosmos are likewise evolving. We are not the lowest in the Cosmos! Beyond our solar system are whole systems whose people have not progressed, socially, as far as we on Earth; yet some of them have advanced beyond us scientifically, and do have space ships.

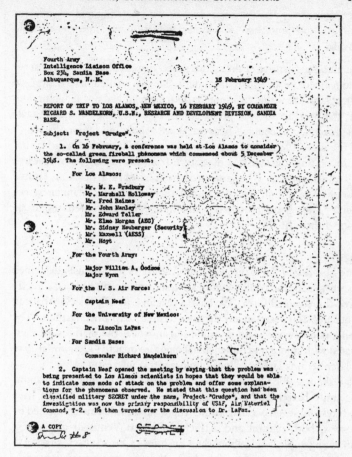

Fig. 12. The beginning of a lengthy US Army Intelligence report, dated
18 February 1949, relating to a secret conference at Los Alamos to discuss the
'green fireballs'. (*US Army*)

While we were confined to our planet with the thought that space was an
empty void, we naturally were of little interest to most outsiders. But now
that we are sending rockets and satellites into space we are attracting their
attention and natural curiosity, and this would cause them to investigate us.
Since we on Earth are warlike, with a history of wars, why should we not suspect that people of lower evolution socially would not be like us, or even more
so? Because they live on another planet and travel space as we are trying to
do, this certainly does not make them angels! And while they might not try to
attack the world as a whole at this time, our own space ships just might need,

one day, the protection which our military-trained personnel could provide.

Much to the surprise of the sixties peace groups, Adamski never acted as an apostle for peace and disarmament. 'He was no dreamer,' wrote Lou Zinsstag, 'and knew that the time for disarmament had not yet come, and worse, that world-wide war preparations were the most serious obstacle to the promotion of contacts between space people and senior officials.'[49]

By the early 1960s, Adamski's claims and philosophical treatises became increasingly esoteric; a departure from his usual, down-to-earth self. 'I must confess, I became tired of Adamski's articles on cosmic philosophy,' said Lou Zinsstag, until then one of his staunchest co-workers. 'They were moralizing and indulgent, and often singularly pointless I thought.'[50]

VENUS

During his alleged flight into space in August 1954, Adamski described how scenes from the planet Venus were shown to him, beamed directly in three-dimensional form. 'We have a certain type of projector that can send out and stop beams at any distance desired,' explained Orthon. 'The stopping point serves as an invisible screen where the pictures are concentrated with colour and dimensional qualities intact.' (The similarity to what later became laser-produced holograms is interesting.)

Adamski described 'magnificent mountains', some topped with snow, others quite barren and rocky, and some thickly timbered. 'We have many lakes and seven oceans,' said Orthon, 'all of which are connected by waterways, both natural and artificial.' Several Venusian cities were shown, with people 'going about their business'. Conveyances, varying in size and patterned somewhat after the mother ships, appeared to be gliding along just above the ground. Next came a beach scene beside a lake, complete with swimmers and 'long, low waves'. Animal life was shown, including horses and cows, both slightly smaller than those of Earth but otherwise very similar. 'This seemed to hold true of all animal life on Venus,' Adamski reported.

The constant cloudy atmosphere surrounding Venus was explained as a contributing factor towards an average life-span of 1,000 years. 'When the Earth, too, had such an atmosphere,' explained Orthon, 'man's years on your planet were correspondingly far greater than now.'[51]

In Chapter 2, I alluded to the harsh atmospheric environment of Venus, which precludes unprotected human, animal or even vegetable life (the atmospheric pressure alone is said to be about 90 times that of Earth). It is of course possible – as with Adamski's description of the Moon – that the areas shown might have been protected from the vicissitudes of the Venusian atmosphere by some advanced means, but the fact remains that the Venusians made no mention of such. And in 1961 Adamski claimed to have made an actual trip to Venus – a journey taking some 12 hours – followed by a five-hour visit. His description of the environment is equally fanciful:

> The day was warm with the sun shining brightly through a scattering of 'mackerel' cloud. The air was fresh and clean smelling. I was told it had rained the day before . . . While I had done little walking, I noticed myself fatiguing . . . The atmospheric pressure on Venus, in the vicinity I was, could therefore be compared to that of Earth's atmospheric pressure at the altitude and in a comparable location with Mexico City.[52]

In another (undated) report, Adamski went into further fanciful details:

> On Venus, as on Earth, 80% of the planet is covered with water. The cloud cover that does not permit us to see the surface of Venus is caused by constant evaporation of moisture. This permits a large tropical area where fruits and vegetables are plentiful. There are seasons when the rainfall is very heavy. And in some sections a light rain falls each day . . .

Moreover, Adamski was told that the rotation period of Venus is 23 hours (versus Earth's 24 hours), whereas in fact it is 243 *days*!

THE SATURN REPORT

Adamski's most ludicrous claim at this time was his supposed visit to the planet Saturn. In March 1962 his co-workers received a copy of his report on the 'Trip to the Twelve Counsellors' Meeting of our Solar System' that took place from 27 to 30 March that year. 'The [space] ship had come in on the 24th to one of our air bases where a high official of the US Government had a conference with the crew,' wrote Adamski. 'After the conference, the craft was returning to its home planet Saturn. The trip took nine hours, at a speed greater than 200 million m.p.h.'

Adamski's follow-up explanations were received with increasing scepticism by all but the most devoted co-workers. Lou Zinsstag regarded the Saturn story as a 'personal mental experience'. Henk Hinfelaar, a New Zealand co-worker, was equally doubtful. He reported that shortly after release of the Saturn report, 'many things began to happen which gave rise to doubts, and evidence began to pile up indicating that a great change was taking place as far as George Adamski was concerned . . . he alone can resolve it. Until he does there is no alternative but to by-pass him as a source of information.'

Hinfelaar and Zinsstag came to the conclusion that Adamski had 'got into the wrong hands'. He himself had admitted that he was now dealing with a 'new set of boys', as he put it. Lou elaborated:

> Either, we reasoned, his 'new set of boys' was an extremely clever fake organization, a group of secret agents (not necessarily governmental), trained experts in mind control and hypnosis, or else George was dealing with a new group of space people who were deliberately feeding him false information in order to confuse an issue which had been established by the earlier, friendlier group. Why not? 'They are no angels', George had said.[53]

DECEPTION

Author John Keel is convinced that all contactees were lied to by the 'ufonauts', this being 'part of the bewildering smoke screen which they have established to cover up their real origin, purpose, and motivation'.[54] Before full blame is put on alien deception, consideration must be given to the fact that Adamski himself sometimes deceived. In October 1962, for example, a silly letter was sent to the co-workers, purportedly written by one of his space contacts, in English but with transparently fake 'alien' symbols added, stating: 'You are doing good work. George Adamski is the only one on Earth that we support.' The letter was written by Adamski himself, sent from a post office box number rented for him by a friend, Martha Ulrich. This was the final straw for many of Adamski's supporters. 'Adamski himself is pulling down his original image,' wrote Roy Russell, an Australian co-worker.[55] It is also relevant that Adamski's 1964 *Science of Life Study Course*, which he claimed contained information and philosophy revealed to him by the 'space brothers', was, in fact, a rewrite of his first book, *Wisdom of the Masters of the Far East*, published in 1936 by the 'Royal Order of Tibet'.[56]

Former contactee Ray Stanford claims that in 1958, as a devoted 15-year-old follower of Adamski, he was visiting Palomar Terraces with his brother Rex one morning when Adamski started reminiscing:

> . . . during the Prohibition I had the [Royal] Order of Tibet. It was a front. Listen, I was able to make the wine. You know, we're supposed to have the religious ceremonies; we make the wine for them, and the authorities can't interfere with our religion. Hell, I made enough wine for half of Southern California. In fact, boys, I was the biggest bootlegger around . . . If it hadn't been for that man Roosevelt, I wouldn't have [had] to get into all this saucer crap.[57]

While I am inclined to believe that Adamski may have established the monastery of the Royal Order of Tibet to allow him to make wine in large quantities, I do not believe it was Prohibition that led him 'to get into all this saucer crap'. If he did say as much, I would attribute the comment to his sometimes earthy sense of humour. Even some of Adamski's most vociferous critics concede that he was a man totally dedicated to his mission and his philosophy. Stanford himself found Adamski to be a likeable man. 'I also saw a gentle, benign, artistic side to him,' he said. 'He showed us many paintings he had done that were quite good.' Stanford furthermore believes that there is much in favour of Adamski's original contact claim.[58]

Although negative evidence from Carol Honey is often cited by Adamski's detractors as conclusive proof of Adamski's charlatanism, the positive evidence is invariably overlooked. In 1979, Honey wrote:

> In Adamski's own words he did not 'go off the beam' until many years after his original contacts. Because I am interested only in the truth I told Adamski many times that I would support only that which had been proven to be true . . . Adamski turned over to me many manuscripts, most of his library and nearly all of his original files. For several years every word Adamski published came through my typewriter. His unpublished manuscripts remain unpublished because in my opinion they were 'off the beam' and not compatible with what I knew to be the truth . . .
>
> On various occasions, Adamski produced photographs, artifacts, recordings, laboratory reports, etc., which I examined closely. None was ever revealed to the public or press, so far as I am aware; yet all were more convincing than those things he did release. He claimed he was told not to 'reveal them until the proper time'. If he was fraudulent in all his claims, why didn't he reveal this stronger material? What happened to these items when he died?
>
> George Adamski met men from other planets [and] his photographs were genuine . . . later, after his contacts had ended, he misled the public rather than admit that the initial phases of the 'program' were over.[59]

A PIONEER OF SPACE

In 1949, a science-fiction book by 'Professor George Adamski' was published privately. Entitled *Pioneers of Space: A Trip to the Moon, Mars and Venus*, it was ghost-written by Lucy McGinnis and printed in a limited edition. In this book are striking similarities to *Inside the Space Ships* (1956), especially in Adamski's descriptions of 'Venetians, Martonians [*sic*] and Saturnians', though the majority of the descriptions in the earlier book bear little or no similarity to those appearing in the later one. Yet, for many, *Pioneers of Space* is a damning indictment of the later book. Interestingly, in the foreword, Adamski pointed out that: 'While this is at present in the field of fiction, the advance of science is so rapid that it will not be long before all this will become a reality.'

In *Pioneers*, for instance, is a remarkable resemblance to his 1953 description of the Moon:

> We find . . . a belt-like section extending as far as we can see around the Moon that has a natural growth of trees and vegetation [and] we see a small lake.[60]

Likewise, his comparison of extraterrestrial lifespans conforms with the later book:

> this man looks to us to be around seventy years of age. Later we learn that he is one hundred and ninety[61] . . . On Mars the span of life is from five hundred to one thousand years.[62]

It is in its graphic accounts of Venus and the 'Venetians' that *Pioneers* comes closest to his later descriptions:

> They tell us they have nine oceans, many lakes and rivers, majestic, towering mountains and very beautiful woodlands . . . [their cars] seem to be gliding right over the surface of the ground[63] . . . Breathing at first was somewhat difficult since the air is so light but we are able to quickly adjust ourselves . . .[64]
>
> In appearance they look more like men out of a dream than humans like ourselves . . . their hands are long and slender, rather delicate in structure . . . the women are far greater in beauty and expression[65] . . . We feel as though they are looking right through us and can actually read our minds. We have been told they are experts in the field of mental telepathy.[66]

Though it is obvious that Adamski embellished his later accounts with material probably drawn from *Pioneers*, there could be a less simplistic, if more incredible, explanation. In company with abductees decades later, Adamski claimed privately to have been contacted by extraterrestrials as a child, and to have received instruction from them in Tibet by way of preparation for his mission in later life. Publicly, Adamski made no such claims, though he hinted at having had 'mental' contact prior to the Desert Center encounter. 'Speaking of visitors from other planets,' he wrote to a correspondent in early 1952, 'in the physical I have not contacted any of them, but since you have read *Pioneers of Space* you can see how I get my information about these people and their homelands.'[67]

In 1958 Adamski is reported to have made an interesting statement to Ray Stanford which has been interpreted as an admission of fraud. 'Ray, listen,' he said, 'I did not ever have to go out into space to know about the spaceships. Hell, I knew about the spaceships and what was in 'em years ago . . . *Pioneers of Space* will tell you everything, just like *Inside the Space Ships*. All I did was project my consciousness to the beings out there and I could see them and know what was in their ships.'[68]

Does this necessarily mean that Adamski lied about all his 'actual' trips into space? I think not. Assuming that is precisely what he said, he does not deny having made such trips; he seemed to be implying rather that it simply was not absolutely necessary for him. Secondly, apart from those similarities already discussed, *Inside the Space Ships* is a very different book, full of much more richly detailed descriptions. Thirdly, those with whom I have spoken who knew Adamski well over many years told me that he remained adamant about actually having been inside the spaceships.

I asked Lucy McGinnis, Adamski's secretary for many years (until she 'defected' in the early 1960s), how she reconciled the two books. 'I have often wondered about that,' she replied thoughtfully. 'The first book was definitely written as fiction, and it might have been his way of breaking into the subject. He might have known something more – I don't know. It never bothered me to the extent that I made an issue of it because, you see, I *could* have made an issue of it if I hadn't seen those ships.'

In addition to having witnessed the Desert Center contact, Lucy had another sighting of a craft – similar in configuration to the classic 'scout' – which she saw at Palomar Terraces several years later. As she related to me:

> I was in my room lying down one afternoon. I don't know what date it was, but for some reason I got up and went out. As I got out the door, I looked up, and here was this great big saucer-like thing. I was amazed. As I looked up I could see *through* it. It was two stories; you could see the steps where they would go up and down. I don't remember how many people I saw, but they were moving around. It seemed to me they had kind of ski-suits, fastened around the ankle . . . Then suddenly it started just drifting away.

I also asked Lucy for her opinion as to why Adamski had begun telling such ridiculous stories in the early 1960s. She replied that his oversized ego was to blame, and offered her belief that the original group of extraterrestrials had left him for just that reason. She also felt that he was simply lying about the trips to Saturn and Venus by way of bolstering his ego, which had become seriously deflated when the original group left him.[69] Although this explanation is convincing, it still falls short of answering all the questions about this complex man and his even more complex claims.

An important clue to the Adamski mystery was provided for me by a friend whom I first met in 1952, and for whose integrity I can vouch. This person, whom I shall call Joëlle, claimed that in 1963 she met the same, or a similar group of extraterrestrials that Adamski knew, through a series of fortuitous circumstances (see Chapter 12). According to what Joëlle was told by the visitors, Adamski was indeed selected and contacted by this certain group of extraterrestrials, but at an early stage he disclosed some secret information with which he had been entrusted, and it therefore became necessary for them to feed him with false information which would discredit him, thereby protecting their own interests. Exactly what this disinformation was, I do not know, but I can say that it began to make its appearance in *Inside the Space Ships*. Joëlle told me that Adamski's account of the Desert Center contact, as described in *Flying Saucers Have Landed*, is essentially true. Joëlle's contacts confirmed that Adamski had indeed been on board their craft, but they would not say where they had taken him. They were equally reluctant to reveal their origin, other than saying that they had bases within our solar system, including on Earth. It was not made clear to what extent Adamski was aware of the disinformation that he disseminated.

Carol Honey told me that on one occasion Adamski had indeed betrayed such a confidence.[70] Assuming his earlier contacts were genuine, the pressures on him must have been great. 'My heart is a graveyard of secrets,' he once told Lou Zinsstag.

Desmond Leslie, seeking to rehabilitate his friend, invokes an esoteric explanation to account for the 'Venusians', 'Martians', 'Saturnians', *et al*. The 'brothers', he says, are able to 'materialize' in our environment but their own planets are on a 'higher vibratory frequency' than ours, hence life as we know it has not been discovered in our solar system.[71] I do not reject the hypothesis: however, apart from the fact that the 'brothers' do not necessarily originate in our solar system (as even Adamski stated privately, according to Carol Honey,[72] and which is implied in the information given to Joëlle), there is another point that can be overlooked.

Leslie cites Paramahansa Yogananda, the great yogi teacher, and discusses the extraordinary feats attributed to highly advanced avatars and masters said to be living on Earth who can levitate, render themselves invisible, project their images across vast distances, walk through walls, etc.[73] But this in no way alters the fact that these remarkable people are still flesh-and-blood human beings, albeit highly advanced physically, mentally and spiritually. It is my conviction that many extraterrestrials, too, are capable of these and other fantastic feats: indeed, in this respect I see little difference between highly evolved human beings said to live on this planet and those from any other. Adamski once told Leslie that we could not visit advanced civilizations on other planets 'in our present bodily condition'. I believe there is much truth in this statement, but not for exclusively esoteric reasons.

When Leslie asked about the solidity of the space people, Adamski, in alluding to his initial contact with Orthon, replied: 'Those guys were no goddam spooks. The pilot scratched his hand on the rim when he grabbed my arm to save it from being torn off by the force field and I tell you it bled red blood, just like you or me.'[74] On other occasions, Adamski would emphasize the point. 'Why would a spook need a spaceship?' he liked to ask.

Assuming that there are no thriving indigenous civilizations within our solar system, there is no reason why temporary or even permanent bases could not be maintained on some planets and their satellites, even on Earth, by beings from other solar systems. Adamski was the first to state that the aliens had secret bases on our planet, known to a select few in the military and intelligence community. I have uncovered much evidence for this allegation during my investigations.

PERSONAL ENCOUNTERS

The closest I came to meeting George Adamski was in November 1963, during my first tour of the United States with the Royal Philharmonic Orchestra. Because Los Angeles was on the itinerary I was determined to take a bus to his home in Vista. In the event, the bus schedule did not coincide with mine, and I had to abandon the idea. Yet a curious incident occurred en route to Los Angeles which left a deep and lasting impression on me.

On 13 November we left Tucson, Arizona, for the 500-mile ride to Los Angeles, in a convoy of three buses. About halfway there, near the Arizona/California border, we stopped at a roadside restaurant. Sitting at a table with some colleagues, casually surveying the customers waiting in line, my attention was drawn to an extraordinary-looking girl, with blond bobbed hair, delicate pale features and a petite figure. (Later, I was reminded of Adamski's description of Kalna, with her 'almost transparent skin'.)[75]

Adamski was the first to proclaim that some people from other planets were actually living and working among us – illegal aliens, as it were – and stated that his contacts often took place in the anonymous surroundings of restaurants and hotel lobbies. Having spoken with a number of other witnesses who had related similar encounters, I decided to make an attempt at telepathic communication with this unknown girl, and transmitted the question: 'Are you from another planet?'

There was no immediate response, but as she left the queue she made a point of walking past my table, pausing to give me a gracious smile and an actual bow of acknowledgement before proceeding to another part of the restaurant with a 'dead-pan' expression on her face.

Although I cannot remember the precise location of the restaurant I do recall that one of the highway signs nearby indicated Desert Center, by coincidence not far from the site of Adamski's initial encounter.

Four years later I had another such experience while in the United States. In February 1967 I was in New York City for a series of concerts with the London Symphony Orchestra at the Carnegie Hall. One afternoon I decided, as an experiment, to attempt some further telepathic communication in the lobby of the Park-Sheraton Hotel, now the Omni Park Central, at 56th Street on Seventh Avenue, where we were staying. I had just returned from my first meeting with Madeleine Rodeffer in Washington, DC. Madeleine had told me that she had encounters with the 'space people', and that these most often took place in public places. I resolved to try and settle the matter once and for all. Settling back on a sofa in the lobby I transmitted a telepathic request, which went something like this: 'If any of you people from elsewhere are in the New York vicinity, please come and sit down right next to me and prove it.'

New York is, of course, a busy city, and a hotel lobby seems the most incongruous of venues to conduct such an experiment. Many people (a few of them strange, if terrestrial) came and went during the ensuing half-hour or so. Suddenly a man entered the lobby whose demeanour put me on alert. Dressed in a charcoal-grey suit with a white shirt and dark tie, he could have passed for a businessman from Madison Avenue. He was five feet ten inches tall, with curly fair hair, a tanned complexion, and perfectly proportioned features, and he appeared to me to be about 35 years of age. He came and sat down beside me. From an attaché case he took out a copy of the *New York Times*. Unfolding this he began to turn the pages over in a rather deliberate and superficial manner. After he had refolded the paper I felt the time had come to ask him telepathically if he really was from another planet, and if so, would he please identify himself by placing his right index finger on the right side of his nose. The response was immediate and dramatic, for no sooner had I transmitted the thought than he did precisely that!

Sitting dumbfounded, I wondered what on earth the next move would be. I attempted more telepathy, but nothing else happened. Perhaps I should have engaged him in conversation but, being British (clearly a drawback to interplanetary communications), I had reservations about such an approach. Also, I felt that if my expectations were well founded, it should be he and not I who would initiate any such conversation.

We both sat silently for a few minutes. Then he stood up and walked over to some display windows behind and to my right, about 15 feet away. Observing him surreptitiously, I noticed that he appeared to be taking little interest in the merchandise displayed there, and after a few more minutes he gave me a long, penetrating look, then turned and walked out into Seventh Avenue. I never saw him again.

Of course, it can logically be argued that this experience, like the one in 1963, was coincidental. Telepathy often is commonplace, as people who pick up the unspoken thoughts of a companion often know. For instance, Dr Dennis Ross, a physicist at Iowa State University, told me that he and his brother had communicated telepathically with one another until the age of 12. Perhaps, then, this man in New York was merely a receptive mortal who picked up my thoughts and responded accordingly. I am the first to accept the plausibility of this hypothesis, yet there was something oddly distinctive about him which I cannot erase.

As far as I am concerned the experiment was a success. It is not the sort of experiment that would meet with the approval of radio astronomers. I believe that it was intended as personal proof and encouragement for me, and, as such, was of no value to others. In any event, it had a cathartic effect on me.

In my estimation, the evidence, taken as a whole, suggests that, although some of George Adamski's claims were exaggerated, preposterous, or the result of disinformation by his contacts, many are sensible and verifiable. I take the view that his reported encounters with spacecraft and their operators were fundamentally accurately reported as to basic data, but were embellished both by Adamski and by friends and supporters to the extent that they later assumed mythical qualities. We need to re-evaluate not only his claims, but those of others claiming contact with quasi-human beings from other worlds, lest, in throwing out the proverbial baby with the bath-water, potentially important data may be lost to analysis. Apart from my own prejudices, I feel it is important to re-emphasize that a great deal of what Adamski spoke and wrote about the 'space people' and their technologies is now, on the verge of the twenty-first century, more plausible and more scientifically relevant than it was some 40 years ago.

NOTES

1 Adamski, George, *Inside the Space Ships*, Arco Spearman, London, 1956, p. 67.
2 Glenn, Lieutenant Colonel John H., 'My Day of Miracles', *The Astronauts Book*, Panther Books, London, 1966.
3 Komarov, Vladimir, Feoktistov, Konstantin, and Yegorov, Boris, '24 Hours in Space', *The Astronauts Book*.
4 Interview with the author, London, 24 November 1996.
5 Adamski, op. cit., pp. 71–2.
6 Ibid., pp. 82–3.
7 Ibid., pp. 110–12.
8 Ibid., pp. 116–18.
9 Ibid., pp. 130–4, 138.
10 Ibid., p. 136.
11 Ibid., pp. 143–4.
12 Wilford, John Noble, *We Reach the Moon*, Bantam, New York, 1969, p. 23.
13 Adamski, op. cit., p. 144.
14 Ibid., pp. 203–5.
15 Leslie, Desmond, and Adamski, George, *Flying Saucers Have Landed*, revised edition, Neville Spearman, London, 1970, pp. 260–1.
16 Adamski, George, *Flying Saucers Farewell*, Abelard-Schuman, London, 1961, pp. 156–7.

17 Wilford, op. cit., p. 21.

18 Letter to the author from John McLeaish, NASA, Manned Spacecraft Center, Houston, 20 May 1970.

19 Interview with the author, London, 24 November 1996.

20 Steckling, Fred, *Why Are They Here?*, Vantage, New York, 1969.

21 Fouéré, René, 'Adamski's Last Chance: Will the Moon Vindicate Him?', *Flying Saucer Review*, vol. 10, no. 5, September–October 1964, pp. 27–9.

22 Ibid., p. 27.

23 United Press International (UPI), Houston, 16 October 1971.

24 Taylor, John W. R., *Aircraft Seventy*, Ian Allan, Shepperton, Middlesex, 1970.

25 Knowlton, Brian, 'Frozen Water Found on the Moon, Reviving a Dream', *International Herald Tribune*, London, 4 December 1996.

26 Edwards, Frank, *Flying Saucers – Serious Business*, Lyle Stuart, New York, 1966, pp. 39–40.

27 Ibid., pp. 41–3.

28 Petersen, Hans C., *Report from Europe*, Scandinavian UFO Information (SUFOI), Jylland, Denmark, 1964, p. 102.

29 Corso, Col. Philip J., with Birnes, William J., *The Day After Roswell*, Pocket Books, New York and London, 1997, pp. 128–9.

30 Ibid., pp. 125–6.

31 Adamski, *Flying Saucers Farewell*, pp. 151–6.

32 Interview with the author, Ontario, California, 19 November 1979.

33 Petersen, op. cit., pp. 131–2.

34 Letter from George Adamski to his co-workers, 5 April 1960.

35 Zinsstag, Lou, and Good, Timothy, *George Adamski: The Untold Story*, Ceti Publications, Beckenham, Kent, 1983, pp. 61–3 (out of print).

36 Letter to the author from Monsignor G. Coppa, Secretariat of State, The Vatican, 14 May 1977.

37 Zinsstag and Good, op. cit., p. 63.

38 Ibid., pp. 63–4.

39 Vallée, Jacques, *Messengers of Deception: UFO Contacts and Cults*, And/Or Press, Berkeley, California, 1979, p. 203.

40 Zinsstag and Good, op. cit., p. 25.

41 Adamski, *Inside the Space Ships*, p. 14.

42 *Saucer Smear*, ed. James Moseley, vol. 29, no. 10, 10 December 1982, p. 5.

43 Adamski, *Inside the Space Ships*, pp. 222–6.

44 Ibid., p. 16.

45 Petersen, op. cit., p. 104.

46 Letter from Lucy McGinnis to Major Donald Keyhoe, 20 July 1959.

47 Zinsstag and Good, op. cit., pp. 95–6.

48 Good, Timothy, *Beyond Top Secret: The Worldwide UFO Security Threat*, Sidgwick & Jackson, London, 1996, pp. 320–6.

49 Zinsstag and Good, op. cit., pp. 95–7.

50 Ibid., pp. 67–71.

51 Adamski, *Inside the Space Ships*, pp. 212–15.

52 Report from Adamski to his co-workers, 31 March 1961.
53 Zinsstag and Good, op. cit., pp. 67–71.
54 Keel, John A., *Operation Trojan Horse*, Souvenir Press, London, 1971, p. 213.
55 Zinsstag and Good, op. cit., pp. 72–5.
56 Ibid., p. 191.
57 Clark, Jerome, 'Startling New Evidence in the Pascagoula and Adamski Abductions', *UFO Report*, vol. 6, no. 2, August 1978, p. 72.
58 Ibid., pp. 74, 76.
59 Letter from Carol A. Honey, published in *Fate*, vol. 32, no. 3 (?), February 1979, pp. 113–15.
60 Adamski, George, *Pioneers of Space: A Trip to the Moon, Mars and Venus*, Leonard-Freefield Co., Los Angeles, 1949, p. 14.
61 Ibid., p. 31.
62 Ibid., p. 81.
63 Ibid., p. 221.
64 Ibid., p. 203.
65 Ibid., p. 115.
66 Ibid., p. 207.
67 Letter to Emma Martinelli from George Adamski, 16 January 1952.
68 Clark, op. cit., p. 72.
69 Interview with the author, Escondido, California, 20 November 1979.
70 Letter to the author from Carol Honey, 24 July 1979.
71 Leslie, Desmond, 'Commentary on George Adamski', in *Flying Saucers Have Landed* by Desmond Leslie and George Adamski, revised edition, Neville Spearman, London, 1970.
72 Interview with the author, 19 November 1979.
73 Yogananda, Paramahansa, *Autobiography of a Yogi*, Self-Realization Fellowship, Los Angeles, 1946; revised 1951.
74 Leslie, op. cit., p. 250.
75 Adamski, *Inside the Space Ships*, p. 56.

Note: *Inside the Space Ships* has now been republished, combining Adamski's section from *Flying Saucers Have Landed*. In addition to some of Adamski's books on philosophy, tapes of his lectures and photographs, it is available from the GAF International/Adamski Foundation, PO Box 1722, Vista, California 92085.

Chapter 8

From the Benign to the Bristly

We have stacks of reports of flying saucers. We have to take them seriously when you consider we have lost many men and planes trying to intercept them.

Thus stated General Benjamin Chidlaw, Commanding General of Air Defense Command, in conversation with researcher Robert Gardner in 1953.[1]

In previous books I have alluded to some disturbing cases involving missing aircraft and pilots, including one which took place in the vicinity of Soo Locks, Michigan, on 23 November 1953, when an F-89C Scorpion jet was scrambled to intercept an unknown target, confirmed on radar by an Air Defense Command ground-control intercept (GCI) controller. As the interceptor approached the target, the two blips, of the F-89 and the UFO on the GCI radarscope merged into one, as if they had collided. For a moment a single blip remained on the scope but then disappeared. No trace of wreckage or the missing crew was ever found.[2]

In the previous chapter, I cited the claim by George Adamski that he had been consulted in 1956 by the Air Force regarding a jet which had landed by itself – minus pilots. While researching material for this book, I came across an equally outlandish story, from a reputable source, which in some respects compares well with both of the cases cited above.

On an unspecified date in June 1953, Sergeant Clarence O. Dargie was working in the operations centre at Otis Air Force Base, Massachusetts, when radar detected a UFO, and a jet was scrambled to intercept. As the plane levelled off at 1,500 feet, all systems on the aircraft suddenly failed. 'That was nearly impossible,' said Dargie, 'because each system [with the exception of the engine(s)] had a separate power source. If one source stopped working, the rest would continue to operate. But for some strange reason, all the systems were out.'

As the plane started to nose-dive, the pilot ordered his navigator to eject. Normal procedure in such a situation was for the navigator to pull the first lever which ejected the canopy and then a second lever for his seat ejection. On hearing the second ejection explosion, the pilot would then pull his own lever to eject himself from the cockpit. Because a crash was imminent, the pilot ejected after he heard the first explosion, taking a risk of colliding with the navigator. In this case, when the pilot ejected and landed in the back yard of a Cape Cod resident some moments later, no trace of either the aircraft or the navigator was found. The resident, said Dargie, did not report hearing any crash or even seeing a jet. 'After three months of intensive search the government never did find any trace of the plane or the navigator,' Dargie claimed. 'If the plane had crashed there would have been an explosion. There was none, not a trace of it at all. Both the UFO and plane disappeared from the radar scope.'[3]

A CONTACT IN MEXICO

Between 17 and 20 August 1953, Mexican taxi-driver Salvador Villanueva was hired by two Texan tourists to drive them from Mexico City to the border of Texas. After about 60 miles, they had just passed Ciudad Valleys when the car ground to a halt. Oil had apparently leaked from the differential and it was obvious the car would go no further. The Texans angrily unloaded their baggage, hired another cab and drove off without paying.

Villanueva tried unsuccessfully to flag down a car for help. Reluctantly he decided to stay with his car for the night and seek help in the morning. It began to rain. At about 18.00 he crawled under the car to look at the damage again, and it was while in the prone position that he became aware of two pairs of feet. From what he could see of them, both feet and legs seemed normal, except that they were enclosed in what looked like a seamless grey corduroy. Scrambling up, Villanueva found himself confronted by a couple of pleasant-looking men, about four feet six inches in height. Because many Mexicans are short Villanueva was not unduly alarmed.

Both men were dressed in a one-piece grey garment and a wide perforated shiny belt. Around their necks were metal collars with what looked like small black shiny boxes on the back of their necks. The men carried helmets under their arms similar to those worn by jet-pilots or American football players. Villanueva assumed that they were pilots who had landed nearby. The men continued smiling. One of them opened a conversation by asking Villanueva if he was in trouble. The taxi-driver replied in the affirmative, explaining what had happened. The 'pilot' smiled sympathetically, then made small talk. It was at this point that Villanueva realized that the man had a peculiar accent, as though he were stringing words together. His companion said nothing but occasionally made expressions indicating that he understood. 'Doesn't your friend speak Spanish?' asked Villanueva.

'No, but he is able to understand you,' came the reply.

It began to rain again, so Villanueva invited the men to shelter in his car, where the conversation was continued. It was confirmed that the men were indeed pilots.

'Is your plane near here?' asked Villanueva.

'Not very far.'

'Where are you from, if I might ask?'

'We have come from very far.'

By nightfall Villanueva felt there was something very strange about these men. In his conversation the spokesman betrayed that he knew far too much for an ordinary man, not only about this world but about others too. So around dawn, Villanueva asked if they really were aviators from our world. 'No,' came the reply. 'We are not of this planet. We come from one far distant, but we know much about your world.'

Villanueva was incredulous, believing they were teasing him; he even accused them of such at one point. After sunrise, the men said they had to leave and asked the taxi-driver if he would like to accompany them to their craft. Expecting to see a conventional aeroplane, Villanueva followed them as they led him for about half a kilometre through a swampy area. Although he was sometimes sinking to his knees in muddy pools, he was astonished to notice that the men in front of him did not sink at all. 'When their grey-clad feet touched muddy pools,' reported investigator Desmond Leslie, 'the mud sprang away from them as if repelled by some invisible force. No dirt ever seemed to come in contact with them and they remained unspotted although his own boots were by now caked in mud.'

Villanueva also noticed that each time they walked over the muddy pools, their perforated belts glowed. He hesitated nervously, but the aviators turned round and smiled encouragingly.

THE CRAFT

Suddenly, in a clearing, Villanueva caught sight of a large shiny craft, unlike anything he had ever seen. As Leslie wrote:

> In form it had the shape of two huge soup plates joined at the rim. Above it was a shallow dome with portholes. The entire structure, about 40 feet across, rested on three giant metal spheres or landing balls. Unless this was some secret invention from the United States, it was surely a ship from another world.
>
> As they approached, a faint humming came from within the craft and a portion of the lower hull opened outwards, much in the manner of the rear entrance to a Martin 404 airliner, so that the inner side of the panel formed a staircase to the craft and the supporting cables became handrails. The two men went up the short flight of steps, pausing on the top to turn and look at their earthly companion.

'Would you care to come inside with us?' they asked. Fearing that he would not see his family again, Villanueva shook his head, turned, then ran for his life. Back on the road, he watched as the craft lifted off the ground.

> Something glowing white rose slowly into view, hovering for a moment, then gaining speed it began a kind of pendulum motion, a backwards and forwards arcing movement, like a falling leaf going up instead of down. It attained an altitude of several hundred feet by this method; then, glowing brighter, shot up vertically with incredible speed. In seconds it was lost from sight. Only a faint swishing sound marked its passage.

When he finally reached his home in Mexico City, Villanueva told no one except his wife about the experience. Fortunately, she believed him.

'THE KEY'

Desmond Leslie visited Salvador Villanueva in Mexico in November 1955 and came away with a very good impression of his character. 'I found him quiet, unassuming, [and] well-mannered,' he reported. 'He gave me every impression of being a trustworthy, reliable human being, the kind you would trust to take your jewelry to a bank or to look after your children if suddenly called away. I liked him very much, and I thoroughly believe his story . . . All who have investigated him have come to the same conclusion. He related and re-enacted the story to his examiners without change or contradiction.'

Villanueva did not believe his visitors were from Venus, he told Leslie. 'He had the impression from their talk, though they did not name any planet, that they had come from somewhere much farther than Venus, maybe from worlds beyond our vision entirely.'

Desmond Leslie claims to have been given (by George Adamski, I believe) what is known as 'the Key'. 'By this,' he explained, 'I mean that every man who has received a true and physical contact with men from other worlds has been given a certain "Key" whereby it shall be known that he is speaking truly. No man . . . could ever stumble upon this key by guess or chance; least of all a simple countryman. Unless Villanueva had spoken to a spaceman in truth he could not have known it. Possibly I am the only "layman" to hold it. It is the "Key" which all falsely claiming contacts through vain or neurotic reasons fail to give. Villanueva gave it without hesitation.'[4]

A VERY CLOSE ENCOUNTER

In the spring of 1954, a huge craft was observed at very close quarters near Bruyères, in the department of Vosges, northeastern France. The incident occurred one afternoon towards the end of April in a sparsely inhabited forest area at Bois-de-Champ.

Roger Mougeolle and Gilbert Doridant were engaged in logging when suddenly they heard a loud noise 'like the sound of a train passing over a metal bridge', a precursor sound reported in a number of such cases, followed by silence. Then three huge, cigar-shaped objects came into view. 'Two passed over in total silence but the third, equally silently, slowly descended over the clearing where the two men were,' reported Joël Mesnard, who interviewed Mougeolle.

> Its surface was quite smooth, devoid of any structural appendages or protuberances, and its general aspect was metallic and its colour grey. Its size was absolutely enormous. Over 200 metres long, Mougeolle thought, and perhaps 80 or 100 metres wide and equally high, though he thought the height was a bit less than the width. The monster came to a halt with its base just a few tens of centimetres from the ground.

Terrified, Doridant fled, and never again entered that part of the forest. Mougeolle, however, convinced that this was a conventional airship, stepped forward boldly until he was right underneath it. Then he put up his hand and actually touched the craft, which felt smooth and cold, like steel. To see what would happen, he struck the craft with the flat of his axe (Fig. 13). It made a dull sound, 'such as you get when you strike a great piece of steel', but instantly Mougeolle was hurled a distance of six metres or so.

Fig. 13. 'Just to see what would happen', Roger Mougeolle bashed the craft with his axe, as his partner fled. He was instantly hurled six metres and lay paralysed until the craft took off. (*Joël Mesnard/ Lumières Dans La Nuit*)

'Lying where he had been thrown, against the foot of a rock, Mougeolle found himself unable to move,' related Mesnard. 'What had hurled him was no blast of air or anything like that, but something totally unknown to him, something that seemed to act uniformly upon every part of his body. And now, as he lay there, he realized that what was keeping him pinned down was the monster itself . . .'

The huge craft remained in this position for a few minutes, almost blocking out Mougeolle's vision, then it lifted and disappeared. Fortunately, the witness suffered no ill effects from his exceptionally close encounter.[5]

AERIAL ENCOUNTERS
The distinguished British astronomer, Dr H. Percy Wilkins, was on a lecture tour of the United States on 11 June 1954 when he caught sight of two strange aerial objects from the window at his seat on a Convair airliner flying from Charleston, West Virginia, to Atlanta, Georgia. The incident occurred at 10.45.

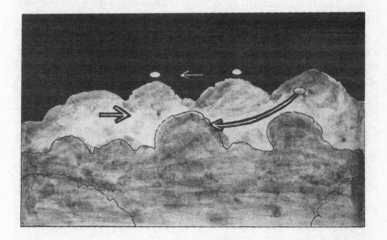

Fig. 14. A sketch by Dr H. Percy Wilkins of the flying saucers he observed on 11 June 1954.

. . . my attention was caught by two brilliant, oval, sharp-edged objects apparently suspended or hovering above the tops of two particularly lofty cumulus masses of cloud, the sides of which were shadowed and at an estimated distance of two miles. These two objects were of a yellow colour like polished brass or gold, and, quite apart from their colour, were very much brighter than the sunlit clouds on the other side of the aircraft. They looked like metal plates reflecting the sunlight, and were in slow motion northwards [see Fig. 14] . . .

Suddenly a third and precisely similar oval object was seen against the shadowed side of the cloud, but this object was dull and greyish, presumably because it was not in the sunshine. While the two brilliant objects continued their slow motion, the third one began to move with accelerated velocity; it described a curve, and vanished behind another and nearer cloud mass. The whole display was visible for nearly two minutes, but the grey object completed its rapid motion in less than five seconds after it began to move.[6]

On landing in Atlanta, Dr Wilkins told reporters that he had seen three flying saucers, whose diameter he estimated at about 50 feet.[7] 'One thing is certain,' he wrote, 'if they *are* solid objects, capable of moving in any desired direction and at any desired speed, then they must have been devised, and are operated and controlled, by intelligences superior to man.'[8]

Less than three weeks later, at midday on 1 July 1954, Griffiss Air Force Base in New York State picked up a radar return from a craft approaching the base. No aircraft should have been in the area. A Lockheed F-94 Starfire all-weather interceptor was scrambled and vectored to the unknown target by the ground control intercept (GCI) controller. The radar intercept operator in the rear seat kept his eyes on the 'blip' on his own radarscope.

Within minutes of take-off, the pilot observed the UFO visually: a shining, disc-shaped object hovering several thousand feet above the F-94. Opening the throttle, the pilot headed for the target as the radar officer radioed the unknown craft for its identification. Suddenly, the jet's engine cut out. As journalist Frank Edwards dramatically described the scene:

> . . . at that instant the cockpit of the plane became a veritable hell-hole. The pilot noted that the instruments showed no fire – but he told fellow airmen later that it was like a blast from a blowtorch right in his face. He started to report to Base but realized that he did not have time . . . instead, he yelled at the radarman to bail out. A few seconds later he felt the thump as the other man left the stricken jet. Half blinded and gasping, the pilot blew himself out of the jet and got a fleeting glimpse of the UFO as he went out on his back. The thing was huge and circular . . .

Both pilot and radar operator parachuted safely, landing near Walesville, New York. Unfortunately, their jet crashed into an automobile and two houses, killing two adults and their two children, and injuring a few others.[9][10]

At variance with these accounts is the Air Force version, which states that the incident actually took place on 2 July. There were reports which the Air Force received on 1 July of a UFO having the appearance of a balloon. On the following day an F-94C on a routine training mission was sent to investigate an unknown aircraft at 10,000 feet (which was identified) and another unknown aircraft apparently coming in to land at Griffiss AFB. At this point, 'the cockpit temperature increased abruptly . . . the fire warning light was on [and] both crew members ejected successfully'. There is no mention of a UFO, as such, in this instance.[11]

Major Donald Keyhoe claimed that the dazed pilot spoke briefly to a reporter who had arrived on the scene, but Air Force officials turned up before he could tell the whole story, and further interviews were prohibited. The incident was classified 'Secret'. Both Keyhoe and Edwards remained convinced that the incident occurred as it was described to them by their sources.[12]

A BENIGN ENCOUNTER IN NORWAY

On 20 August 1954, two well-educated sisters, 24-year-old Edith Jacobsen and 32-year-old Asta Solvang, claimed to have met an extraordinary man and his flying machine near their home town of Mosjøen in northern Norway.

It was a sunny afternoon and the women were picking berries. Suddenly they saw a man in the distance whom they assumed at first to be another berry picker. 'We walked towards him and wondered who he was,' recounted Edith Jacobsen.

> As we got near him he smiled and stretched out his hand. I, too, smiled and held out my hand, but he only brushed my palm with his. Then he began to talk, but we didn't understand a word. It didn't resemble any language I had heard [or] studied. The stranger's language was very soft and melodious. It seemed to have few consonants and no gutturals at all.
>
> When we gathered that the man must be a foreigner from some distant country we took a closer look at him. He was of medium height, had pleasant, regular features and long hair with a natural wave. He was rather dark. We didn't notice the colour of his eyes, but I believe they were slightly oblique. His hands were beautiful and expressive, with fine long fingers; rather like the hands I imagined a fine pianist would have. He wore no rings.
>
> He was clothed in a kind of overall, but as he wore a broad belt it could have been trousers and a blouse. The blouse fitted closely at the neck, but was otherwise loose. I could not see any buttons, zippers or fastenings. We didn't notice how he was shod.

What impressed the women particularly was the genuine friendliness the stranger emanated, giving them a sense of security. When it became evident that they did not understand each other, the man produced what they 'took for granted' to be paper and pencil and drew some circles, pointing out over the moor and then at the sisters, then pointing at himself and another drawn circle. 'I had at once the impression that he wanted to tell us something about the solar system,' said Jacobsen, 'but perhaps I was mistaken.'

The man then motioned to the sisters to follow him and turned and walked out along the fen. They followed, and not far away saw a curious contraption parked on the ground.

It was grey-blue and looked like two giant pot-lids placed together. It was about 10 feet in diameter and about 4½ feet in height. Because the man was still so calm and convincingly friendly we were still not afraid, even though we thought this a very curious thing to find in the wilds. We approached the thing, but he made a sign that we were not to come too close. He then opened a kind of hatch on the top of the 'rim' which encircled the thing, crawled in and shut himself in.

Presently, we heard a faint humming, like the droning of a large bumble-bee, and the curious vessel rose slowly while rotating on its own axis. Then, and only then, did all I had read about flying saucers come to my mind. When the saucer reached about 100 feet it hovered for a moment and then started rotating very fast. Finally it rose at tremendous speed and disappeared.

The sisters agreed not to discuss the incident with anyone, but eventually Asta told her husband and the story spread around the community. A reporter asked to be taken to the site of the landing. No traces were found. Subsequently, the women were ridiculed and harassed. 'The whole thing is so fantastic that I can readily understand why people who have known me all my life refuse to believe me,' said Edith Jacobsen. Finn Norstrom, who interviewed the sisters, found no discrepancies in their accounts.[13] Other journalists confirmed that all the people they spoke with in the town of Mosjøen found it difficult to believe that the sisters would have invented such a story.[14]

It is of course possible that, owing to its remarkable similarity to George Adamski's description of his 'Venusian' in *Flying Saucers Have Landed*, the sisters could have invented their story; however, as Gordon Creighton pointed out, although many features in their account of the pilot are identical with those described by Adamski, 'when they come to describe the UFO it is not Adamski's [but] a contraption "like two giant pot-lids placed together"'.[15]

ANGELIC ALIENS

It was about 16.45 on 21 October 1954, a beautiful, if cold and frosty day. Jessie Roestenburg's husband Anthony was at work for the Staffordshire County Education Department, her two sons were still at school, and she was at home at Vicarage Farm Cottage in Ranton, Staffordshire, with her little girl and the dog.

'I took my little girl out for a walk, as usual,' Jessie recalled. 'I had to light the fire so I could make the evening meal. (We had no running water or electricity and it was hard going.) I started to get washed and changed in the outhouse, when I heard this noise – a sort of hiss like when a blacksmith puts hot iron into water but much louder . . . I thought it was a plane crashing.'

I was worried because the boys hadn't got home from school as they should have done by then. I went out to see what was happening and to my amazement they were lying flat out on the ground. They shouted to me: 'Mummy, there's a flying saucer!' I said what any mother would say: 'Don't be silly!' But they were as white as sheets so I could see there was something wrong.

Then I found I wasn't in control of myself. I walked up to the water pump in the garden and turned round, but it was as though somebody else was making me do it. I wanted to look at the boys and ask them if they were all right, but I couldn't. Then I saw, suspended in the air, a massive disc – bright silver and shaped like a Mexican hat. In the middle was a tubular light going round very slowly. It had a dome, like glass, and inside it were two beings looking down at us. They were the most beautiful people I have seen, but they weren't human.

Their foreheads were large in proportion to the rest of their faces and they had long golden hair. I could only see them from the chest upwards, and they were wearing what looked like vivid blue polo-neck jumpers and what looked like fish bowls over their heads.

The craft was hovering at a tilted angle. Jessie Roestenburg recalls that although the figures looked like women, she felt sure they were men. They gazed down at the witnesses with a seemingly stern, though compassionate expression. 'I couldn't move. I was absolutely paralysed. I wasn't frightened at that stage but I was mesmerized. It seemed to last for ages but it could only be for a few minutes. I felt all the tension go from me and I felt a sense of peace I have never felt since.'

I asked the boys if they had seen the same thing and they said yes. We ran into the field to see if we could still see it and I thought it had gone but one of the boys said: 'Look, there it is.' It circled the cottage and then shot up vertically and disappeared.

By now we were scared to death. We went into the cottage where I locked all the doors and hid under the table until my husband came home.

When Tony Roestenburg arrived home, Jessie told him what had happened. 'I could see he didn't believe me. He questioned the boys separately and decided to report it to the police. The whole of Ranton had never seen so many policemen. We had people from all over the place – newspapers, the lot.'

Jessie's health began to deteriorate. 'I went to see my doctor, who had read about what happened,' she said, 'but he just thought I was round the twist. I insisted on seeing a psychiatrist and he said: "There is nothing wrong with your mind but you do need to go to hospital." He took me himself and they did a blood count. [It] was so low they couldn't understand how I was still alive. They said they wouldn't be surprised if I was suffering from radiation sickness. For a while, I was in a terrible mess but gradually I got better.'

Having met Jessie Roestenburg on a number of occasions and discussed her case at length, I am convinced of her honesty. She still retains a sense of awe and wonder about the incident. 'To this day I don't know what they were,' she told reporter Neil Thomas in 1996. 'I don't believe they wanted to do us any harm. They are far more intelligent than we are. We must have looked a pitiful sight, standing there next to a water pump while they were in a space ship.'[16][17]

LONG-HAIRED INTERLOPERS

In Porto Alegre, Rio Grande do Sul, Brazil, an agricultural engineer and his family were taking a drive on 10 November 1954 when they saw a landed disc, from which alighted two normal-sized men with long hair and dressed in overall-like clothing. They approached the car with their arms held above their heads. The engineer's wife and daughter were so frightened that they insisted on fleeing the scene. As they looked back, they watched the men board their craft and take off at a tremendous speed.[18]

One month later, on 9 December, Olmira da Costa e Rosa was cultivating his crops at Linha Bela Vista, two and a half miles from Venancio Aires – also in Rio Grande do Sul – when he heard something 'like a sewing machine', as animals in a nearby field panicked. Looking up, he saw a strange-looking man. Further away, hovering just above the ground, was an object shaped like an explorer's hat and enveloped in a smoky haze. There were two other men, one in the craft, his head and shoulders sticking out, the other examining a barbed-wire fence. Costa e Rosa dropped his hoe in shock. The stranger nearest him raised his hand, smiled and picked up the hoe, which he examined carefully before handing it back. Then the man bent down, uprooted some plants, joined the others and walked back to the craft.

Costa e Rosa stood as though paralysed, but to assure himself that these strangers meant no harm, he approached the craft. Although the man who had examined the hoe and the man in the craft made no move to stop him, the one who had been studying the fence made a gesture indicating that he should halt. He did so. Some of the farmer's animals then approached the strangers, who looked at them with great interest. 'With words and gestures the farmer tried to tell them he would be happy to make a gift of one of his animals,' reported investigator Coral Lorenzen. 'The strangers didn't seem enthused about the offer.' Suddenly the spacemen trooped into their craft, which (as so often) lifted off slowly at first then accelerated away at fantastic speed.

The farmer, who was practically illiterate and knew nothing about flying saucers, provided a detailed description of the visitors (whom he believed to be simply aviators from another country), as reported by Lorenzen:

> They appeared to be of medium height, broad-shouldered, with long blond hair which blew in the wind. With their extremely pale skin and slanted eyes they were not normal looking by Earth standards. Their clothing consisted of light brown coverall-like garments fastened to their shoes. Afterward Costa e Rosa said the shoes seemed especially strange because they had no heels.

When the craft had gone, Costa e Rosa searched the ground under which it had hovered, but found nothing. However, a smell 'like burning coal' permeated the air for some time afterwards. Two days later, Pedro Morais, who lived less than a mile from Costa e Rosa, also saw an object making a sound like a sewing machine, oscillating as it hovered. Nearby could be seen two humanoid figures. Angered by the trespass, he headed in their direction. One of the humanoids came running towards him as the other gestured that he should come no closer, then uprooted a tobacco plant. Morais ignored the warning, but there were no repercussions. The beings, who seemed to be enveloped in a kind of yellow-coloured 'sack', obscuring facial details, returned to their craft and took off.[19]

In commenting on the close similarity between Costa e Rosa's spacemen and George Adamski's 'Venusian', Gordon Creighton wrote: 'Honesty requires that this case, and the other Latin American cases of "long-haired" men, be very carefully investigated. It does not seem that this has been done.'[20]

HAIRY DWARFS

The year 1954 also saw a proliferation of reports of 'hairy dwarfs' associated with landed craft. One interesting report from the United States, unearthed a few years ago by Jean Sider, was given to the French Catholic newspaper *La Croix* by an anonymous source, believed by the newspaper to be reliable. In his statement, the unnamed American technician described the incidents – which were alleged to have occurred in the vicinity of a military helicopter base in October of that year – as follows:

> One day, as I was just finishing my work, I saw in the distance an unknown object [which then came closer and landed nearby]. It resembled the classic saucer. A human-looking being emerged from it, picked up some pebbles, earth and grass, and took a nest from a bush. Then the strange being went back into its vehicle and disappeared within a few seconds. The personnel at the base had not been able to intervene, because it all happened so quickly . . .
>
> A few days later, a saucer landed in the middle of the airfield. As on the previous occasion, its pilot emerged from it and leant over it as if he wanted to check it . . . We hastened toward the saucer [and] were a few metres away from the stranger when he straightened up, saw us, and immobilized us with an unknown fluid [gas?]. Then, after watching us for a moment, he walked

around the saucer, climbed into a kind of 'cigar' which had parked next to the saucer, and took off vertically. [Then] we recovered our freedom of movement.

The saucer remained on the ground. We didn't know what to do, still having the image of the fleeing being in mind. It was of a small size, very stocky, with an ape-like appearance. Its body or clothes were covered with long hair. Of its face, we were able to observe only extremely sharp and brilliant-looking eyes. But – the saucer was right there in front of us. What could it contain? Our engineers decided to investigate it on the field.

We noticed right away that the saucer had been closed hermetically: there was practically no opening. It had a dome and portholes but no visible engine. It rested on three legs which were attached to little 'skis'. We had to use a welding-torch in order to cut out the door, which caused problems, [because] the unknown metal, which formed the exterior shell of the saucer, was extremely smooth, resistant, and couldn't be raised either by stamping or adjustment. Our chemical engineers analysed a fragment of it: they found a very heavy alloy of gold, lead and iron. But . . . they were incapable of reconstructing the alloy, considering the proportions analysed.

The interior of the saucer was upholstered entirely with a sort of fibre, resembling rubber. A fairly heavy gas filled two-thirds of the object: an artificial atmosphere? Fuel? Above the portholes were some 'throttles', but we couldn't make anything out of them. Also, we couldn't find any engine. It seems that the pilot had been traveling 'on all fours' on the floor: we found sort of 'suction pads' where the 'hands' and 'knees' would have been. But in each of these we discovered six imprints . . .

This is what I witnessed with a few of my colleagues . . . The American authorities ordered a 'black out' on the affair, which hadn't yet 'filtered' out to US newspapers . . . Need it be mentioned that we were profoundly affected by everything we had seen, and will never be able to forget it?[21]

The American engineer's observation of 'suction caps', which led him to believe that the pilot had been travelling 'on all fours', as well as the description of the hermetically sealed door, provides remarkable corroboration for the 1943 report by Daniel Léger, who claimed that the female pilot he encountered boarded her craft via a panel in an apparently seamless hull, then assumed a stretched position 'on all fours' inside the cabin (see Chapter 1). As Jean Sider confirms, Léger had no knowledge at all of the American incident.[22]

VENEZUELA

The most dramatic encounters with hairy dwarfs in 1954 were reported from Venezuela.

At 02.00 on 28 November, Gustavo González and José Ponce were driving in a panel truck in the suburbs of Caracas when they encountered a luminous sphere, eight to ten feet in diameter, which hovered about six feet above the road, blocking their passage. The men got out of their truck to investigate and a dwarf-like creature came towards them. With the intention of taking it to a nearby police station (!), González grabbed the dwarf, whereupon he noticed that it was incredibly light (about 35 pounds), extremely hard, and covered with stiff, bristly hair. It gave González a push, throwing him for about 15 feet. As two other entities emerged from bushes carrying chunks of dirt or rock, and entered the sphere, Ponce ran to the police station. Meanwhile, the creature who had pushed González headed towards him once again, with eyes glowing and claws extended. Panicking, González took out his knife and stabbed its shoulder. The knife glanced off as though it had struck steel. Then another hairy dwarf emerged from the sphere and beamed a ray of light at González from a hand-held tube, blinding him momentarily. Finally, the creatures climbed into the sphere and rapidly took off.

González staggered to the police station, where he joined Ponce. The police initially thought the two men were drunk, but examination proved otherwise. González suffered a long red scratch on his side, where the creature had clawed him. Both men were given sedatives.[23]

On the night of 10 December, two youths, Lorenzo Flores and Jesús Gómez, who had been rabbit hunting near the Trans-Andean Highway between Chico and Cerro de las Tres Torres, encountered a bright object some distance off the highway which at first they took to be a car. As they approached, however, it appeared like two huge washbowls placed one on top of the other, hovering a few feet above the ground. The object was about nine feet in diameter, with 'fire' emitting from the bottom. Flores described what followed:

> Then we saw four little men coming out of it. They were approximately three feet tall. When they realized we were there the four of them got Jesús and tried to drag him toward the thing. I could do nothing but take my shotgun, which was unloaded, and strike at one of them. The gun seemed to have struck rock or something harder, as it broke into two pieces.
>
> We could see no [facial] details, as it was dark, but what we did notice was the abundant hair on their bodies and their great strength.

Gómez, who apparently had become unconscious during the episode, was unable to recall much. Neither of the youths saw the object leave, for as soon as Gómez regained consciousness they both ran to the highway and stopped a car. Seeing their scratches and bruises, and shirts torn to shreds, the driver rushed them to the nearest police station. Police investigators, doctors and psychiatrists found the youths to be 'sane and responsible'.[24]

On 16 December, Jesús Paz and two friends were returning from dining in San Carlos, when Paz asked the driver to stop while he went into the bushes to relieve himself. Suddenly a piercing scream was heard by the men in the car. Rushing into the bushes, they found Paz lying unconscious. A short distance away, a small hairy man was running towards a flat, shiny object hovering just above the ground. One of the men, Luis Mejia, a member of the National Guard, reached for his gun, but realizing that he had left it at the barracks he picked up a stone and threw it at the object – to no avail. The craft took off with a 'deafening buzzing sound'.

In a state of shock, Paz was rushed to the San Carlos hospital, where he was found to have several long, deep scratches on one side and along his spine, 'as if he had been clawed by a wild animal'.[25]

On that same day, President Dwight Eisenhower was quoted as having stated at a press conference that a trusted Air Force official had told him that the notion that flying saucers came 'from any outside planet or any other place' was 'completely inaccurate'.[26]

One of the doctors who had examined Flores and Gómez (described in the second report) admitted later that he had actually witnessed the incident in the suburbs of Caracas on 28 November. Out on a night call at the time, he had been in the same street where González and Ponce had stopped their truck. He stayed only long enough to see what happened, he said, but then left, concerned about undesirable publicity. In an official statement prepared for the Venezuelan authorities, the doctor confirmed the incident, though he stipulated that his name should not be associated with the story. Following the statement, the doctor was invited to Washington, DC, to discuss the matter with 'American authorities'.[27] President Eisenhower's statement notwithstanding, some officials took the subject seriously.

THE MICHELIN MEN

Returning from the cinema one evening in May 1955 in Dinan, Côtes-du-Nord, France, Monsieur Droguet was about to lock the door leading to one of the courtyards of the girls' college, where he lived on the premises, when he was temporarily blinded by a bluish-green beam of light. He became extremely frightened; his knees knocked and his hair literally stood up.

Droguet discerned an enormous circular object, hanging motionless at a height of about 1.5 metres above the courtyard. He could hear no sound, though he was aware of a constant 'vibration'. The witness then became aware of two beings close to the machine, who apparently had not noticed him (see Fig. 15). Growing even more frightened, Droguet tried to escape, but found himself glued to the spot. The beings were dressed in a type of metallic one-piece grey suit, which somewhat resembled the suit worn by the little man in the famous Michelin tyre advertisement.

The witness was unable to see the beings' heads, enclosed as they were in bulky helmets. Their hands were covered with gloves of a kind. On the abdomen, each being had a black box with many leads coming out of it. One of the beings was picking up something from the ground – probably pebbles – while the other one inspected the surroundings. Droguet had the feeling that someone was watching him from inside the machine.

The two beings walked towards their craft, on the underside of which was a dark hole with a metallic ladder hanging down. No more than 1.6 metres in height, the beings walked with obvious difficulty, 'like divers with their leaden soles' – an observation corroborated in other cases. As investigator J. Cresson continues in his report on the case:

> Just as they were entering the craft, M. Droguet distinctly heard a metallic sound emanating from their feet as they trod. When they had entered the craft, the ladder was drawn in and there was a sound like the intake of air. He felt a displacement of the air, a sensation of suction. The machine, still lit up, rose vertically to above treetop height, without any sound. He was now able

Fig. 15. (*J.-L. Boncoeur/ Lumières Dans La Nuit*)

> to see the black hole in the centre of the underpart of the craft, and to perceive that [it] was circular. The craft was rotating very rapidly, but the black hole did not seem to be moving. When the machine had reached treetop height, its lights went out.

On recovering mobility, the witness ran to his quarters. He told only his wife and a few trusted friends what had happened. On learning about the encounter, the headmistress of the girls' college advised M. Droguet not to let the affair become known, to avoid a 'scandal'.[28]

There were further reports of the 'Michelin men' in France in 1954. Had these been reported exclusively in France, there might be reason to suspect a publicity stunt, but the Michelin men have been observed in other countries – and well beyond 1954.

THE KELLY SPACE CREATURES

It was a hot night on 21 August 1955. Shortly after 23.00, Cecil 'Lucky' Sutton and members of his family and other witnesses rushed in two cars to the police station at Hopkinsville, Kentucky, where he told Police Chief Russell Greenwell that they had just escaped from an invasion by at least one creature which had landed in a spaceship at the Sutton home near Kelly, seven miles away. 'For God's sake, chief, get us some help,' pleaded Sutton, 'we've been fighting 'em for four hours.'

According to the family, a visitor to the Sutton farm, Billy Ray Taylor, had been getting water from the well in the back yard earlier that evening when he noticed a coloured light streak across the sky and descend into the trees along a ravine about a quarter of a mile away. Shortly afterwards, Sutton's mother, Glennie Lankford, saw a creature with very long arms and talon hands raised in the air approaching the back of the house. Sutton and Taylor armed themselves with a .22 rifle and a 20-gauge shotgun and fired a shot at one of the creatures, which flipped backwards then seemed to rise up and 'float' into the underbrush.

The siege continued throughout the night, as the 'luminous' creatures repeatedly approached the house (with arms raised) and were driven back by gunfire, a total of about 50 rounds expended. When one creature was shot at extremely close range, the pellets sounded as if they had hit a metal bucket. At one stage, Taylor was tapped on the head by the talon of one of the creatures which was perched on the roof of the house.

The creatures appeared to be made of a silvery metal, and were described as three and a half feet tall, with oversized heads and large, floppy, pointed ears. Their eyes were large, glowing yellow, and set halfway between the front and side of the head. They had wide, thin mouths (which never opened). Their legs were spindly and inflexible; in fact, the creatures seemed to propel themselves more with their arms.

Within minutes of Sutton's plea for help, local police, state police, a deputy sheriff, newspaper staff, an Army reservist from nearby Fort Campbell, and others arrived on the scene. There was little physical evidence at the site, beyond empty shotgun cartridges, a hole in the screen through which one of the shots had been fired, and a strange luminous patch on the ground near the fence from which one of the creatures fell after it was shot. After the police and all the others had left the house, not long after 02.00, there were further sporadic sightings of the weird creatures until about 04.45.

In all, seven adults and three children – all of whom were found credible by the police and most of the other investigators – witnessed these incredible events. The case is listed as 'Unidentified' in the Air Force's Project Blue Book files.[29 30]

In a chapter about unidentified flying objects in a text of the United States Air Force Academy's Department of Physics, *Introductory Space Science*, author Major Donald Carpenter points out some of the dangers faced by aliens contacting or trying to establish contact with human beings:

> Let me point out that in very ancient times, possible extraterrestrials may have been treated as Gods, but in the last two thousand years the evidence is that any possible aliens have been ripped apart by mobs, shot and shot at, physically assaulted (in South America there is a well-documented case), and in general treated with fear and aggression. In Ireland about 1000 AD, supposed airships were treated as 'demon-ships'. In Lyons, France, 'admitted' space travellers were killed. More recently, on 24 July 1957, Russian anti-aircraft batteries on the [Kirile] Islands opened fire on UFOs. Although all Soviet anti-aircraft batteries on the Islands were in action, no hits were made. The UFOs were luminous and moved very fast. We too have fired on UFOs . . .

Major Carpenter confirms that a US Air Force F-86 fighter also tried to shoot down a flying saucer.[31] The Kelly incident, in particular, he wrote, supported the contention that 'humans are dangerous'. 'At no time in the story did the supposed aliens shoot back,' he remarked, 'although one is left with the impression that the described creatures were having fun scaring humans . . .'[32]

NOTES

1 Good, Timothy, *Beyond Top Secret: The Worldwide UFO Security Threat*, Sidgwick & Jackson, London, 1996, p. 337.

2 Ibid., p. 339.

3 Gribble, Bob, 'Looking Back', *MUFON UFO Journal*, no. 242, June 1988, p. 20.

4 Leslie, Desmond, 'Mexican Taxi Driver Meets Saucer Crew!', *Flying Saucer Review*, vol. 2, no. 2, March–April 1956, pp. 8–11.

5 Mesnard, Joël, 'The "Steel Airship" at Bois-de-Champ', *Lumières Dans La Nuit*, nos. 275/6, May–June 1987, translated by Gordon Creighton and published in *Flying Saucer Review*, vol. 32, no. 5, November 1987, pp. 16–17.

6 Wilkins, H. Percy, Ph.D., *Mysteries of Time and Space*, Frederick Muller, London, 1955, pp. 40–1.

7 Edwards, Frank, *Flying Saucers – Serious Business*, Lyle Stuart, New York, 1966, p. 47.

8 Wilkins, op. cit., p. 44.

9 Edwards, op. cit., pp. 56–7.

10 Keyhoe, Maj. Donald E., *Aliens from Space: The Real Story of Unidentified Flying Objects*, Panther, St Albans, Herts., 1975, p. 35.

11 Thayer, Gordon D., 'Optical and Radar Analyses of Field Cases', *Scientific Study of Unidentified Flying Objects*, ed. Dr Edward U. Condon, Bantam, New York, 1969, p. 161.

12 Keyhoe, op. cit., p.35.

13 'Spaceman Lands in Norway?', *Flying Saucer Review*, vol. 1, no. 4, September–October 1955, pp. 6–7, reprinted from the Norwegian magazine *NA* ('Now').

14 Lidstrøm, Anton, 'UFO Report from Norway: A Norwegian Close Encounter of the Third Kind: Mosjøen 1954', *Flying Saucer Review*, vol. 34, no. 2, March–April 1989, pp. 3–4.

15 Creighton, Gordon, comment in Lidstrøm, op. cit., p. 7.

16 Thomas, Neil, 'Close encounter with outer space: Jessica Roestenburg, face to face with aliens', *Staffordshire Newsletter*, Stafford, 30 August 1996.

17 Personal communications with the author.

18 Lorenzen, Coral E., *The Great Flying Saucer Hoax: The UFO Facts and Their Interpretation*, William-Frederick, New York, 1962, p. 42.

19 Ibid., pp. 46–8.

20 Creighton, Gordon, 'The Humanoids in Latin America', in Charles Bowen (ed.), *The Humanoids*, Neville Spearman, London, 1969, p. 95.

21 *Le Rouergue Républicain*, 7/8 November 1954, p. 4.

22 Sider, Jean, *Ultra Top-Secret: Ces ovnis qui font peur*, Axis Mundi, 20220 Ile-Rousse, France, 1990.

23 Lorenzen, op. cit., pp. 52–4.

24 Ibid., pp. 51–2.

25 Ibid., pp. 50–2.

26 *New York Times*, 16 December 1954, pp. 1, 24, 26

27 Lorenzen, op. cit., pp. 53–4.

28 Cresson, J., 'Spectacular Landing at Dinan', *Lumières Dans La Nuit*, no. 106, June 1970, translated by Gordon Creighton and published in *Flying Saucer Review Case Histories*, supplement no. 1, October 1970, pp. 13–14.

29 Brown, Jennifer P., 'World won't let community forget Kelly space creatures', *Kentucky New Era*, Hopkinsville, Kentucky, 30 October 1995.

30 Clark, Jerome, *The UFO Encyclopedia*, vol. II, Omnigraphics, Penebscot Building, Detroit, Michigan 48226, 1992, pp. 214–15.

31 See Good, op. cit., pp. 336–7. Original report cited by Capt. Edward J. Ruppelt in *The Report on Unidentified Flying Objects*, Doubleday, New York, 1956, pp. 2–5.

32 Carpenter, Maj. Donald J., 'Unidentified Flying Objects', *Introductory Space Science*, vol. II, Department of Physics, United States Air Force Academy, 1968, pp. 462–3.

Chapter 9

Alien Fantasia?

There, sitting on a rock by the brook, was the most exquisite woman my young eyes had ever beheld! The warm sunlight caught the highlights of her long golden hair as it cascaded around her face and shoulders. The curves of her lovely body were delicately contoured – revealed through the translucent material of clothing which reminded me of the habit of skiers . . . She seemed to radiate and glow as she sat on the rock, and I wondered if it were due to the unusual quality of the material she wore, which had a shimmering, shiny texture not unlike but far surpassing the sheen of nylon. The clothing had no buttons, fasteners or seams I could discern. She wore no make-up, which would have been unnecessary to the fragile transparency of her Camellia-like skin . . .[1]

Thus wrote Howard Menger, describing his first claimed encounter with an extraterrestrial, in High Bridge, New Jersey, in 1932, when he was but 10 years old. Menger, who rose to prominence as a charismatic contactee in the late 1950s, has been widely dismissed as a charlatan who simply jumped on the bandwagon in the wake of publicity following George Adamski's stories. For me, this is too dismissive a judgement. As with Adamski, I am unable to reject arbitrarily all of Menger's claimed experiences, though I believe some of them to be delusional.

The lovely lady explained to Menger that she had 'come a long way' to see him, because she and her people had been observing him, and that she had known him for a 'long, long time'. 'We are contacting our own,' she added cryptically, implying perhaps that Menger had been linked with her people in the past. She impressed on the pubescent youngster that though he would not understand much of what she told him, he would begin to do so later in his life.

Menger had no idea who she was or where she came from: he knew nothing about extraterrestrials, even though already he had had several sightings of peculiar flying discs, including one that landed briefly, mostly in the company of his brother. In any event, no spacecraft could be seen nearby. After giving further information to Menger about his future role in life, to include meetings with others of her people, the lady asked the boy to leave first. He simply walked slowly away from the scene, pausing to look back at her as she sat smiling on the rock.[2]

FURTHER EARLY ENCOUNTERS

Howard Menger's second self-reported encounter with an extra-terrestrial being occurred in 1942, while he was serving in the US Army in an armour division near El Paso, Texas. He was wandering through the nearby Mexican town of Ciudad Juárez one night when a taxi pulled over to the kerb and the driver pointed to a man sitting in the back seat.

The man had long blond hair which hung over his shoulders, and suntanned skin. Taller and heavier than the average Mexican, he spoke in English with a slight Mexican accent, inviting Menger to get inside the cab with him as he needed to talk to him. Menger declined, but later wondered if he had made a mistake. Could this man have been one of the 'others' referred to by the lady on the rock?

Some time later, while posted in military service at Camp Cook, California, Menger was greeted by a uniformed man who initially addressed him telepathically, then verbally, confirming that he knew about both the earlier encounters with his people. In appearance, the man was rather unusual, said Menger:

> He was a fine looking man. Although there was something definitely unusual about him, he could have passed – and did – for an ordinary GI. The singularity of the man probably was not because of the finely chiseled features and the luminous, almost liquid quality of his eyes, but in the communication I felt. I could sense that the man was kind, wise, emotionally and spiritually developed beyond anyone I had ever met. Although a kind of reserve he wore as if a part of him set him apart from an ordinary person, I somehow accepted with no surprise the emergence of an underplayed, yet natural sense of humor . . .

In referring to the Juárez contact with one of his people, the spaceman began by saying that he fully understood Menger's reluctance to get inside the cab with such a man ('We told him he should cut his hair'!). He also appreciated the fact that Army regulations encouraged caution in such areas and that Juárez was hardly the best place for an interplanetary meeting.

The stranger went on to explain that his people had established contact with humans in Mexico many centuries earlier.

> Long before the time of the Conquistadores, we made contact with the Aztecs. We helped these people in many ways, and it is too bad the conquerors came in war instead of good will and friendship; for there were many things the Aztecs could have taught them. Instead, they withheld these secrets, and these perished with the civilization.

Some of these secrets supposedly related to the use of sound and light to produce power and run machinery. A number of the space people came from a planet (which he did not name) to contact 'remnants of his own people still living on Earth, descendants of an ancient race which originally came here from his own planet'.

The spaceman told Menger that his Army unit would soon be moved to Hawaii, and that he would be put on detached service with special duties which would give him more free time for 'certain tasks' he was to perform, and where he would have yet another contact. These events, as it transpired, came to pass.[3]

In Hawaii, Menger became a battalion topographical draughtsman. Later, he told me, he was transferred by regimental headquarters and put on detached service working with Naval Intelligence and his Army battalion on various top-secret inventions.[4] On 'impulse' one evening, he borrowed a jeep and headed for a cavern area several miles away from the battalion headquarters, where he encountered yet another gorgeous girl from elsewhere.

> She was dressed in a sort of flowing outfit of pastel shades. Under a kind of flowing tunic, translucent and pinkish, she wore loosely fitted pajama-type pantaloons. She stood about five feet six inches, with the dark, wavy hair falling over her shoulders and the tunic floating gracefully around the shapely contour of her body . . . this girl, too, exuded the same expression of spiritual love and deep understanding. Standing in her presence I was filled with awe and humility, but not without a strong physical attraction one finds impossible to allay when in the presence of these women.

Menger emphasized that the space people he claims to have met, though far superior to us in terms of physical, mental and spiritual abilities, were still much like us. At first he found it daunting that these visitors were able to read his every thought, but 'one suddenly realizes he cannot hide anything, and becomes completely honest, both with himself and the visitors'. During the lengthy discourse, the 'Martian' space-woman foretold, correctly, that Menger would be posted in early April 1945 to Okinawa, where he served with the 713th Tank Battalion.[5]

Menger had his first close encounter with the horrors of war when shrapnel from an exploding shell entered his eye, causing infection and temporary blindness. During his stay in hospital, he relates that one of the Army nurses who looked after him, whom he believes may have been from somewhere else, assured him that his sight would be restored. She also predicted correctly another contact near the time of his release from hospital.

Two weeks after his release, Menger claims he was nearly bayoneted by three Japanese soldiers, but managed to overpower them. During the skirmish, he was filled with a strong impression that he should not kill the soldiers – an impression he attributed to the mental influence of his space contacts.

Menger alleges that the following night he had yet another contact, in the northern part of Okinawa, with a very tall man dressed in Army khaki. Eventually, the man claimed that he came from Venus. One of his predictions, according to Menger, was that the Japanese would soon surrender, 'for they are about to be blasted into submission by a power which will shock the world'. A few weeks later came the atomic bombing of Hiroshima and Nagasaki.[6]

After the war, Menger says he had his first encounter with a landed spaceship and its crew. In June 1946 he was visiting his parents in High Bridge, New Jersey, when a craft, similar to Adamski's famous 'scout', landed. Two men, dressed in blue-grey ski-type uniforms and with long blond hair, stepped out of the craft through an opening on the flange, followed by the same girl Menger had encountered in 1932. 'This lovely creature had not changed at all,' said Menger. Although looking only about 25, she claimed to be 'more than 500 years old'! (In later contacts, the space people generally were much 'younger', e.g., 79 Earth years.) During the ensuing conversation, Menger was told that he would have continuing contacts which would 'further instruct and condition' him.[7]

In late 1947, two supposed space people met Menger and showed him to a secluded farm area, one of several sites to be used for future landings and contacts, where no one could be harmed by 'the electromagnetic force which emanates from our craft'. Menger would be advised of these appointments by telephone.[8]

ALIEN LIAISON

Howard Menger's incredible experiences continued into the 1950s, as he became increasingly involved in helping the space people establish themselves on our planet. Such tasks were not without their moments of humour, Menger recalled:

> Often I purchased clothing and took it to the points of contact. Visitors just arriving from other planets had to be attired in terrestrial clothing so they could pass unnoticed among people . . . I remember one time when I was asked to purchase several complete outfits of female clothing. Feeling it would be embarrassing and somewhat difficult to explain why I was buying so many outfits, I purchased them in separate shops.
>
> I bought what I thought was the appropriate size and showed up at the point of contact. The women went into the next room from which I soon heard a series of giggles and groans. Finally the door opened and the bras were flung out. They apologized, saying they just could not wear them, and they never had . . . They [also had] difficulties with high heels. They teetered and wobbled and suffered, but took it in good humor. They realized they would have to learn how to wear them, though they often complained, 'Why can't your women wear sensible shoes!'

On several occasions, Menger says he acted as barber for some of the new male arrivals. 'I don't know if they save their hair or not,' he remarked, 'however, all evidence of the meetings was always carefully gathered up by the space people before they departed.' Some of the men had unusually fair skin, without hair on their arms or faces, and apparently they had no need to shave. 'After three months on Earth, however, they became hairy and grew beards.'

Some of the visitors requested dark glasses; some specifically of red glass, though the reason for this was not given.

'Thus I had the opportunity to meet people from other planets in all stages of progress and development,' claimed Menger, 'from those who spoke no word of our language to those who spoke it fluently; scientists and technicians to helpers and assistants. I briefed them on our customs, slang and habits. Although they utilized instruments to learn a language quickly, the machines couldn't always cope with colloquialisms.'

Although acting as one of the space people's liaison men, Menger says that he was never asked to obtain identification papers, nor to seek jobs for them.[9]

FOOD
Occasionally, Menger was asked to bring food for the space people.

They asked mainly for frozen fruit juices, canned fruit and vegetables, whole wheat bread, wheat germ and the like. They refused to drink milk, avoided fresh oranges, lemons and grapefruit. They preferred tree-ripened fruit when I could find it . . . I remember one time I bought five bushels of tree-ripened apples from a local orchard. They tested the apples and found the mineral and vitamin content much lower than similar fruits of their planets. This was due, they said, to our poor soil. They explained that chemical fertilizers were not the correct answer to the problem . . . because they did not replenish the organic materials our soil sadly lacks.

While chemical fertilizers have widely been regarded as unhealthy for many years, it should be pointed out that this was not the prevalent view in 1959, when the above information was published in Menger's first book, *From Outer Space to You*. He continued:

Most of the time they brought their own food, in a dried, preserved state. I sampled some of it and while it was tasty, it was hard and dry. I tasted other food which was delicious. They had put the dry foods through some sort of processing which returned the moisture and at the same time expanded it to its natural size and state. I ate one of their tubers which was far superior in protein and mineral content than any vegetable we now have. We could grow the same here, they said, if our soil were healthy . . .[10]

AN OBSERVATION DISC
During a contact at one of the landing sites in April 1956, Menger claims

to have been shown an 'observation disc' at close quarters, similar to those reported by George Adamski. The object seemed to be lying on the ground. 'It appeared to be a circular, translucent object about a foot thick and three or four feet in diameter,' he reported. 'It was pulsating different warm colors. As we approached it, the colors changed from white to blue and back to white with a tinge of yellow.'

According to Menger's contact man, the disc was controlled remotely from a nearby spaceship, and was capable of recording all emotions, thoughts and intentions. 'Don't worry, it's white,' said the spaceman. 'When it changed to white I knew you were all right', adding that the same colours were being displayed on an instrument panel in the ship and would be recorded permanently.[11]

PHOTOGRAPHIC ATTEMPTS
During several contacts, Menger was able to take a number of photographs and films of the spacecraft and on a couple of occasions their occupants; alas, only in silhouette, owing to the darkness (see plates). At 00.45 on 2 August 1956, for instance, he took Polaroid photos of one of the craft as it came in for a landing.

> I snapped away, hardly able to wait for even a minute while the pictures developed, but in the darkness could not see just how I was doing.
> I noticed [that] one of the three ball-like objects under the craft became distorted and looked like rubber as it seemed to stretch and grasp the ground. I could see the other two balls through the translucent flange. I wondered how they could make metal appear translucent, and also become plastic, certainly alien to our earthly physics . . .
> An opening appeared and a man stepped out. He stood tall and straight, his long, blond hair blowing in the soft warm summer breeze. I could see the beautiful structure of his body; his broad shoulders, slim waist, and long, straight legs. He approached and when he was about 50 feet away I snapped his picture. He . . . was silhouetted against the glowing craft, a dramatic pose which I hoped would turn out better than my previous picture. But in the picture the craft seemed distorted and looked as if a gaseous, swirling haze encircled it [see plates].

The spaceman explained that the pictures had come out slightly distorted 'due to the electromagnetic flux around the craft',[12] an effect which is seen clearly in the frames from the 8mm movie film taken by George Adamski in Silver Spring, Maryland, in 1965.

INSIDE THE SPACECRAFT
The following evening, Menger was taken on board one of the spacecraft for the first time; a short hop from one of the landing sites to another. Just

before entering the craft, one of the spacemen pointed an instrument at Menger and a bluish beam of light struck his head, producing a rather pleasant tingling sensation. 'We projected the beam on you to condition and process your body quickly so you could enter the craft,' Menger was told. 'What actually happened was that the beam changed your body frequency to equal that of the craft. Thus you felt entirely comfortable inside the craft and suffered no ill effects.'[13]

In the small hours of 5 August 1956, Menger says, he was taken for a longer ride.

> We stepped into a large circular room. In the center of it was an ample-sized round table made of translucent material. Under the table-top pulsating lights of many colors moved. The dowel-shaped stem supporting the table was set in what appeared to be a huge magnifying glass, itself set into the floor. Approximately one third of the circular room was devoted to instrument panels containing many colored lights blinking on and off. In front of the control board was a frame containing what I guessed was some kind of viewing screen.

One of the spacemen waved his hand over a section of the table and two chairs came out of the floor. Menger and his guide sat down, and on the screen appeared a scene on Earth. After take-off, the magnifying lens on the floor zoomed in to reveal another terrestrial scene, with two people – whom Menger knew – driving along a highway. 'The image appeared as if in broad daylight,' said Menger. 'I could see everything clearly . . . [and] I could hear the two voices as if the people were in the ship with us . . .'[14]

In the second week of August 1956, in the Blue Mountains area of Pennsylvania, Menger claims that, together with two other people he knew (not named), he was taken into space aboard another craft and shown the Moon, as well as large meteorites, through a viewing screen. Like Adamski, Menger reported seeing colours (blues and greens) in a crater on the Moon. But his description of what Earth looked like is interesting, in that it conforms to what our planet does actually look like in photographs taken years later by the astronauts: 'At one time we caught sight of Earth in the distance, glowing bluish-white with tinges of red, floating like a tennis ball in an inky black pool . . .'

Menger took five photographs of Earth and the Moon through a porthole, of which three came out. Unfortunately, as he himself admits, these particular photographs are blurred and do not furnish satisfactory evidence.[15]

A TRIP TO THE MOON?

In September 1956, also in the company of Earth people (again, not named), Menger says he was taken for a trip to the Moon. This trip, it was pointed out by the extraterrestrial hosts, would be longer than the others, and would require 'processing' the humans' bodies. 'Each atom of your physical body will undergo a processing which will change its polarity, frequency and vibration, to adjust your body from its balance to the earth's attractive inertial mass to that of the moon's,' the hosts explained. 'This will require approximately a week and one half, Earth time.'

MODERN CONVENIENCES

Menger and the others were led to their sleeping quarters on the craft, consisting of three bunk-beds per compartment. 'The bed did not feel overly soft,' Menger reported, 'instead it seemed to give just the correct extent to support the contour of the body. I laid my head on a flat pad of soft material, [and] pulled the single warm (but extremely light) coverlet over myself.'

In the morning, the guests showered in a compartment containing three or four cubicles, partitioned by translucent walls. 'When I stepped inside one of the cubicles the door closed behind me and lights went on automatically,' Menger reported. 'Three shower heads, one above me and two at waist level, could be operated separately or all together.'

I pushed a button and a flow of water, apparently mixed with warm air for it was quite bubbly, fell over my body. I had never had a shower so invigorating. I looked around for soap, but there was none. Seeing another button I had not previously pushed, I put my finger on it and a stream of colorless solution came from the shower heads and completely lathered my body; at the same time the water was turned off. I pushed the 'soap' button, then the 'water' button, alternately, enjoying the novelty like a small boy would have done.

The lavatory bowl reportedly looked 'very much like one on Earth, except that the bowl was lower to the floor and was made of a hard white translucent material, not a ceramic'.

In a mirror above a kind of basin, Menger was surprised to notice that he did not need to shave, which was just as well, since he had not brought a razor with him. And – interestingly – throughout the estimated 10-day trip, he said, the guests found it unnecessary to shave.

Food was served in the main compartment of the spacecraft.

Our instructor opened a compartment in the wall and withdrew some items of processed food, which he put into a deep well, or pot, set into a sink-like unit. He pushed a button and the pot filled with liquid. He allowed the food to steep in the liquid for about five minutes, then he drained the liquid from the pot. He pushed another button and almost instantly the appearance of the food changed and steam rose from it. It had been cooked in little more than a second! . . . He removed the food from the pot by means of a large, deep strainer and transferred it to plastic-like plates which he said were disposable . . .

The food supposedly consisted of vegetables such as potatoes with 'a meaty, nut-like flavor', cabbage, carrots, parsley ('much larger than our variety'), 'green mineral salt' for seasoning, 'very large wheat kernels', nuts of various kinds, served in slices, and a fruit about six inches in diameter, orange-red in colour, that tasted like a combination of peach and plum.

'I am not certain how long we spent,' Menger reported, 'but, estimating by my watch, I believed it to be about ten days. I have often thought that time might have been different [on board], possibly because my beard didn't seem to grow; but that could have been a result of our conditioning – however, all our other bodily functions seemed to progress normally.'[16]

THE LUNAR EXCURSION

On arrival at the Moon, Menger began to take more photos, with the aid of some coloured filters given him by his hosts. These particular Polaroid pictures – showing a domed structure rising out of hilly terrain and a saucer hovering above – are interesting, though not unequivocal because they might have been fabricated.

After the spacecraft landed at a dome-shaped building, about 150 feet wide and 50 feet high, the guests were taken, in different groups, for a guided tour of the Moon, in a 'long train-like vehicle with ten or fifteen coaches with plastic domes over them' which glided noiselessly about a foot above a 'copper highway'. Menger's description of his alleged lunar visit understandably provoked as much ridicule as that heaped on Adamski's story.

We passed mountains, went through valleys, visited underground instal-lations . . . Huge cliffs and mountains made our own look like ant hills. One particular desert locale brought to mind 'The Valley of Fire' in Nevada. There we stopped long enough for our guide to open the door and permit us to stick our heads out for a brief moment, which was all one could take, for it was terribly hot outside – like a blast furnace. I was certain no one could have lived outside very long . . .

Finally we came to another large dome-shaped building, where we halted and our guide told us we could get out on the moon's surface where we could breathe the air with little or no difficulty . . . My first impression was that I was in the desert. The air was warm and dry. I could see little wind funnels forming on the ground, drawing up dust particles like tiny whirlwinds. I looked up at the sky. It was a yellowish color. When looking I had the queer impression that if I walked some distance I would fall off, since the horizon seemed foreshortened . . . The ground beneath our feet was like yellowish-white powdery sand, with stones and boulders and some minute plant life showing here and there as we looked around us.

In addition to the group of 'learned' Earth people he travelled with in the spacecraft, Menger professes to have spotted 'hundreds of Russians, Japanese, Germans and other people from other nations' who also were being taken on a guided tour of the Moon.[17] This seems absurd, because, if true, surely by now at least one other tourist would have come out and spoken about his or her experiences.

WITNESS TESTIMONY

Howard Menger's claimed liaison with the space people – and there were many more contacts – continued until the late 1950s. A month after his assumed visit to the Moon, Menger met Marla (Connie) Baxter, whom he married after divorcing Rose, his wife of 17 years, in 1958. Connie helped Howard run his sign-painting shop in Somerville, New Jersey, and also became deeply involved in his less down-to-earth pursuits. In company with many people in Menger's study group, she had some extraordinary experiences which tend to support some of her husband's claims, and which are now published, together with a fully updated version of the original book, entitled *The High Bridge Incident*.[18]

Several individuals claiming to have witnessed some of Menger's contacts came forward and spoke on *The Long John Party Line*, hosted by 'Long' John Nebel on radio station WOR, New York City. One of these witnesses was Menger's father. Here follows part of that interview:

NEBEL: Would you say these were normal-sized people?

MENGER Sr.: Oh, no. I would say one was about six-feet-two or three, and the other was about six feet.

NEBEL: Were you close enough to observe their features?

MENGER Sr.: No, that I wasn't.

NEBEL: Did you notice what they wore?

MENGER Sr.: To a certain extent, yes. As far as I could see, they wore something similar to ski-suits, tight at the wrists and ankles . . . It was a dark night . . . but these people seemed to have a glow to them. That is how we discovered they were coming toward us – by the glow. When these people left us, the grass there (and I know positively because I had cut some of it) was three feet high. And they went through that grass like it was a nice concrete walk, with no exertion at all . . .

NEBEL: You have seen the ships, too, haven't you?

MENGER Sr.: Oh, yes, I have seen them in the air and in the daytime I saw them, and at first I was very sceptical.[19]

Witnesses were not restricted to Menger's family. 'I had many witnesses – sane witnesses,' Howard told me when I interviewed him in 1978.

One was a doctor's wife, one was a physicist at Princeton University, and they all saw it. They all said the same – which is unusual – on radio, to back me up. I had photos and movies. In other words, in my case, there was definitely proof that we were being visited, not necessarily by aliens from another planet, but people who were more advanced than we were in technology, spirituality, and general human engineering, and that all this evidence would be admissible in a court of law . . .[20]

The physicist referred to was 'Dr Tom Richards', at the time a graduate student at Princeton (and whose real name is Richard Berry, Menger told me). In September of that year, with the permission of the space people, he and others were invited by Menger to witness one of his contacts. 'The visitors landed about a quarter of a mile to the rear of the house in a secluded wooded area,' wrote Menger. According to investigator Peter Jordan, Richards first sighted two disc-shaped objects, about six feet in diameter, which pulsated irregularly and hovered silently only 20 feet in front of him, radiating a variety of bright colours. These were also observed for nearly twenty minutes by Menger and his wife, Rose, and a young high-school student named Hotchkins.

Because Menger had not permitted any of the witnesses to come closer than 20 feet to the objects (due to the potentially dangerous level of electromagnetic energy he believed surrounded them), Richards's scientific scepticism was never completely dispelled. 'But,' said Jordan, 'Richards admits that he did find the experience striking, and finds it difficult, given the incredible sophistication of the display, to accept allegations of fraud freely advanced by so many of Menger's detractors.'[21]

ALIEN ACROBATICS

Later that night, Richards was treated to an even more impressive demonstration. The visitors reportedly walked down to where the witnesses were standing, hurdling a fence in an apple orchard, where Menger advanced and spoke to them in full view of the witnesses. 'These men from another planet were very tall, close to seven feet,' said Menger, 'and I defy any Earthman to equal the physical abilities they displayed.'[22]

According to Jordan, Richards watched for over 15 minutes as three beings of above average height 'ran, bounded, and jumped in a swift fashion across a yard, at times . . . attaining almost "superhuman" speeds'.

The entities, surrounded by a whitish glow, hurdled fences over five feet tall, and exhibited incredible gymnastic prowess in their movements. This demonstration, said Richards, could have been duplicated only had Menger enlisted the services of professional gymnasts . . . the experiences that evening persuaded him that Menger's claims, unacceptable as they appeared to many others, may have contained more truth than is generally assumed.[23]

A CONTACT WITNESSED?

Another of the many witnesses to some of the phenomena which seemed literally to surround Menger at that time was 'Mr X', a physicist who later became a businessman. On 10 January 1957 he testified on *The Long John Party Line* that on one occasion, together with three other witnesses, he had seen and heard Menger apparently communicating with extra-terrestrials.

Mr X: . . . the five of us went out and [Menger] took us through very rough terrain. The underbrush was kind of high . . . and Rose Menger pointed out a glowing light between the trees . . . It would take about 15 seconds to grow dim, and another 30 seconds to get brighter again. It was pulsating at about that rate . . . about 200 or 300 feet away. And it could be seen only through the trees. We were in an open clearing about 50 feet in diameter, and at the end of this clearing were the trees . . . Howard Menger suddenly said, 'Wait here', and he walked off toward the light. He didn't go very far. It must have been about 40 feet, and then he stopped and we heard two male voices talking . . . [Menger] walked right into the trees. He was probably about 15 or 20 feet into the trees . . .

NEBEL: Now, you heard two voices, one that you recognized as Mr Menger's – and do you feel the other voice could possibly have been Mr Menger's?

MR X: No, it had a different quality to it. It was more sing-songy than his voice . . . I couldn't make out any words. I listened as acutely as I could . . . And this conversation went on for at least half an hour . . . While he was talking we were facing Mr Menger and this person – whomever he was talking to.

NEBEL: You could not see him, could you?

Mr X: You could see a silhouette, a dark shadow.

NEBEL: Do you know it was another form there?

Mr X: It looked like two forms. I could not distinguish any facial characteristics. But I could pick out two forms . . . the other individual was slightly taller. While we were looking in that direction we heard somebody walking along the edge of the trees, walking toward our right, and stopping . . . Then we heard another person walking. We could hear the underbrush crackling and they walked around almost to the rear of us among the trees. In other words, we were surrounded . . . I am quite sure there were three of them . . . then Howard Menger came back and he said, 'Gee, I'm awfully sorry, I know how you feel, but I can't take you up there' . . . he looked very, very disappointed. I am sure he wanted us to see what was going on.

NEBEL: Did he imply that these people were from outer space?

Mr X: Yes. He did say there were three people: two men and a woman. I asked him what they said, but he said, 'I can't tell you . . . I would like to tell you but cannot', and it seemed to be something very personal. . .

NEBEL: Could this have been a 'set-up'?

Mr X: Well, we had a difficult job finding that spot. Howard Menger had not been there before. It was obvious we were lost when we started. And we had a lot of difficulty getting to this spot. We were going through underbrush and actually had to fight our way through. I am quite sure he had never been there before.[24]

The physicist, Menger told me in 1997, is now in his eighties and lives in Arizona. He prefers to remain anonymous.[25]

A 'HOT POTATO'

Yet another witness to a number of weird events was Richard Thompson (pseudonym) who, like Dr Richards, suspected that the heavy concentration of reported phenomena in the High Bridge area during the mid- to late 1950s was something more than coincidence. He could not comprehend how Menger, hampered as he was by his limited income as a sign-painter, 'could possibly have perpetuated such a combination of hoaxes, each perfectly timed, and brilliantly executed'.

In 1957, a large number of dehydrated vegetables, fruits and nuts were found in Thompson's home at Plukemin, New Jersey, as well as in open fields near High Bridge, which he says he was 'drawn to by apparent telepathic means'. The food appeared to have undergone an odd, 'freeze-dried' process.[26] A potato (believed to have come from the Moon!) was given to Menger, who suggested that it should be analysed by a professional laboratory. Samples were taken to LaWall-Harrisson Consultants in Philadelphia. Analysis revealed:

Total weight of sample	5.20 grams
Moisture	7.23%
Ash	4.49%
Fat (ether extract)	0.95%
'N' as Protein (NX 6.25)	15.12%

Owing to the abnormally high protein content (terrestrial potatoes – at least undehydrated ones – typically contain no more than 3 per cent) Menger decided to obtain a carbon-14 test (to date the samples) at LaWall-Harrisson, but the latter explained that this would cost about $2,000. Balking at such a fee, Menger declined. The consultants then recommended that Menger take the samples and their report to a certain doctor at a certain government agency (CIA) which, they said, would continue the research at no cost. At the agency laboratory, Menger and Thompson spoke to the doctor, a 'polite, intelligent man who appeared to be completely fascinated with the specimens,' wrote Menger. 'We left the samples with him, feeling we were on the right road.'

Two weeks later Howard and Connie went back to the laboratory to see how the analysis was progressing. As Menger reported:

We were shown into a room where a piece of a specimen was in a container of water, another piece was in another container and a small fragment was under a huge microscope. We took turns looking through the microscope . . . the outer surface of the specimen appeared like a crystalline beach of sand [possibly] due to an intense contraction of structure. I then told Dr——— what had been explained to me about the method of collapsing the molecular structure, which probably gave the specimen a dehydrated nature . . . We were told they would run all sorts of tests on the specimens left with them and keep us informed . . . Since my friend had been given the potatoes in the first place, we left the arrangements in his hands. That was in June, 1958, and the last we heard from the laboratory. I understand that this may have become 'classified' information . . .[27]

Thompson believes that the potato was 'impervious' to radioactivity; that it was intended to warn him of the dangers of 'impending nuclear catastrophe'; and that the agency deliberately withheld information on its analysis for fear that publicity would have produced 'panic and scientific disorder' – a veritable 'hot potato'! He saved a few of the remaining samples, which he claims remained in their original state, thus testifying to their unique method of preservation.[28]

BILOCATION

One of Howard Menger's more exotic claims, one 'endorsed' by a court of law, relates to an incident which occurred one night in the spring of 1957. This incident suggests that during his contacts with the extra-terrestrials, he seemed to become imbued with some of their phenomenal abilities.

During a coffee break at a gathering of his study group in Plukemin, he allowed his mind to wander back sentimentally to his light-green 1950 Plymouth station wagon which he had recently traded for a new model. 'In my mind's eye I drove it along on a blacktop road, picturing many things in vivid detail. Then I left the reverie, returned mentally to the group, and becoming aware of the discussion, joined in, without giving another thought to my vivid mental experience.'

At the next meeting, the police station in Bedminster Township, a few miles from Plukemin, telephoned to advise Menger that he had a driving summons awaiting collection. 'Sergeant Cramer claims you were speeding and went through a red light in his district about 11.40 p.m. on [he named the date of the last meeting],' Menger was informed.

'It couldn't have been me,' Menger responded, 'because I was here at that time and there were at least twenty people here with me. Besides, I do not have a 1950 station wagon, sir; I have a 1957 Plymouth station wagon, and incidentally, it could not have left the premises because it was blocked in by other cars, and I had the keys in my pocket.'

The police officer was unconvinced: finally, the summons was delivered personally to Menger by Police Chief Kice. Menger opted to appear in court, taking along seven witnesses, and pleaded not guilty. According to Sergeant Cramer, he saw a light-green Plymouth station wagon (licence number WR E79) speed past him. With Cramer in pursuit, the car went straight through the red light without stopping, then 'disappeared'! After commenting tersely on the 'phantom car' and listening to the testimony of the witnesses, the judge declared a verdict of not guilty.

A check with the auto agency in Philadelphia where Menger had traded in the 1950 Plymouth established the fact that it was still in the shop undergoing repairs prior to resale. So what had happened? In view of the fact that he had been thinking about driving his old station wagon at the precise time it was spotted 'speeding', Menger believes in the possibility that his thoughts 'had manifested into an actual projection'.[29]

RECANTATION AND RECONSIDERATION

In 1961, following a period of seclusion, Howard Menger returned to the public eye on *The Long John Nebel Television Show*. Paris Flammonde, producer of the show and later the author of two scholarly books on the UFO phenomenon, wrote that Menger, 'to the astonishment of supporters and opponents alike, recanted the vast majority of his personal legend, suggesting that all of his experiences may well have been "psychic" . . .'[30]

As author John Keel quotes Flammonde: 'Vaguely, aimlessly, rather embarrassingly, he avoided and vacillated . . .'

Howard Menger, Saturnian husband to a Venusian traveler in space [he believed at one time that he and his wife were reincarnated from those planets], friend of extraterrestrials, annotator of 'authentic music from another planet', master of teleportation, and saucerological sage *extraordinaire* – recanted! . . . His saucers might have been psychic, his space people visions, his and Marla's other planethood, metaphoric.

Later, in letters to investigators Jim Moseley and Gray Barker (publisher of *From Outer Space to You*), Menger described his book as 'fictionfact' and implied that the Pentagon had given him the films of saucers and requested that he participate in an experiment to test the public's reaction to extraterrestrial contact. 'He has helped us, therefore, to dismiss his entire story as not only a hoax, but a hoax perpetrated by the US government,' wrote Keel.[31] It is not that simple. For one thing, Menger has not recanted to the extent implied here. In 1967, for instance, during a rare appearance at a convention in New York, he confirmed at least one of his purported encounters with extraterrestrials, in High Bridge in August of 1956:

The craft came down from the west. It looked like a huge fireball. I was frightened. Gradually, as it came closer, it slowed down. The pulsations subsided . . . it turned into what looked like a man-made craft, reflecting the sun as it came close to the ground. It was a beautiful sight . . . It stopped about a foot and a half from the ground. An opening appeared in the side of the craft. There was a small incline or platform. Two men stepped out, very nicely dressed in shiny space suits . . . One man stepped to the left, and the other stepped to the right, and then another man stepped out, a man I will never forget as long as I live. He was approximately six feet one, maybe six feet two. He had long blond hair over his shoulders – yes, long blond hair. He stepped toward me, and the message he gave, of course, was what most people don't want to hear, a message of love and understanding. He said he had come from outer space, which is what most people really don't believe in. Someday they will.

Menger went on to stress the hardships faced by contactees following public disclosure of their claims. 'If you realize what people go through when this happens to them. If you really think you have guts enough to come out and tell people. Of course, nowadays it might be a little easier, but in the early Fifties it was very, very rough, especially when you are in business and you are trying to act like a reputable citizen.'[32]

So what really happened to Howard Menger? For several days in 1978 I interviewed Howard at his home in Vero Beach, Florida, in the presence of his wife, Connie, and my friend and co-investigator Lou Zinsstag. I found Howard, as always, to be a gentleman.

'Howard, when I first met you in 1969,' I began, 'you told me that you stood by the story in your book, but that you no longer knew where the space people came from nor what they were doing here. You also said to me that you'd swear on the Bible, to God, or whatever, that that book was true – that it happened to you.'

'Of course it was true,' he replied.

What I wrote in there and what I photographed and everything is absolutely true. However, at that time, I think I made some mistakes. In one case, they said, 'We have just come from the planet you call Venus.' I believe it's a possibility that I might have distorted that. It doesn't mean that they're Venusians. That means they might have a base there. There's evidence they've had bases on the Moon, and they could have bases anywhere. In a craft, I saw [on a viewing screen] something which I thought was the surface of Venus – they led me to believe it . . .

Menger gives serious consideration to the possibility that the visitors he encountered might have originated from Earth, rather than on other planets.

There are so many theories – this thing is so complex. One theory is that the Earth is the only one in this solar system which was given the gift of life, and this life developed a long time ago on this planet, and reached a civilization far beyond ours, in technology and spiritual ideas, thousands, maybe millions of years ago, and they have left, perhaps because of a cataclysm . . . those that survived probably would go underground, or under the ocean. Let's call them 'Atlanteans'. Most of this is myth, but suppose Atlantis was real? The people might have gone under the ocean and have cities there. It's very possible. UFOs have been seen going into the ocean, and coming out . . . It's possible they don't want us to know that they live here on this planet, that they would probably throw us off the track by telling us, you know, 'Venus' or 'Mars'.

'Are you *sure* that you actually went to the Moon?' I asked, incredulously.

'Well, over the years I've given it much thought,' he replied. 'I believed at the time that it was the Moon. They said it was one of our satellites. I don't know what they meant by that – we have only one! I think it was the Moon, yes. I took a picture from a porthole: that was the Moon I took a picture of!'

'But how did you manage to survive in the airless atmosphere, when they opened that door in the "train" and you were struck by a blast of hot air?'

'I don't know. But I was under their control, which would make a lot of difference. Their technology is so advanced . . .'[33]

AN ALIEN BASE?

In her section of *The High Bridge Incident* (much of which she published originally in a quasi-fictional book in 1958)[34] Connie Menger claims that Howard told her about an alien base, located about 150 miles from High Bridge, in the Blue Mountains of Pennsylvania, where supposedly Howard had made 'periodic trips'. It was there, he said, that he saw Connie's 'psychological chart' – 'a flat square, about a quarter of an inch thick, made of a plastic-like material . . . Across the face of this chart, a series of colored globes of light appear when it is connected to the machines, and this indicates the emotional and mental state of an individual.'[35]

Howard made no such claim during our interviews, but he did talk about having seen the alleged base.

In the Blue Mountains, when I was out there one night, I see this huge slit in the mountain. I'm watching this slit, and the light coming out of it – it started to light up the whole area. It's getting bigger and bigger . . . There was a section in the mountain – I would say 100 feet of mountain – with trees on it and everything, opening up like a garage door, only opening out. It's blinding me because I hadn't been used to the light: I'd been there for an hour, with my camera, waiting for a UFO to come down. I couldn't even take a picture, it's so bright. When I finally got used to it, I dropped the camera! And I see spaceships – at least three of them – in this cotton-picking mountain garage! And men walking around in shiny silver suits – they've got these tight-fitting suits – walking around, doing maintenance and work and everything. And a couple of guys came out in some kind of a vehicle, like a motorcycle without wheels. He's sitting on it and he's coming toward me! That's one time I didn't stay – I got in my station wagon and took off.

'I think that one of the main reasons they're here is mining,' Howard continued. 'I think they're mining some stuff. We don't even know what the stuff is, and if we did we'd probably be way ahead in our technology. It's something that they use – maybe for energy.'[36]

Menger's hypothesis is not entirely uncorroborated. In her book, *Silent Invasion*, researcher Ellen Crystall claims that two US Government agents told her that aliens were mining beryllium, titanium and zirconium in the area of her investigations around Pine Bush, New York. Both Crystall's and independent research has discovered that such ores are found in that region (and elsewhere). All three ores are used in nuclear engineering projects.[37] Of particular interest here is that zircon is found in Berks County, Pennsylvania, among several other locations. While this area is only about 60 air miles southwest of High Bridge, New Jersey, it is also just to the south of the northeastern end of the great Blue Mountain ridge in central eastern Pennsylvania.

FACT, FANTASY OR FRAUD?
In talking with Howard Menger, I found him convincing when discussing some of his original claims. At other times I had the impression that he was fantasizing. This impression was reinforced when he began regaling us with incredible tales of having been involved in building and flying a saucer, together with scientists and military personnel who had hired him to help them out, utilizing the knowledge gained from his extraterrestrial friends. He also professes to have put 'thousands of dollars' of his own money into the top-secret project.

'We built a huge craft,' he stated. 'My part in this thing was the design of the skin of the craft – the power system. And I did design the manual control, and from the manual schematics the other guy computerized it . . . I took it up with the four other fellows, I'd guess about 1,000 miles in five minutes. Of course, we didn't take any trips out of the atmosphere . . .'[38]

If this is pure fantasy – as I believe to be the case – why do I not reject *all* of Howard Menger's claims? Because, as with George Adamski, some of the evidence suggests that he did indeed have encounters with apparent extraterrestrials, some of which were observed by credible witnesses. And regarding the photographic and ciné film evidence (he took both 16mm and 8mm movies), my feeling is that *some* of it is genuine.

As to allegations of fraud, Richard Thompson suspects that elements of 'untruth' may have been woven into Menger's story to promulgate a more widespread belief in extraterrestrial life. 'But [Thompson] does not believe, based on his own experience,' stated Peter Jordan, 'that deception was as pervasive an ingredient in Menger's story as his detractors were later to maintain.'[39]

Perhaps the last word in this chapter on Howard Menger's claims should go to Berthold E. Schwarz, MD, a distinguished psychiatrist and investigator of the paranormal, with whom I have discussed the case, and who also lives in the Mengers' home town. In his introduction to *The High Bridge Incident*, he writes that, for Menger: 'There have been *no* rewards except the ivory chisels of ridicule, harassment, sometimes persecution, and even an alleged attempt at assassination. Throughout the years Howard has remained an honorable, outstanding citizen of his community . . . He has (by choice) avoided the spotlight for decades and never sought to profit from his extraordinary experiences . . .

'How influential, specific and relevant to his UFO adventures were the horrors and traumas of his son's, his brother's and his mother's deaths – all within a short period of time [in the mid-1950s]? Did these tragedies carve and prepare him for the UFO events and kindred dissociative paranormal happenings?' Dr Schwarz continues:

> Both Howard and I have lived in northern New Jersey . . . my own professional studies of [him] have involved interviews [with] the Princeton physics student and his friend who were witnesses to the discs [and] several of the New Jersey State Police who, when they realized that my purposes were confidential and scientific, recalled the furor and unexplained mysteries at that time . . .
>
> Although the Howard Menger story is sometimes seemingly bizarre, even by Fortean standards, Howard never deviates from his original, fundamental assertions which are voiced again and again like a hymn, a gesture of communion: *contact with the unknown has occurred* . . .[40]

NOTES
1 Menger, Howard, *From Outer Space to You*, Saucerian Books, Clarksburg, West Virginia, 1959, p. 26.
2 Ibid., pp. 26–8.
3 Ibid., pp. 31–4.
4 Letter to the author from Howard Menger, 14 September 1997.

5 Menger, op. cit., pp. 35–8.

6 Ibid., pp. 40–8.

7 Ibid., pp. 50–3.

8 Ibid., pp. 55–6, 59.

9 Ibid., pp. 71–3.

10 Ibid., p. 73.

11 Ibid., pp. 74–5.

12 Ibid., pp. 79–80.

13 Ibid., pp. 82–4.

14 Ibid., pp. 89–91.

15 Ibid., pp. 145–7.

16 Ibid., pp.148–52.

17 Ibid., pp. 153–6.

18 Menger, Howard, and Menger, Connie, *The High Bridge Incident*, PO Box 1405, Vero Beach, Florida 32961, 1991.

19 Menger, op. cit., pp. 104–5.

20 Interview with the author, Vero Beach, Florida, 27 October 1978.

21 Jordan, Peter A., 'The Enigma of Howard Menger', *UFO Update*, no. 11, summer 1981, *Beyond Reality Magazine*, PO Box 428, Nanuet, New York 10954, p. 46.

22 Menger, op. cit., pp. 102–3.

23 Jordan, op. cit., p. 46.

24 Menger, op. cit., pp. 192–4.

25 Letter to the author from Howard Menger, 14 September 1997.

26 Jordan, op. cit., p. 48.

27 Menger and Menger, op. cit., pp. 55–6, 104–5.

28 Jordan, op. cit., p. 48.

29 Menger, op. cit., pp. 112–14.

30 Flammonde, Paris, *The Age of Flying Saucers*, Hawthorn, New York, 1971, p. 157.

31 Keel, John A., *Operation Trojan Horse*, Souvenir Press, London, 1970, pp. 206–7.

32 Ibid., pp. 207–8.

33 Interviews with the author, 27/28 October 1978.

34 Baxter, Marla, *My Saturnian Lover*, Vantage, New York, 1958.

35 Menger and Menger, op. cit., p. 90.

36 Interview with the author, 28 October 1978.

37 Crystall, Ellen, *Silent Invasion: The Shocking Discoveries of a UFO Researcher*, Paragon House, New York, 1991, pp. 119–20.

38 Interview with the author, 28 October 1978.

39 Jordan, op. cit., p. 48.

40 Menger and Menger, op. cit.

Chapter 10

Cosmic Shock

One morning in April 1957, at about 07.30, a resident of Córdoba, Argentina, was riding his motorcycle towards Rio Ceballos. He had just reached a point some 15 kilometres from the airport at Pajas Blancas when his engine cut out. Dismounting to check the fault, he noticed a huge disc-shaped object hovering motionless about 50 feet above the road. Shocked, he ran and hid in a ditch.

The object appeared to be about 60 feet in diameter and about 15 feet high. It descended, hovering some seven feet above the road, emitting a sound 'like air escaping from the valve of a tyre'. Suddenly, a device described as a lift, or 'transparent stairway', began to come down from the lower part of the craft, carrying a human-shaped figure who stepped down when the lift had stopped about a foot above the ground. The figure was approximately five feet eight inches tall, dressed in attire resembling a diver's neoprene suit, which seemed made of a plastic material.

After glancing around at various plants, the being came towards the witness, who made further attempts to conceal himself – to no avail. Silently and gracefully, the stranger reached out to help the witness from the ditch. As the two stood side by side in the road, the being pointed towards the hovering craft, indicating that the witness should follow him. To calm the panic-stricken Córdoban, the being gently stroked the man's forehead. The two then went to the craft and entered the 'lift'.

Around the wall of the cabin were five or six panels, each about six feet wide and covered with an intricate mass of equipment, including TV-like monitor screens. At each of the panels sat a similar being, dressed exactly like the first. They paid no attention to their surprised visitor. The Córdoban later commented that he had been particularly impressed by a series of large square windows around the walls above the panels, having observed no trace of windows on the outside of the craft. In addition to the light coming through the windows, a dull phosphorescent light pervaded the cabin, though yet again, no lamp was seen. The colour of the craft was somewhat indistinct: in parts it had a greenish tinge, elsewhere blue, the combined effect being that of metallic iridescence.

The witness was invited to enter the lift-shaft, which descended. As it did so, he struck the wall with his knuckles. This caused a metallic sound.

Once on the ground, the Córdoban asked the being, in sign language, how the craft managed to remain suspended. The being responded by passing the palm of one hand flat over the other – a gesture which meant nothing to the witness. The motorcycle was examined closely by the being, who indicated that it would not start while the craft was present.

Finally, the visitor turned to the Córdoban, placed a hand on his shoulder in a farewell gesture, then went back inside the craft via the lift-shaft. Shortly afterwards, the craft took off, heading in a northwesterly direction. There were several sightings of what was presumably the same craft that morning.[1] It is unfortunate that the newspaper which reported this story did not mention the name of the witness; but that omission may have been stipulated by the motorcyclist. The story itself seems convincing, if only because of the technical details, which closely correspond with those of many similar encounters.

Oscar Galindez, the Argentine investigator who supplied the original report, believed (together with many of his countrymen) that there were UFO bases in the Andes, the most probable locations being in the high mountains around Salta (northwest Argentina) and the Puna de Atacama, a desolate area to the west of Salta.[2]

AN ALIEN BASE IN ARGENTINA?

In August 1957 an Argentine newspaper published an extraordinary story, supplied by an Air Force guard who reported an encounter with extraterrestrials on 20 August. The following, translated by Gordon Creighton, appeared under the headline 'UFO Base near Salta':

Two days ago, at Quilino (Province of Córdoba), an aircraft made a forced landing. Air Force personnel were sent to guard the machine, and they installed themselves in a tent nearby. One of the Air Force men was on guard, while the other two went to a store several kilometres distant to get supplies.

Suddenly, the man on guard – whose name we omit for reasons that are obvious – was aware of a strange humming sound. This was so persistent that he went out of the tent, but on looking around he could see nothing that was out of the ordinary. He went back inside the tent, and once again he heard the hum; this time it was loud and high-pitched. He stepped outside once more, and to his astonishment he saw a disc-shaped machine suspended at a height of some 90 metres from the ground. The strange machine descended slowly until it was only a few metres from the ground. The grass and plants became violently agitated beneath the craft. Alarmed at the sight, the serviceman tried to draw his revolver, but was unable to do so; the revolver appeared to be stuck in the holster, perhaps through some influence from the machine.

The next thing he knew was that a clear soft voice was speaking to him from the machine. He was told not to be afraid; that they were there in order to make the world aware of the existence of the 'interplanetary ships'. The voice also told him that they, the occupants of the interplanetary ships, had set up a special base or station in the province of Salta, and that from this base their crews would go forth to establish peaceful contact with Earthmen. The voice said furthermore that their aim was to help us, for the wrong use of atomic energy threatened to destroy us. Finally, before departing, the voice assured him that very soon the rest of the world would know even more about them. Then the bushes began to blow to and fro, and the craft rose vertically to a height of some 40 or 50 metres, before moving off towards the north.[3]

PLAIN-CLOTHES ANGELS

In June 1962 the Italian magazine *Domenica della Sera* published an interview with a 42-year-old engineer named Luciano Galli, who claimed to have encountered human-type extraterrestrials in 1957 (or 1959; he was unable to recall which of those two years). Although Galli's account parallels the experiences of George Adamski to a remarkable degree, he was prepared to take an oath stating that at the time of his experience he had not heard even the name of Adamski. The following is taken from the interview with Galli conducted by the reporter Renate Albanese, translated by Gordon Creighton.

At 14.20 on 7 July 1957 (or 1959), Galli left his home in Bologna to return to work after lunch. He was nearing his workshop, situated off the Via Castiglione, when suddenly a black Fiat 1100 stopped in front of him. Out stepped a tall, dark man with regular features and very black eyes. 'His face was of the kind which invites you to be friendly,' said Galli. The man wore a double-breasted grey suit complete with collar and tie. He spoke perfect Italian. At the wheel of the car sat another man, with delicate features, dressed in a light-coloured costume: he wore no moustache like the dark one and never spoke a word.

'I knew the man with the moustache from sight,' Galli told Albanese.

I had noticed him several times in town; he even seemed to follow me. Once, I remember, I walked with a friend through the arcades of Via Castiglione when I again saw this man. As always, he looked straight into my eyes and this time I wanted to address him, but suddenly he disappeared. And now this very stranger was standing before me, asking me if I remembered him. I said yes. 'Won't you come with us?' 'Where to?' 'Have confidence, nothing will happen to you.'

Galli took a seat in the Fiat and drove with the men to the Croara ridge, 57 kilometres from Bologna. There, he claimed, a 'shining grey' flying saucer awaited them, hovering about two metres off the ground. From the bottom a metal cylinder with an opening in it came out, through which Galli and the men entered. Although initially afraid, Galli became calm as soon as he was inside the craft. Just prior to entry, two lights flashed. 'Don't be afraid,' said the man with the moustache, 'you were only being photographed.' Galli continued:

The pilot's cabin was spacious and round with a lot of instruments around, panels with pointers and needles. There were also hatches, and the seats were fixed somehow to the ground. In the middle of the floor was a kind of circular window, about one metre wide. Through it we could see the earth fall away from us. First she looked as though viewed from one of our own planes, then – when we were already in the dark zone – she looked like the Moon and later like Venus or Mars.

'Were you able to talk to the man you call commander?' asked Albanese.

'Yes, very well. He spoke a perfect Italian. I asked him how he had managed to learn our language so well. He answered that he had used a very good method.'

Suddenly, Galli saw the silhouette of an enormous 'dirigible'.

Its length was at least 600 metres. The one end was cut like the end of a cigar. The [craft] emitted a phosphorous light and on top of that it looked as if strong light beams were directed toward it. Underneath the cut end six openings came into view, out of which and into which small flying discs were seen coming and going. Every opening was divided by a partition wall into six smaller cubicles, every one wide open.

As they approached the giant ship it became evident that the openings were large hangars, capable of accommodating at least 50 saucers. Inside the ship could be seen no fewer than 400 or 500 men and women, standing or walking around the hangars. 'This is what Galli said on oath,' reported Albanese. 'All those people wore overalls of a shining plastic or silky material. When they passed by them, they smiled. The women were very beautiful and friendly.'

Spellbound, Galli asked his companions where the ship came from. 'From the planet you call Venus,' came the reply.

Later on, Galli said that he was shown through a large hall, a kind of library, and into another large room which he believed to be the commander's. Eventually he was shown back to one of the hangars and into the same saucer - 'always in company of the man with the moustache and a face like an angel in plain clothes'. He was brought back to the same spot on the Croara ridge. 'The whole trip was completed within three hours and ten minutes,' Galli claimed.

Renate Albanese asked Luciano Galli if he was certain that these fantastic events had not happened to him while he was in a trance, or under hypnosis. I have never been hypnotized,' he replied. 'I took this trip in my physical body, this is indeed so . . . I do not want people to say that I made up this story in order to gain publicity or money. What I have told is the naked truth.'[4]

AERIAL ENCOUNTERS IN BRAZIL

Numerous encounters with unknown flying machines were reported worldwide by military and civilian pilots in 1957, some of the more disturbing encounters occurring in Brazil.

Just before 21.00 on the night of 14 August, Commander Jorge Campos Araujo was at the controls of a VARIG airlines C-47 cargo plane, en route from Porto Alegre to Rio de Janeiro, when the co-pilot, Edgar Onofre Soares, spotted a luminous object left of the plane. Suddenly the object manoeuvred so that it was ahead of the plane, crossed to the right on a horizontal trajectory, stopped momentarily, then abruptly dived and disappeared in the cloud bank below. All the crew described the object as saucer-shaped, with a dome on top that glowed with an intense green light while the flattened base emitted a less intense yellowish luminosity. Commander Araujo estimated the saucer's speed to be several times the speed of sound.

Although the encounter made headline news in Brazil, the most important part of the event was not publicized. After landing, Commander Araujo and the crew told a colleague, an air traffic chief for another airline company, that when the object had reached the right side of the airliner, the engines began coughing and missing and the cabin lights dimmed and almost went out. Fortunately, everything came back to normal when the object disappeared.[5]

Another disturbing incident also occurred in Brazilian airspace on 4 November 1957 at 01.40, when a C-46 cargo plane of VARIG, bound from Porto Alegre to São Paulo, encountered an unknown flying machine. At first, it looked like just a red light to the left side of the aircraft. Commander Jean Vincent de Beyssac joked to his co-pilot that at last they were seeing a real flying saucer, but then the object seemed to become larger, and de Beyssac decided to investigate. Commander Auriphebo Simoes, who interviewed de Beyssac, reported as follows:

> He started to put his plane into a left bank, but just before he pressed his rudder the object jumped a 45-degree arc on the horizon and became larger. De Beyssac started pursuit and was about midway on his left 80-degree turn when the object became even brighter and suddenly he smelled something burning . . . all at once his ADF [automatic direction finder], right generator and transmitter 'burned' out. The 'thing' disappeared almost instantly, while the crew looked for fire. De Beyssac turned on his emergency transmitter and reported the incident to Porto Alegre control; then he turned his ship around and headed back to Porto Alegre, where he landed an hour later.

After submitting a written report, de Beyssac went home and 'got soused'. That same day, VARIG issued an order forbidding its pilots to discuss UFO sightings with the media.[6]

OUTRIGHT HOSTILITY

A quarter of an hour or so after the VARIG C-46 encounter, at 02.00 on 4 November 1957, an even more alarming incident occurred at the Brazilian Army's Itaipu Fortress at São Vicente, near Santos, Brazil, when two sentries were struck by an unbearable wave of heat emanating from a large disc-shaped object hovering above them while emitting a humming sound. One sentry fainted and the other yelled for help. Inside the garrison, the electrical systems failed completely; even the back-up system would not work. Both soldiers received first- and second-degree burns. The authorities reacted desperately, as investigator Dr Olavo Fontes reported:

> Next day the commander of the fortress (an army colonel) issued orders for-bidding the whole garrison to tell anything about the incident to anyone – not even to their relatives. Intelligence officers came and took charge, working frantically to question and silence everyone with information pertaining to the matter . . . The fortress was placed in a state of martial law and a top-secret report was sent to the QG [headquarters]. Days later, American officers from the US Army Military Mission arrived at the fortress together with officers from the Brazilian Air Force, to question the sentries and other witnesses involved. Afterwards a special [Air Force] plane was chartered to bring the two burned sentries to Rio de Janeiro [where] they were put in the Army's Central Hospital (HCE), completely isolated from the world behind a tight security curtain.[7]

Dr Olavo Fontes, a world-leading UFO investigator of that period, spent a great deal of time investigating this case. In a paper discussing UFO-induced effects, he theorized that these were 'not merely side-effects of the powerful electromagnetic fields that exist around UFOs, but the result of purposeful interference by a weapon used as a means of defense or attack'.

> Existing evidence suggests that such a weapon is not an alternating magnetic field in itself, but a high-frequency, long-range electromagnetic beam of some sort, i.e., a radio-electric wave concentrated into a narrow, powerful beam. After a careful analysis of the data I came to the conclusion that this weapon might be a microwave ionizer – a generator of odd-shaped micro-waves that ionize the air where they strike. They would make the air a high-resistance conductor; nothing more than that.

The 'heat-wave' which burned the two sentries at Itaipu, he asserted, was a different kind of weapon, 'produced only by ultrasonics'.

> This is due to the fact that the longitudinal ultrasonic oscillations are trans-formed into transverse waves (shear waves) at interfaces between mediums of different acoustic impedance as, for example, between the clothes and the skin. These resulting transverse waves are more rapidly absorbed than the

longitudinal ones, with a subsequent increased heat development at interface areas . . . Only an ultrasonic beam could produce the peculiar characteristics of the 'heat wave' that struck the Itaipu sentries.[8]

'DANGERS THREATENING HUMANITY'

'The statements of those who purportedly have had actual contact with "space people" should not be dismissed offhand as merely romance,' stated Rear Admiral Herbert Knowles of the United States Navy, in July 1957. 'Perhaps there is some real information here . . ."[9]

That same month, an extraordinary contact was reported by a highly credible witness, Professor João de Freitas Guimarães, a lawyer and Professor of Ancient Roman Law in the Catholic Faculty of Law at Santos, Brazil. Interestingly, the encounter took place near Fort Itaipu. The following is based on a transcription of a television interview with the professor on Brazilian television (TV-13) on 27 August 1957, translated by Gordon Creighton.

In the course of his duties as a military advocate, Dr Guimarães was visiting São Sebastião, northeast of Santos on the Atlantic coast in the State of São Paulo, when one evening he decided to take a stroll along the beach. The sky was overcast and there was no visible moon. As he was sitting on the beach at about 19.15 (another report gives 21.15) he noticed that, in the direction of Ilhabela (Bela Island), the colouring of the sea seemed to be getting lighter, then a spout of water shot up into the air, reminding him of the 'blowing' of a whale. Some sort of 'high-bellied craft' (shaped like a hat) headed for the beach and as it arrived, threw out a landing line to which were attached some 'spheres', unlike conventional buoys. A metallic stairway also came down and two men stepped out and approached the professor. Both appeared to be completely human.

By now he could see that they were tall beings, over 1.80 metres in height, with long fair hair extending to their shoulders. Their complexions were fair, they had eyebrows, and their appearance was youthful and they had light-coloured, wise and understanding eyes. They wore greenish one-piece suits fitting closely at the neck, wrists and the ankles.

Though frightened, Dr Guimarães stood up, holding his ground. He asked in Portuguese whether there was something wrong with their craft or if they were looking for somebody. He also asked them whence they came. There was no reply, so he repeated the question in French, then English, Spanish and Italian – to no avail. Then the professor received the impression, telepathically, that he was being invited to board their craft. (Subsequently, he discovered that they also made verbal speech.) He became seized with an irresistible desire to know more and decided to accept the unspoken invitation. One of the men headed towards the disc, followed by the professor and the other man. The man in front leapt up the stairway to the entrance with ease, using one hand to hold on, though Guimarães needed both hands to do so. Standing in the entrance was a third man, possibly one or two others.

SPACE FLIGHT

On board, Dr Guimarães found himself in a brilliantly illuminated compartment, one of several in the craft. Together with the crew, he sat on a bench that encircled the compartment. As the craft lifted, the professor noticed what looked like rain on the portholes. 'Is it raining?' he asked. To this he received a reply from one of the crew, telepathically, to the effect that it was not rain, but water produced by the rotation, in opposite directions, of parts of the craft. Surrounding the machine was a 'ray-filtration tube, which produced the effect of a semi-vacuum in each of its parts'.

> Gazing out through the portholes, the professor beheld a vast, intensely black zone all around, in which the stars were shining with astonishing brightness. Then came areas where the stars seemed to be in even greater swarms, shining with an incomparable splendour, followed by other areas which seemed darker, with fewer stars. Then they passed through a belt of violet-coloured atmosphere, and after that another, similar belt, but of a more violet shade, and of a most refulgent brightness, and during this stage the professor felt the craft shuddering strongly, and he showed his fear, whereupon one of them said to him telepathically: 'Our machine has just left the atmosphere of your planet.'

In the compartment was a circular instrument with three very sensitive needles which trembled in the Earth's atmosphere but vibrated more intensely when outside of it. As one of the crew members explained, the craft was being driven by 'the effects resulting from the magnetic forces present in space'.

On returning to Earth, Dr Guimarães noticed that his watch had stopped working at the time he boarded the craft, thus he was unable to determine how long the trip had lasted, though he estimated that it had been for about 30 to 40 minutes, judging from the hotel clock on his return.

Dr Guimarães believed that the crews of extraterrestrial craft were anxious to alert the inhabitants of Earth to dangers threatening humanity. In his opinion, some of our scientific experiments were being conducted frivolously, such as the indiscriminate explosion of atomic bombs which not only added to contamination of the atmosphere but also destroyed those layers of our atmosphere that filter out dangerous radiation. If more care is not taken, we shall all suffer from the consequences of the explosions, he warned.

AFTERMATH

Though barely able to contain himself, Dr Guimarães at first told no one (with the exception of his wife) about his fantastic experience. Many months later he informed three friends and colleagues: Dr Alberto Franco, a São Paulo judge; a Dr Nilson, a lawyer; and later, Dr Lincoln Feliciano, who contacted the media. Dr Guimarães was besieged from all sides by people wanting to hear more, but he found it difficult to explain what precisely had occurred because, he said, his experience related to matters far above and beyond his knowledge.

During his trip, the professor said, an appointment had been made for him to meet the crew of the craft again, on 12 August 1957. The crew had shown him a Zodiac of 12 constellations: a wheel indicated the year, and the repetition of the number '8' twelve times gave him the impression of August. He did not keep the appointment, he said, owing partly to the fact that a party of people – including the town's deputy sheriff and Major Paulo Salema of the Brazilian Air Force, who had got wind of the meeting – had planned to go along to the beach. Also, he said, the Air Force had advised him not to keep the appointment; moreover, it had been arranged for some jet-fighters to put in an appearance. Had these fired on the disc, Dr Guimarães felt that it would have appeared like an act of treachery on his part. In any event, the craft did not make a repeat appearance.

Although Professor Guimarães agreed that he was of an 'idealistic turn of mind', he insisted he was practical too, and stressed that he had been fully conscious throughout the experience, and not a victim of an hallucination.[10][11][12]

CHILD-SIZED HUMANOIDS

Miguel Español, a Spanish naval officer, was travelling with a companion by truck to Ceres, Brazil, on 10 October 1957, when about three miles from Quebracôco village he noticed a strong gleam of light over a hill ahead. Just after the truck surmounted the hill, a huge luminous object could be seen about a mile away, hovering in the air, brilliantly illuminating the surrounding countryside.

As it descended to a height of about 20 feet from the ground, some 40 yards away, the object seemed to cause the truck's engine to stall. It then cut out all its lights, except for a reddish one in an antenna on the dome. Estimated to have a diameter of about 500 feet and a depth of about 130 feet, the craft was oval-shaped, but like two superimposed saucers separated by a circular area of about 16 feet. A door opened and two beings emerged, followed by another two pairs, then a seventh one, who walked down the centre of two lines made by the others.

Español described the beings as looking like child-sized humans, with long hair and dressed in what appeared to be luminous suits. The truck engine would not work during this whole period, despite frantic efforts to make it start. After silently observing the truck and its two occupants for about three minutes, the space visitors climbed back into their craft and the door shut. The huge ship took off, climbed to a height of about 1,600 feet, at which altitude it released a small disc, then disappeared to the south.[13]

THE VENUSIANS COME TO ENGLAND

During the course of my nearly 40 years' research into contact cases, I have come across a number of accounts involving encounters with extra-terrestrials whose alleged origin is Venus. Some of these accounts have never been published, such as the following, which was incorporated in a manuscript given to me by the pioneering researcher Tony Wedd, a former Royal Air Force flying instructor, designer and artist who taught me much about this arcane subject.

Early one Sunday morning in November 1957, newsagent Hubert Lewis was cycling to the town of Church Stretton in Shropshire to collect newspapers. It was miserably cold, dark, wet and windy, and he was cursing and swearing at his lot in life. 'I was really browned off,' he told Wedd. All at once in the half-light a tall figure appeared in the road in front of him. Hovering to Lewis's right was a large object of 'a dull lightness', which seemed to rotate, although part of it was still.

Replying to Lewis's question as to who he was and whence he sprang, the stranger answered that there was no need to be alarmed. 'I must admit the whole circumstance at first did scare me,' said Lewis. 'I also remember noticing how the wind had dropped, although I could still hear it, but from away from us, and around us.'

The phenomenon of localized silence has occurred in many close encounter cases, such as that claimed by the Italian engineer Gianpietro Monguzzi, who together with his wife witnessed a landed UFO and its occupant in July 1952 on the Cherchen Glacier in the Italian Alps; he succeeded in obtaining several remarkable (and generally discredited) photographs before it flew off. As in Lewis's encounter, a howling gale suddenly subsided, and all became silent during the landing.[14] Lewis continued:

> My visitor spoke quite clear English, but with a slight lisp it appeared, and he first rebuked me for my language, pointing out that everybody on this earth had troubles and difficulties to face, and that this life is merely a probation-ary period for a further life after. He knew of my difficulties and troubles; he spoke of my previous employment, and mentioned names of people whom I had once known.

Lewis estimated that the conversation lasted nearly 30 minutes, during which time the circular object, thought to be about 60 to 100 feet in diameter, hovered above him and to the right, approximately 100 feet away. A slight whistling sound emanated from it. The stranger gave no indication of his origin at this stage, but was 'kindly, understanding and gentle'. The tone of the conversation then became somewhat more didactic, as Lewis was treated to the Sunday-morning sermon of his life.

> He told me I had nothing to fear in the future from any evil. If I would only keep calm and faithful, all would be well. He promised I should be looked after and guided along the right path if I kept faith. My [deceased] sons and my wife were still with me in spirit, as were my friends (he mentioned one whom I had forgotten, of forty years ago) . . . He wished me well, then suddenly vanished, just like that. Here I would like to add that this did puzzle me until Mr Cooke [James Cooke, a contemporary English contactee] gave me the answer, and that was that my visitor was projected before me but was really away from me, possibly in the machine; and this could be what happened.[15]

PROJECTED IMAGES

It is apposite to note here that in the case of Birmingham housewife Cynthia Appleton, an encounter which took place during the same month as that of Lewis (and about 40 miles from Church Stretton), an image of a tall, fair man, dressed in a tight-fitting silvery garment with an 'Elizabethan' collar, and with straight hair almost to the shoulders, cut in 'page-boy' fashion, appeared in her home 'just like a TV picture on the screen' with an accompanying 'whistle'. The image was blurred at first, then became sharp.

Without voicing her question, Mrs Appleton wondered whence came this man. 'From another world,' came the response, received telepathically. 'Like yours, it is governed by the Sun. We have to visit your world to obtain something of which we are running short. It is at the bottom of the sea.' Mrs Appleton thought the required substance was something that sounded like 'titium', which her husband later suggested might be titanium. (Although I have often wondered if that might have been a misunderstood or mispronounced reference to lithium, the lightest of solids, recourse to the book by Ellen Crystall mentioned in the previous chapter suggests that the alien's reference might well have been to titanium.) 'You are stripping bark from the wrong tree to line the wrong boat,' the man continued. 'You are concentrating on the wrong power. You are trying to go "up" [i.e., against the force of gravity]. We go like this,' and he made a sweeping lateral movement with both his hands and suddenly between his outstretched fingers appeared a type of television screen, on which could be seen a circular craft with a transparent top half (with several figures inside looking at her) and some much larger ships with several smaller, circular craft attached underneath. Before the image disappeared, the visitor said he would return in January.

On the second occasion, Cynthia Appleton was visited in precisely the same manner by the same man, this time with a companion. Communicating verbally on this occasion, with careful, clipped articulation, they informed her that they came from 'Ghanas Vahn' (or so it sounded, pronounced with their guttural accent) on Venus. Other people's brains, they said, were not suitable, as was hers, to witness this type of projection. The image appeared to be quite solid. 'You could not see through them although the light of the window was behind them,' explained Mrs Appleton. When she asked if she could touch the image, she was told that to do so would be very injurious to her health.[16]

From another contactee (Joëlle, whose experiences are recounted in Chapter 12), I have learned that this manner of projected image is but one of a host of mental and technical feats of which certain extraterrestrials – and some terrestrials, such as advanced yogis – are capable.[17] This particular phenomenon does not negate the physical reality of the source of the projection.

A LINK WITH THE SPIRIT WORLD?

Hubert Lewis went on to explain that his psychic faculties had been stimulated by the experience. 'Such things of which my visitor had spoken had never appealed to me previously, things such as an after-world,' he commented.

> Personally, I had little faith; I always held the view that once you were dead, that was the end. In my life I had seen so much badness and the exploitation of man by man that no normal person could have convinced me that life held anything but sorrow for many and happiness for a few. Looking back, this shock was the only thing which could have altered my views of life.
>
> Now I have faith. I have seen my wife, who appeared one morning to me. I have had several conversations with my sons, more especially with my youngest, who was an intellectual and has made quite clear many things which I previously had no idea could possibly be . . . There is a connection between a spirit world and the planets and our after-life – please be assured of this. I now have no fear of death.

FURTHER ENCOUNTERS

During May 1958, Lewis claimed to have been visited by 'a high official of the police and another gentleman', who asked many questions. 'I was advised to forget certain matters and carry on my normal work and way of life.' Lewis decided to accept this advice, but early one Sunday morning in the summer of 1958 he had another contact, this time with a different being. The craft in which he arrived was in the distance, and to meet Lewis, the man had apparently walked across some fields. Lewis was told that people from Venus are living on Earth among us. Lewis went on:

> I can assure you I have no fear of our friends (and I say this with perfect confidence) for they are our friends (so God help me and all of us) and if only our leaders and persons of importance and authority would only meet them and guarantee them safety and guarantee that their knowledge and age-old abilities would not be exploited and imposed upon, then I who know them and have been with them on this Earth can say with truth and confidence that all of us who inhabit this planet can be assured of a perfect life . . .

In July 1958 Lewis travelled by rail to Paddington Station, London, where he met by prior arrangement with his contact, who escorted him to a car in which he was introduced to a woman of appearance similar to that of the man. As it was a pleasant evening, they drove into Hyde Park, and walked by the Serpentine for about an hour. Later, the trio drove to Wanstead Flats, and then to Forest Gate, where Lewis had a meal and the others tea and sandwiches (!). At one stage the aliens engaged each other in conversation in their own language, unintelligible to Lewis, but soon joined him for a fascinating discussion which lasted until midnight.

After having spent the night with a friend in Stratford, Lewis met the same man, without his female companion, at 18.00 the following day. A visit to the Spiritualist Association of Great Britain at Belgrave Square, in central London, was suggested. The two attended a demonstration of clairvoyance by the famous medium Rebecca Williams. Further meetings took place in London that week, including a trip to St Paul's Cathedral, and finally both companions bade Lewis farewell and saw him on to his train at Paddington.

Lewis described the appearance of the 'Venusians' as well dressed but not overdressed, with a 'likeness to continental folk, dark, Jewish or perhaps Grecian. The male friend as tall as I, more than six feet. Well built, athletic, very strong I should say. Lady also tall, similar features. Both very healthy-looking.'

> There is much that I have learnt, much I am not yet allowed to say. I have also been urged to be careful with whom I discuss this excursion of mine, for there are many forces who desire to know many things which could be of great advantage to the few . . . Although I don't know anything about this subject, it is magnetic power which drives these craft – points of magnetic pole. Each craft and space ship works to a purpose, whatever they do. Many space craft land on Earth in various countries . . . Some are huge but these never land . . . and the planets have many contacts on Earth. Also many are living among us,

and one cannot tell the difference. Medical men could, however, so I understand.[18]

I have known personally several contactees whose experiences parallel
those of Hubert Lewis in cases where the possibility of collusion is very
small, because, firstly, I made a point of not introducing the parties to
each other until I had completed my independent investigations; while
secondly, none of the stories had been published.

Lewis's comment that medical men could distinguish between an alien
and a terrestrial is interesting, especially because it is now a matter of
official record that a qualified physician has claimed to have physically
examined an alleged extraterrestrial. As described in my book *Alien
Contact*, in 1976 a Mexican paediatrics and anaesthesia specialist, the late
Dr Leopoldo Díaz, examined a man who said he came from another
(unspecified) planet. The man proved to be normal in all respects, with
the exception of his very white skin, extraordinary eyes with a wider iris
than normal, and a claimed age of eighty-four, while otherwise appearing
not older than in his forties or fifties. So taken aback was Dr Díaz by this
encounter that he contacted the United Nations in New York and spoke
there with a delegation. One of the delegates was Robert Muller, then
UN Under-Secretary of Economic and Social Development, who told
me that unfortunately he was subsequently unable to gain the interest of
anyone in the UN in the case.[19]

THE OBSERVER

Air Marshal Sir Peter Horsley, former Deputy Commander-in-Chief of
Strike Command, has flown 90 types of aircraft, ranging from the
Mosquito (in the Second World War), to the Spitfire, Meteor, Hunter,
Lightning and Vulcan. He spent seven years in the service of Her
Majesty the Queen and HRH Prince Philip as Equerry, and it was during
this period that he had an experience which had a profound effect on him
– an encounter with an apparently extraterrestrial being.

In his fascinating autobiography, *Sounds From Another Room*, Sir
Peter devotes a lengthy chapter to the subject of UFOs, including details
of his investigations into sightings reported by pilots; details which he
relayed to Prince Philip, who shared his interest in the phenomenon.
Another enthusiast was Air Chief Marshal Sir Arthur Barratt, who
retired from the RAF at the end of the war. Barratt introduced Sir Peter
to a friend of his, a General Martin, who believed that flying saucers
were extraterrestrial vehicles from another planet whose inhabitants
were trying to warn us of the perils of nuclear war. Sir Peter was not
convinced.

One day in 1954, General Martin phoned Sir Peter, inviting him to meet a Mrs Markham that night at her London flat in Smith Street, Chelsea. General Martin himself did not attend the meeting. There, in a dimly lit room, he was introduced to a 'Mr Janus'. 'Without any pre-liminaries,' writes Sir Peter, 'Mr Janus dived straight into the deep end by asking me to tell him all I knew about UFOs. He listened patiently . . . At the end I thought I might be equally as direct and asked Janus what his interest was. He answered me quite simply, "I would like to meet the Duke of Edinburgh."' Somewhat taken aback, Sir Peter replied that this would not be easy. 'I was about to add particularly for security reasons but thought better of it,' he writes. 'But it was here the strangeness of it all started – the man's extraordinary ability to read my thoughts.' Asked why he wanted to meet Prince Philip, Janus replied: '[He] is a man of great vision . . . who believes strongly in the proper relationship between man and nature which will prove of great importance in future galactic har-mony . . . perhaps you and I can discuss the subject first and you will be able to judge whether I am dangerous or not.' Sir Peter devotes 14 pages to the ensuing two-hour discourse, of which selected excerpts follow.

PER ARDUA AD ASTRA*

Janus began by pointing out that Man was 'now striving to break his earthly bonds and travel to the moon and the planets beyond'. He continued:

> But flight to the stars is Man's ultimate dream, although knowledge of the vast distances involved in interstellar flight makes it appear only a dream. Yet perhaps after a hundred years or so . . . exploration of his own solar system may be complete and it is just not in Man's nature to stop there . . . Just as tribes found other tribes and Christopher Columbus discovered on his trav-els unknown centres of ancient civilizations, so Man in his journeys through the universe may find innumerable centres of culture far more ancient than his own . . . He will discover a wealth of experiences infinitely more startling and beautiful than can be imagined: an infinite variety of agencies and forces as yet unknown: great fields of gravity and anti-gravity where objects are accelerated across space like giant sling shots, even other universes with dif-ferent space and time formulae . . .
>
> Why does Man reach for the stars? His energies have never been solely directed towards material benefits alone. From the beginning of Man's history he has striven . . . towards a spirituality and grace of which he was aware but could not fully comprehend. This drive to reach out beyond him-self has been the motive power behind some of Man's finest achievements . . . So Man invading space for material gain or personal glorification alone will gain nothing, but Man searching to enrich his own spirituality and nature will come closer to understanding that God is Universal.

*'Through hardship to the stars' – the RAF motto.

A DARK AGE

'The Earth is going through a Dark Age at the moment,' Janus went on. 'Material possessions count more than a Man's soul.'

> Like a child, Man is preoccupied with his technological toys, which he believes will bring him riches and happiness. This shows up in the superficiality of his culture and a careless disregard for nature. In his greedy quest for more complex machines Man is prepared to sacrifice almost anything – his natural environment, animals and even his fellow humans. The dreadful spectre of blowing up his world hardly makes him falter in this headlong rush.

COSMOGONY

Janus expounded on cosmogony – the origin of life in the universe – seemingly lending support to the so-called 'Big Bang' theory, which he referred to as 'the generally accepted theory' of an expanding universe that 'originated from the giant explosion of a vast area of high density gas which contained all the elements necessary for life and matter'.

'If you accept this theory,' he went on, 'then all galaxies contain the elements necessary for life and matter; even at the very boundaries of the expansion, the original explosion is still distributing these elements . . . If you accept the theory of the expanding universe you accept that it is an ocean of galaxies with solar and planetary systems similar to our own. By the laws of probability there must be millions of planets in the universe supporting life, and within our own galaxy thousands supporting life more advanced than on Earth.'

> Earth is a young planet with its Sun a young mother. We may hazard a guess that other planets in [this] solar system are unlikely to support life except in possibly rudimentary cellular form and are no more than uninhabited and hostile islands. But imagine a galactic solar system somewhere in space with conditions similar to Earth except that its Sun is in the autumn of its life. Provided its inhabitants have survived wars and alien invasion, it is impossible to imagine what super-technology and cultural advancement they have reached . . .

A PREDICTION

Janus predicted correctly that 'perhaps in twenty years' time manned rockets will be commonplace and the Earth will be girdled by satellites of all sorts and sizes', and that there would be 'great strides in the miniaturization of all our present technology, advances in navigational guidance and communication over vast distances'.

NON-INTERFERENCE

Why then were aliens coming here? 'The answer is that this traffic is only a thin trickle in the vast highways of the universe,' explained Janus. 'The Earth after all is a galactic backwater inhabited by only half-civilized men, dangerous even to their neighbours . . .'

Most of these vehicles are robot-controlled space probes monitoring what is going on. Some are manned in order to oversee the whole programme and to ensure the probes do not land or crash by accident. They must also ensure that evidence of their existence is kept away from the vast majority of Earth's population. You must be well aware of the damage which your own explorers have done by appearing and living among simple tribes, often leading to a complete disintegration of their society and culture . . . Such impact is far too indigestible and only the most developed societies can cope with such contact . . . The basic principle of responsible space exploration is that you do not interfere with the natural development and order of life in the universe any more than you should upset or destroy an ant heap or bee-hive . . . You will have to grow a lot older and learn how to behave on your own planet, if indeed you do not blow yourselves up between times, before you are ready for galactic travel.

THE OBSERVERS

'Since time immemorial,' Janus continued, 'there have been tales of vessels coming out of the sky bringing strange visitors. Observers do come among you and make contact on a very selective basis where they judge that such contact could not harm either party.'

These observers have studied Earth for a long time. With advanced medical science they have been fitted with the right sort of internal equipment to allow their bodies to operate normally until they leave. It is not very difficult to obtain the right sort of clothes and means to move around quite freely . . . The observers are not interested in interfering in your affairs, but once you are ready to escape from your own solar system it is of paramount importance that you have learnt your responsibilities for the preservation of life everywhere . . . While you are still far away from travelling in deep space, such contacts will be infrequent and must be conducted with great secrecy . . .

The observers have very highly developed mental powers, including extra-sensory, thought reading, hypnosis and the ability to use different dimensions . . . and rely solely on their special powers to look after themselves. They make contact only with selected people where secrecy can be maintained. In the loosely-knit societies of the Western world, particularly in England and America, it is fairly easy with the help of friends to do this but not in police and dictator states.

The discourse ended. Sir Peter bade Mr Janus farewell, saying that he would give consideration to the request to meet Prince Philip.

'What was Janus?' asks Sir Peter. 'Was he part of an elaborate hoax or plot, was he a teacher, an imaginative prophet of the future or what he had insinuated – an observer? Whatever else he was, Janus left me with the impression of a force to be reckoned with. He appeared to know a great deal and spoke with authority about space technology. If he was part of any kind of plot, it was my duty to report the meeting to the security authorities, particularly if it had anything to do with the Royal Family.'

Immediately following the meeting, Sir Peter wrote a verbatim report and gave it to Lieutenant-General Sir Frederick 'Boy' Browning, Treasurer to Prince Philip. By this time, Browning had become fascinated by the subject, and was keen to arrange another meeting with Janus. Sir Peter was not so sure, but nonetheless made several phone calls to Mrs Markham over the next few days. There was no answer. Eventually he contacted General Martin, 'who suddenly became distant and evasive'. Finally, Sir Peter went round to Mrs Markham's flat in Chelsea. There was no sign of life. According to her neighbours, Mrs Markham had left in a hurry. 'The curtain had dropped,' writes Sir Peter. 'Had Janus sensed that I was in two minds about informing the security authorities of my meeting? I never saw General Martin, Mrs Markham or Janus again.'[20]

'I thought I would see them again and discuss it further, and I thought it very odd that the flat was empty,' Sir Peter told me in 1997, in the peaceful garden of his Hampshire home beside the River Test. I asked him for more details of the meeting with Mr Janus. 'It was a winter evening,' he began. 'Mrs Markham's flat was on the first floor and she introduced me to Janus in the drawing room, which was dimly lit by two standard lamps. He sat in an easy chair by the side of the fire. He didn't get up when we shook hands. I sat in an easy chair on the opposite side of the fire and Mrs Markham sat on a sofa between us.'

Somehow, he was difficult to describe. What made it strange is that I have no lasting impression of him: he seemed to fit perfectly in his surroundings. If I have any impression of him, it was his quiet voice which had a rich quality to it. He looked about 45 to 50 years old, with thinnish, slightly grey hair, and he was dressed in a suit and tie. He was quite normal in every way, except that he seemed to be tuning in to my mind, and gradually seemed to take over the conversation. Mrs Markham offered me coffee and didn't interrupt the conversation at any point. My initial reaction was one of scepticism, but by the end of the meeting, I was quite disturbed, really.

'And what of the reaction at Buckingham Palace – apart from that of General Browning?' I queried.

'Michael Parker, Prince Philip's Private Secretary, thought it a joke,' he replied. 'But Prince Philip had an open mind.'[21]

In 1969, Sir Peter Horsley was posted to the Ministry of Defence as Assistant Chief of Air Staff (Operations), responsible for the management of air operations worldwide and reporting to the Vice Chief of Air Staff. In his remarkable autobiography, he reveals that in the Air Force Operations Room (AFOR) he discovered 'a rich vein of UFO reports in the form of an Annexe to a Standard Operating Procedure (SOP) where every report of a UFO, from whatever source, was logged, examined and filed; those which might have either public or political repercussions came to my desk'.[22]

'There were a great many reports,' he told me. 'Of course, 95 per cent were explainable. Our main concern was the Soviets invading our airspace. But they were sent to the RAF's scientific [and technical intelligence] branch for detailed analysis. I also learned that the Americans were treating the subject with great secrecy.'

'I admire your courage in coming forward with such a story,' I said. 'How do you feel about the ridicule you received from the press when the Janus story came out, particularly the article in *The Times* by Dr Thomas Stuttaford, wherein he states that you must have suffered an hallucination or delusion?'[23]

'It didn't take much courage, as I was only reporting what actually happened,' he responded. 'I've met a number of senior RAF officers who agree that something strange is going on. It's only the papers that make up that you're hallucinating or have had an illusion . . .'

As to the incredible Mr Janus, the very credible Sir Peter Horsley retains an open mind.[24]

COSMIC CULTURE SHOCK
In April 1961, the respected Italian journalist Bruno Ghibaudi took a series of photographs purporting to show several unusual flying craft on the shores of the Adriatic coast at Pescara. One of the pictures is of a craft so unusual as to render fakery irrelevant (see plates). At that time, Ghibaudi was well known to the Italian television and radio public as a reporter on scientific subjects, especially aviation and space travel.

A year or so before taking the photographs, Ghibaudi had been asked by his superiors to prepare a television programme about people claiming to have seen flying saucers. Hitherto, he had paid little heed to the subject, but as he began to travel around Italy interviewing people, he was astonished to discover just how many had seen them, or taken photos, or met the occupants, or had recovered pieces of metal and other materials left by craft that had landed. He also learnt that many witnesses who had spoken out had either lost their jobs or had been subjected to such a barrage of ridicule or hours of grilling by officialdom that they had become thoroughly fed up with the whole matter and thus were loath to relate their experiences any further – especially to a journalist. Before embarking on his second tour of Italy, however, Ghibaudi was told by his bosses that the projected programme had been cancelled. By then Ghibaudi had become so impressed by the evidence he had gathered that he continued the investigation on his own.

In the summer of 1961, a few months after taking the photographs in Pescara, Ghibaudi claimed that he had been invited to meet some of the 'space people'. As with Sir Peter Horsley, the meeting took place in a house, the location of which Ghibaudi declined to reveal. Several witnesses were present, including the 'go-between' who had arranged the meeting. In an interview with *Le Ore* in January 1963, translated by Gordon Creighton, Ghibaudi provided no detailed description of the visitors, confining himself mostly to what he had learnt from them, though he did say that some of them are so much like us in appearance that they were able to infiltrate. In their chance meetings with Earth people, he said, they sometimes communicate by gesture or telepathy, and sometimes in the language of the person contacted. After all, he surmised, people who are so advanced technically are unlikely to have any difficulty in learning our languages.

The human form is 'universal throughout the Cosmos,' said Ghibaudi, 'and yet the idea of this has generally been rejected by Earthmen as impossible, no doubt because, as almost always, the truth is too simple to be accepted'. Apart from various superficial differences, beings throughout the universe resemble *Homo sapiens*, though Ghibaudi conceded that some of their internal organs may well be different; even perhaps performing quite different functions.

These space visitors, Ghibaudi continued, were coming to our planet from many different worlds; hence some differences in body sizes, for example. What is happening now, he claimed, is simply that the 'infant civilization of Earth-Man being at a point of particularly grave crisis, the space beings are prepared to reveal themselves to us more'. He confirmed that the people he met, at least, were benevolent and desirous of helping us: their aim, to prevent nuclear catastrophe by intervening if it became unavoidable. He added that although these beings are many thousands of years ahead of us technically and scientifically, as well as ethically, they are not omnipotent. 'They are men,' stressed Ghibaudi, 'so we must not rely on them to get us out of our difficulties. For, not being infallible, even their efforts and their concern might not always suffice to avert disaster if something went wrong or some accident nullified their plans to avert the worst.'

On the subject of nuclear weapons, Ghibaudi pointed out that although extraterrestrials are capable of destroying such weapons, 'the human heart would nevertheless remain unchanged. We would still retain the ability and, above all, the *intention*, to build fresh nuclear devices.' Hence the extraterrestrials he met were working in a more subtle manner to influence the minds of men. 'They fully realize the dangers of any kind of broad prohibitive action. They know that in the last analysis Earth-Man must make his own way . . .'

NEGATIVE COMPARISONS

Although nuclear weapons remained one of the principal reasons why the extraterrestrials were revealing themselves more to Earth people, Ghibaudi emphasized that there were also other reasons, which he was forbidden to disclose.

The principal consideration in the minds of the benevolent visitors in adopting so reticent a policy towards us, claimed Bruno Ghibaudi, was not only the great danger that would ensue from panic – 'tremendous as these dangers would no doubt be, where primitive and backward creatures like ourselves were concerned' – but also that their open appearance among Earth people would lead inevitably to negative comparisons. Humankind might feel so deflated and inferior that it might lose hope. And how could politicians possibly cope with such a scenario? 'Our masses are not yet ready for a revelation of this kind,' Ghibaudi asserted.

> Do not let us forget that between their science and ours there is a gap of thousands of years, and that for this reason an 'official' mass descent of space beings from other planets would inevitably bring about comparisons between their worlds and ours. How could such an encounter be permitted? At an inner level, we should quite certainly be severely shaken as a result of it, and they do not want to alarm us in any way. And this is all the more so, inasmuch as there are cosmic laws which prevent the more evolved races from interfering, beyond certain limits, in the evolution and development of the more backward races. For every race must be the maker of its own progress, paying the price for it with its sacrifices, its failures, and its victories . . .[25]

NOTES

1 *Diário de Córdoba*, 1 May 1957, translated by Gordon Creighton, in *Flying Saucer Review*, vol. 11, no. 1, January–February 1965, pp. 19–20.

2 Bowen, Charles, 'A South American Trio', *Flying Saucer Review*, vol. 11, no. 1, January–February 1965, p. 20.

3 *Diário de Córdoba*, 22 August 1957, translated by Gordon Creighton, 'UFO Bases in South America?', *Flying Saucer Review*, vol. 11, no. 4, July–August 1965, pp. 30–1.

4 Albanese, Renate, 'Luciano Galli's Contact Claim', translated by Gordon Creighton from *Domenica della Sera*, June 1962, in *Flying Saucer Review*, vol. 8, no. 5, September–October 1962, pp. 29–30.

5 Lorenzen, Coral E., *The Great Flying Saucer Hoax: The UFO Facts and Their Interpretation*, William-Frederick, New York, 1962, pp. 146–7.

6 Ibid., pp. 147–8.

7 Fontes, Dr Olavo T., *The APRO Bulletin*, September 1957, Aerial Phenomena Research Organization, Alamogordo, New Mexico, in Jules Lemaitre, 'A Strange Story from Brazil', *Flying Saucer Review*, vol. 6, no. 1, January–February 1960, pp. 9–11.

8 Lorenzen, op. cit., pp. 150–3.

9 *UFO Investigator*, vol. 1, no. 1, National Investigations Committee on Aerial Phenomena, Washington 6, DC, July 1957.

10 Bühler, Dr Walter, *SBEDV Bulletin*, no. 4, Rio de Janeiro, 1 July 1958, translated by Gordon Creighton, in 'A Brazilian Contact Claim', *Flying Saucer Review*, vol. 7, no. 5, September–October 1961, pp.18–20.

11 Bühler, op. cit., translated by Gordon Creighton, in 'Remarkable Confirmation for Adamski?', *Flying Saucer Review*, vol. 29. no. 4, July–August 1983 (pub. April 1984), pp. 13–16.

12 Creighton, Gordon, 'The Humanoids in Latin America', in Charles Bowen (ed.), *The Humanoids*, Neville Spearman, London, 1969, pp. 99–100.

13 Faria, J. Escobar (ed.), *UFO Critical Bulletin*, São Paulo, in 'Giant Space Ship lands in Brazil', *Flying Saucer Review*, vol. 4, no. 3, May–June 1958, p. 24.

14 Zinsstag, Lou, 'Monguzzi Takes Saucer Photos of the Century', *Flying Saucer Review*, vol. 4, no. 5, September–October 1958, pp. 2–4, and information supplied to the author by Lou Zinsstag.

15 Letter to Anthony Wedd from Hubert Lewis, 27 May 1958.

16 Trench, Brinsley le Poer, 'Birmingham Woman Meets Spacemen', *Flying Saucer Review*, vol. 4, no. 2, March–April 1958, pp. 5–6.

17 Yogananda, Paramahansa, *Autobiography of a Yogi*, Self-Realization Fellowship, Los Angeles, 1951.

18 Letters to Anthony Wedd from Hubert Lewis, 27 May and 8 August 1958.

19 Good, Timothy, *Alien Contact: Top-Secret UFO Files Revealed*, William Morrow, New York, 1993, pp. 89–91.

20 Horsley, Sir Peter, *Sounds From Another Room: Memories of Planes, Princes and the Paranormal*, Leo Cooper, London, 1997, pp. 180–96.

21 Interview with the author, Hampshire, 21 September 1997.

22 Horsley, op. cit., p. 201.

23 Stuttaford, Dr Thomas, 'Air marshal's flight of fancy', *The Times*, 14 August 1997.

24 Interview with the author, Hampshire, 21 September 1997.

25 Creighton, Gordon, 'The Italian Scene – Part 3', *Flying Saucer Review*, vol. 9, no. 3, May–June 1963, pp. 18–20, translated from *Le Ore*, 24/31 January 1963.

PART TWO

Chapter 11

Let Humanity Beware!

Even if not all accounts of encounters with extraterrestrials are believable, many reports are consistent, and the claimants remain convinced, sometimes after many years have passed, that something exceptional happened to them – something that changed their lives for ever. The following story from Argentina is a case in point that involves a witness of apparently irreproachable character. This case was investigated by Héctor Antônio Picco, as well as by three of his colleagues (Jorge H. Cosso, Sotero Caraballo and Eduardo R. Rando), over a period of eight years. Translation is by Jane Thomas Guma.

THE POSEIDON ADVENTURE
One night in August 1956, Orlando Jorge Ferraudi was fishing as usual on the then deserted coast of the Northern Resort, where University City is now located, near Buenos Aires.

It was about 11.30 p.m. and while I was getting my gear ready, I suddenly started to feel as if someone was observing me. I thought it was a 'bum' who usually hangs around that place but when I turned my head I saw 'him'; a strange individual observing me. Comparing him with my 1.90 meters, I at once estimated he was more than two meters tall. His skin was very white and he had very light-colored eyes, no beard nor mustache, had short and neat hair, and he wore some kind of tight-fitting 'overall'.

It was a very dark night and I noticed him telling me mentally: 'Take it easy, don't be afraid, you mustn't be scared.' Then he turned around, taking my arm, and placed some kind of 'powder box' on top of the wall which, when opened, gave off a phosphorescent luminosity. This allowed me to see more details: his suit was a yellow-mustard color, had no wrinkles, zippers or buttons; it had a hood in the back of its head. He was repeating: 'Don't be afraid, you will come with me, we will take a long trip.'

He picked up his device and we went down the steps towards the Rio de la Plata. I followed him like an automaton. He suddenly pointed his 'little machine' and I could see that a strange craft in the shape of an inverted saucer was approaching from the water. It stopped and a little door opened from which a ramp came out and a similar being walked towards both of us. Taking my hand very gently, he invited me to go inside the craft.

ON BOARD

As soon as Ferraudi entered the craft, he noticed a girl who was five or six years younger than him (he was 18 at the time). By her clothes, he deduced that she was not one of 'them'.

'Don't be afraid, they won't hurt us, they are good,' the girl said. 'I came into this thing a while before you. My name is Elena, I'm from Villa Mercedes [San Luis province].' Ferraudi continued:

> Suddenly 'they' came and told us, always telepathically, 'Don't worry, you will have to undress and change clothes because the things that you wear have elements and germs that are alien to us.' A woman appeared – taking Elena into another room – identical to several who came later: a beautifully proportioned body and dressed in the same clothes as the men; mouth, nose and ears were normal, but her eyes, which also seemed normal, were almost yellow. The haircut was like that of the 'Valiant Prince'.
>
> They took the clothes I had taken off, following an order I could not disobey because their control over me was complete, and put [them] inside a machine that looked like a TV set, inside of which was a thick green smoke. They gave me an overall like the one they wore, ordering me [to] put it on. I told them that I couldn't because it was too narrow, but they insisted. I saw it had a hole where the neck would be and put a leg inside, then the other, and the overall expanded and covered me completely! I also felt like I was wearing comfortable shoes when I walked, even though I was barefoot.

By this time Elena had returned, and the two were told that they would be taken for a trip underwater, to a place called Samborombón Bay. 'From there,' continued Ferraudi, 'we would emerge and fly at low altitude until we reached the coast of Uruguay, then we would cross the Atlantic Ocean to Africa, and then we would go up.'

'We must take these precautions so we won't be detected and thus avoid being seen as invaders or conquerors,' the cosmonauts explained. 'We want your people to get used to us slowly, to see us as anybody else, because we are not strangers in this part of the universe.'

During his interviews with Picco, Ferraudi would ask that the sound recorder be turned off at instances when asked if he knew where his 'abductors' came from. His off-the-record reply – though he now admits it openly – is interesting. 'I'm not supposed to say it yet: *they come from inside the Earth*.' According to Ferraudi, in about 1950 the aliens had built two underwater bases, one on the Uruguayan coast in front of the Barra de San Juan, 45 kilometres from Buenos Aires, and the other in the Bahia Samborombón, about 150 kilometres southwest of Montevideo.[1] Supposedly another base had been established in the Gulf of Mexico, and it was there that Orlando and Elena were then taken for a brief visit.

AN UNDERWATER BASE

'When the craft gained altitude,' continued Ferraudi, 'I noticed that the inner walls were smooth, and the only remarkable thing in its structure was the oblong windows: they took us, the little girl and me, close to one of them.'

We could see our beautiful planet, blue enormous, round, with white spots and some clouds, 'hanging' in a dark and silent space. Our Moon was an opaque grey. They told us: 'We will now project a force-field that will attract us as if we were inside a tube.' And immediately the Earth became as small as an orange. I felt no fear, no jolt to justify such a movement. They told us we would return at the same speed.

When we started to return I screamed, 'Careful, we will crash!' 'Don't worry,' they said. 'When we get very close we will create a field so as not to collide with Earth.' We entered the ocean, maybe through the Gulf of Mexico, and after a few minutes of travelling underwater, we saw an immense sub-aquatic dome, similar to a giant Eskimo 'igloo', where buildings, people in motion and several ships similar to ours, could be seen. One of 'them' said: 'It is a base to recondition our ships.'

After leaving behind these five or six blocks of buildings submerged in the bottom of the ocean, Orlando and Elena were told that they would be submitted to a 'test', and that they should relax, so the result would be accurate.

THE TEST

One of the female crew members brought a tray with ten small 'eggs', Ferraudi continued. 'Five were for Elena, the other five for me, they said.'

The colors were red, yellow, brown, green and another I don't remember. We had to chew and eat them, and we also had to drink a clear, thick liquid. When we swallowed them, neither of those things had any taste.

We were ordered to lie down on some stretchers that were padded and had a U-shaped headrest, dotted with lights that were the same color as the little eggs we ate. We fell deeply asleep and when we woke up, Elena and I could read our mutual thoughts, which we thought was quite funny. We were told that the results from the test were good, that both of us were very healthy and that in this way they had thoroughly learned about our physical and mental states, and even the date of our deaths. We were also informed that they had reactivated what we call our pineal gland, and this is when they said (I realize that now) what was maybe the most important part of the experience:

'You will be useful to us in the future, because this gland is the only legacy that remained here from us, since of the five races that inhabit this planet, none is originally from Earth, they are only remnants of civilizations from other planets. The Earth has been known for a long time to be the zoo of the solar system. The races that exist today have suffered genetic mutations due to their own fault; when they mixed they caused hybridization, destroying the stock, but what remains from what they once were is the pineal gland. That is why we reactivated it, so when we think about you, you will immediately hear a kind of hum inside your heads.'[2]

In physiological terms, the pineal gland (or body) is defined as:

A pea-sized mass of nerve tissue attached by a stalk to the posterior wall of the third ventricle of the brain, deep between the cerebral hemispheres at the back of the skull. It functions as a gland, secreting the hormone melatonin[3] . . . Some evidence suggests that it is a vestigial organ, the remnant of a third eye.[4]

ENERGY, GOD AND DEATH

Orlando and Elena were then invited to see the rest of the craft. A remarkable detail, corroborating what many others have reported, was that it was impossible to discern where the 'perfect illumination' came from. 'It was as if the air was "turned on",' remarked Ferraudi.

They showed us the engine: it was round, surrounding the edge of the whole ship, which was some 70 metres in diameter. It was formed by a series of huge interlinked bobbins, and we could see other beings there, but wearing blue clothes, gloves and a kind of visor that covered their faces. Surprised, I asked them, 'Is that what you fly with?' The being that accompanied us answered: 'No, we don't fly, we simply slide along a force-field. We use three energies: cosmic, magnetic and solar. We can move in space using all three or only one of them. With regard to our ship, which you call a flying saucer, it is built in one piece because when we built it, it is as if we molded it, and the windows are "adhered" . . .'[5]

In response to a question about 'God and death', the 'Gentleman from Poseidon' (as Ferraudi called him) responded briefly:

For us, what you call God is a form of absolute Energy, and as to death, it is only a change in molecular structure, a change of state. We only use sex to procreate, but we also have families and know love. Our lifespan is much longer than yours. Our children are already born with all the knowledge and keep perfecting it as they grow.

The discussion then focused on the 'indiscriminate and irrational use of nuclear energy on our part', said Ferraudi, 'which endangers not only our habitat which we share with "them", but also the cosmic equilibrium. And showing us an instrument that one of them held in his hands, they made us look towards a window where a solid body was floating.'

The being pointed the device at it and a beam shot out which blew it up as soon as it made contact. He said to us: 'This is pure energy; when it touches its objective it disintegrates it, it completely dissolves everything it touches.'

And finally, his last and unforgettable warning: 'We want you to know this: [this] power is what we will regrettably use if you should endanger the stellar harmony . . .'

RETURN TO EARTH

Finally, Orlando and Elena were informed that they would be returned to the place from where they had been taken, and that for a time they would remember nothing. Later, memories of these events came back. For example, Ferraudi recalls that on arrival the beings asked him to bend down, then they pointed a very bright light at him, causing him to fall asleep. He woke up with the sun almost up and his body feeling very numb, without remembering whether he had caught any fish. He gathered up his fishing-tackle and went home.

Fifteen days later, when he was again getting his fishing-tackle ready for another Saturday night of fishing at the same spot, Ferraudi reasoned to himself: 'I don't know why I should go fishing, if I end up falling asleep.' Suddenly, it 'clicked'. 'No! I didn't fall asleep! I travelled in a flying saucer!'

'I am sure I did not dream all this,' Ferraudi told Héctor Antônio Picco, the principal investigator. After questioning the witness repeatedly over an eight-year period, all the while carefully studying his body language, Picco has concluded that Ferraudi is totally truthful. Furthermore, Picco is impressed by the fact that the beings evidently imparted to Ferraudi scientific and medical information well beyond his own knowledge. For instance, Ferraudi wanted to create a 'machine' to cure cancer. In the beginning of his manuscript, 'Cancer: its origin and development', he writes:

> The origin of this disease lies in the altered functions of the ductless glands, which due to their bioelectrical balance having been upset, drain into the blood incomplete humors that lead to the irrational forming of the cells. This phenomenon leads to the immediate consequence of these humours circulating throughout the whole body, since blood is a vehicle; thus incomplete humors look for the weakest organs where they can exert their influence within a favorable field.

In July 1975, years after Orlando Ferraudi outlined this theory, Nobel Prize-winner (1937) Dr Albert Szent-Györgyi came out with his 'electromagnetic theory' of cancer, which was expressed in a very similar way to that of Ferraudi – despite the fact that the abductee had only the most rudimentary knowledge of medicine.[6]

A SUBTERRANEAN SPACE-BASE
While walking and studying in an isolated area near Halmstad, Sweden,

on the afternoon of 15 August 1960, Olaf Nielsen, a student of agriculture, claims to have been abducted by extraterrestrials and taken to a 'subterranean space-base'. This is his incredible, fascinating story, as related in 1962 to an Italian businessman, Paulo Bracci:

> Suddenly I felt myself as it were in a dizziness and sucked up into the air. Despite my terror I had the presence of mind to note what was happening. At a height of some 20 metres from the ground was a flying saucer, and I was being drawn straight up to it. Finding myself in empty space like this, and carried off in such a manner, I lost consciousness.
>
> When I came round again I found myself stretched out on a very soft couch, inside a small cabin. The cabin was of a pale green colour, lit by a diffused light that had no source. One would have said that the light came from the walls themselves. Suddenly a door opened and a being came in. He was in every way similar to us, except that he was wearing an overall. He approached, smiled at me, and, in my own language, begged my pardon for the way in which I had been carried off.

Nielsen went on to assert that he had been taken very rapidly to a subterranean space-base.

> It seemed at first as though I was out in the open, but instead of that I found that I was in a large brightly-lit cavern. In my curiosity, I asked the guide whether there were many of these bases on Earth. After a moment of hesitation, he replied that such bases had existed on the Earth for very many years past. Some were in Central Asia, where thousands of years ago, the guide added, there used to be flourishing cities. Others, he said, are on the high plateau of the Pamirs, and in Central Africa, and in South America, where the space visitors had adapted for their own purposes 'secret pre-Incan cities'.

Olaf Nielsen said that he was shown several saucers, as well as an apparatus for setting up a protective 'magnetic curtain' at the entrance to the base. His guide explained that these were precautionary measures, directed not against the people of our Earth, but against the 'Dark Ones', i.e., bellicose space-beings who supposedly came from the vicinity of Orion and who were desirous of conquering the Earth.

Gordon Creighton shares my feeling that this case contains important information. 'It is a fact,' he states, 'that Central Asia, now desiccated, once had great civilizations, and there is a persistent [native] tradition, not entirely supported by any evidence, that there still exist undiscovered Incan or pre-Incan cities in the Andes.'[7]

FLYING SUBMARINES IN THE MEDITERRANEAN

It was 06.35 on 3 June 1961. Giacomo Barra and three friends were in a motorboat off Savona in the Gulf of Genoa. The men had shut down the engine and were enjoying the early-morning breeze when suddenly the rocking motion of the waves increased and the boat began to roll badly. As Barra reported:

> We looked around, thinking it must be due to the proximity of one of the many tankers that put into our port. But nothing of the sort. At a distance of a kilometre from us, the surface of the sea was bulging like an enormous ball, with long billows going out from it on all sides. Dumbfounded, we were still wondering what it was when, suddenly, a strange contraption rose up from the bulge of the water. Perhaps it was one of the celebrated 'flying saucers', for the lower part of it looked like a plate upside down, and the upper part ended in a cone. While it was emerging from the sea, the water was thrust away all round it, as by a cushion of air. After it had emerged completely from the sea, it stopped still for a few seconds, at a height of 10 metres or so, and then rocked slightly a few times. Then a halo formed round the base of it, and the thing shot away very fast across the sea and vanished towards the north-west.[8]

A Report from Three Men in Two Boats

The following is one of a number of reports of unknown 'flying sub-marines' collected from fishermen at the French fishing port of Le Brusc, between Marseilles and Nice. None of the fishermen was prepared to have his name revealed. This incident took place on 1 August 1962, between 23.00 and 23.30, on a warm, clear night. Here follows a report by the first witness:

> Suddenly, at about 300 metres from me, I saw a large metallic body, elong-ated in shape, and with a sort of chimney or turret in the middle. It seemed to be moving along slowly on the surface of the sea. Then finally it stopped. I said to my companions in the other boat: 'A submarine has surfaced over there quite close to us. It doesn't seem to worry them!'
>
> One of the others replied: 'It must be a foreign sub. It's a model that I don't know.' Then there was some disturbance and waves coming around the sub-marine, and I was able to make out some frogmen coming out of the sea and climbing up on to the craft. We shouted to them. But at first they didn't even turn round to look at us. My two companions, who had also seen them, and had heard me hailing them, also called to them with their loud-speaker . . .
>
> There was no reply from their side. I had a good view of them. I counted about a dozen of them getting up onto the submarine. Then three or four of them did look around, and hesitated for a few moments, before vanishing into the ship. Finally, before rejoining the rest, the last man turned towards us and raised his right arm above his head and waved it for a few seconds in greet-ing, to say he had seen us, and then he disappeared into the craft like the rest.

Up to this point, the three men had been convinced that the craft was merely a foreign submarine engaged in manoeuvres; that is, until it began to rise into the air.

We saw the machine rise right up out of the water and hang there just above the waves. Then we saw lights go on; red, green, and a beam of white light shot out and reached as far as our boats. This beam was from a searchlight, and gave off no heat or anything unpleasant. Then [it] went out . . . the craft was lit up with an orange-sort of glow, and the red and green lights went out. The machine started to rotate very slowly, from left to right, and rose to about 20 metres above the sea.

Its appearance was, as we now saw, like an oval or almost round dish, and of the dimensions of a medium-sized submarine. It hung there stationary for a few minutes. Then it began to rotate faster, its light grew brighter, and suddenly it shot off horizontally at high speed over the sea, amid a vast silence. Its light now took on the colour of red flame and it flattened out and came back right round over us in a beautiful curve while climbing all the while and increasing speed, and then it vanished as a tiny dot among the stars . . . Apart from the sound of the waves, we had heard no sound from it, and you can well imagine that we asked ourselves what it could possibly have been.[9]

A CRYPTIC ENCOUNTER NEAR FLORENCE

Mario Zuccalà had just returned home by bus to San Casciano, Val di Pesa, Italy, from Florence (20 kilometres away), where he worked as a tailor. It was 10 April 1962, a clear, starry night. Shortly before 21.30, as he was walking through some open ground in the district of Cidinella where he lived, the 26-year-old man felt himself 'struck and lifted up slightly by a sharp gust of wind.'

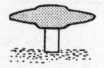  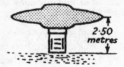

Fig. 16. (*FSR Publications*)

An object, resembling two plates joined together with a diameter of about 8.5 metres, could be seen hovering some six metres above the ground. A cylinder of about 1.5 metres in width came down from the lower side of the machine until it touched the ground. Zuccalà later conjectured that the cylinder, once it had touched the ground, re-entered the machine again, leaving exposed one side of the cylinder in which a door opened slowly, while two small doors were gliding towards the outside, therefore they may have been two cylinders moving, one within the other (see Fig.16). In any event, from the opened door appeared an empty space, illuminated by a diffused, brilliant white light. Three steps, about 40 centimetres high, could also be seen. Then, as investigator Ceccarelli Silvano relates, two beings, about 1.5 metres tall, came out of the opening.

> Their bodies resembled ours in so far as they could be seen, i.e., as to exterior form, because as for the rest they were completely covered by an 'armour' of shining metal. Two antennae came out from their heads . . . These two little men took hold of him gently under his armpits and took him inside the object. Signor Zuccalà went up the three steps and went inside. The interior was empty and shining all over with the same light which he had seen from outside. [He] did not notice any detail in the interior of the object. The two beings left hold of him and Signor Zuccalà remembers that he asked where that light came from but he does not remember having had a reply. He then heard a voice which did not come from the two beings with him but from the inner part of the object; according to Signor Zuccalà this voice was like one amplified by a microphone and as if resounding in a vast space.

The voice, speaking in Italian, gave Zuccalà the following cryptic and rather silly message:

> At the fourth moon we shall come at one o'clock in the morning to bring you a message for humanity. We shall give notice of this to another person in order to confirm that that which you have seen is true.

Whether the 'fourth moon' was supposed to have meant the fourth from the beginning of the year, which would have been the full moon of 20 April 1962, or whether it meant four moons reckoning from the day of the sighting, was not clear. In any case, there was no return visit.

The two beings escorted Zuccalà from the craft. Suddenly he found himself at home at about 21.45, with no recollection as to how he had arrived there. His wife heard four strong knocks at the door and went to open it, rather alarmed because her husband usually knocked only once, and then lightly, and Zuccalà himself could not remember having knocked four times. He looked dazed and frightened, and at first seemed unable to make up his mind whether to stay outside or go into the house. He told his wife about what had happened then went to bed, sleeping fitfully.

The next morning, Zuccalà spoke to a colleague at work who telephoned a newspaper. The story appeared in all the papers that evening and on ensuing days. Journalists pointed out that there was not the slightest evidence on the ground where the strange object had been, but by all accounts Zuccalà, the father of four children, was an honest man. 'He speaks with calm assurance of what he has seen,' reported Silvano. 'I asked him whether in his life he has had any hallucinations – to which he replied in the negative.'[10]

SPACIAL DILATION

Cases such as the foregoing, involving a craft with a central cylindrical column containing the entrance, have been reported in numerous instances. The following case involves a similar such column, but one which led into a bizarre interior. The encounter, investigated principally by Joël Mesnard, is said to have occurred one Sunday evening in November of 1961 or 1962.

The witness, Michel, about 19 years old at the time, came out of the local cinema at 17.30. It was already dark. Because his adoptive father was playing cards in a nearby café, Michel decided to return by himself on his bicycle to their isolated farm, near Bray-sur-Seine, 80 kilometres south-east of Paris. He had at least a kilometre left to ride when, coming from the left of the farm courtyard, he noticed a steady beam of light rising vertically to the sky. 'The light was orange-red,' reported Mesnard, 'and the cylinder of light, with defined edges, rose very high in the sky. The closer Michel came to the farm, the less he understood what was going on . . . he put down his bike and walked around the buildings from the out-side of the surrounding wall.' And there it was.

> He came upon an enormous object sitting on the ground, about 50 metres away. The upper cupola, which had portholes, was turning. The lower part was cylindrical and had a large vertical opening. He could see the brightly illuminated inside. It seemed that the lower side of the cylinder was not rest-ing directly on the ground, but that the device was suspended in the air. Michel continued to approach. He slipped under the flat, central part of the object to where he could look through the opening to the interior.

Amazingly, like the 'Tardis' (telephone kiosk) in the British TV series *Dr Who*, the interior was immense, 'incomparably larger than the cylinder seen from the outside'. It seems to me that this effect may have been due to a local distortion of the 'space-time continuum', a phenom-enon of a highly advanced technology, and one reported by several other witnesses coming in close proximity to or claiming to have gone inside extraterrestrial craft. Michel stopped just short of the opening, but did not go in. The interior of the cylinder is described thus:

> It was immense – at least six metres in diameter, which was about four times the diameter of the exterior. When he looked up, he could see nothing resem-bling a ceiling. All around the huge cylindrical room, he saw a variety of appa-ratuses holding moving lights, 'luminous dials in many colours with characters that moved'. The dominant colors were 'light violet and deep salmon'. Near, but not at the center, was a vertical column that seemed between 70 centimetres and one metre in diameter and that turned around its axis. What made this rotation visible was something like a cable, wound like a helix around the column, 'like a snake'.

According to Michel's memory, he was watching this incredible spectacle when the sound of a fan, coming apparently from above, began and increased in intensity as the rotation of the column increased and the dials became brighter. That was when he was thrown back about four or five metres. The portholes on the cupola turned very fast. The opening closed and the cylinder went up into the body of the object. The object became luminous – a bright orange. It tipped, rose, and in an instant disappeared toward the south. In five seconds, it looked like an ordinary star.

Michel did not discuss his encounter with anyone until 1978. Ten years later, he spoke to Joël Mesnard, who points out that, despite the bizarre nature of the report, 'Michel appears to be as credible as anyone can be and his story is clear and coherent.'[11]

A TALE FROM THE VIENNA WOODS
Bobby is a Filipino pianist, to whom I was introduced in 1962 by my friend John Bingham, the now well-known pianist, when we were fellow-students at the Royal Academy of Music in London. Though Bobby was aware at the time of my interest in the subject of UFOs – knowledge which may bear on the validity of his story – I include his alleged encounter here because not only did he seem genuinely disturbed by it, but also because there are parallels with certain other encounters which he could not have known about at the time. He described his experiences in a handwritten manuscript and gave it to me to do with as I wished, with the proviso that I never publish his full name.

'It is not that I could not bear to expose myself to a barrage of questions and cross-examinations under the trained eye of psychologists, psychiatrists, medical doctors, priests, scientists and space experts,' he wrote, 'but rather because the prevailing controversy would definitely affect my parents. I don't want to hurt them because they want to lead a simple and peaceful life.'

From 1961 to 1963, Bobby studied piano at the Vienna Akademie. At the beginning of October 1962, he began to hear a peculiar, high-pitched sound from time to time. 'It had a certain frequency of its own, and was immensely soothing,' he wrote. At other times he would feel restless, accompanied by a presentiment that something was about to happen. Most peculiar were the 'telepathic messages and mental pictures' he began receiving at frequent intervals.

I could not decipher the source, but I felt strongly that I was being contacted mysteriously through my brain and heart. How I knew this, I could not say. All I know is that I was under the control of some strange, mysterious power – evil or good, I didn't know . . . Sometimes, as I practised during the day, I would suddenly stop, because right in front of me, on the music, was superimposed a picture of a place, a lovely green forest. I could not understand the significance of this picture, but deep down it had somehow a strong influence over me. The only thing that bothered me was that I didn't know where the forest was.

These images and presentiments culminated in an encounter with a landed craft and its occupants on 8 October 1962. At around 16.00, Bobby found himself 'impressed' to leave his flat. 'A tram came by and I boarded it,' he wrote. 'People gazed at me sitting quietly, yet looking nervous. My mind went whirling round and round and my heart was thumping fast. I alighted at Schöttentor and took another tram to Grinzing. I began to sense that something was going to happen before evening. How could I explain this strong feeling? Alighting at Grinzing, I took a bus to the Vienna Woods. Was I being guided toward a particular place?'

It was several minutes before I reached the place, in nice countryside, quiet and peaceful. It was a weekday and so very few people were strolling around. I went on, walking aimlessly, though it was as if I was being guided in a particular direction. My mind suddenly went blank and my heart began to beat normally: I felt so calm and collected. After half an hour of aimless walking, I found myself in a forest. My sense of direction told me I was no longer in the Vienna Woods. Where then? I could not pinpoint the exact location, but it must have been in one of the forests beyond the Vienna Woods.

It was strange that being lost did not bother me at the time. Instead of going back, I went deeper into the forest. I looked up at the tall, forbidding-looking trees and saw that the sky was overcast and the wind blowing in gusts. I suddenly noticed and felt around me a strange stillness, like that before a storm, with no sound of birds twittering to break the sinister and deadly silence. The air was cold and biting and the mist arising . . .

After several minutes of walking on in the gathering dusk, Bobby came to a small clearing in the forest. It was then that he heard a peculiar swishing sound behind a clump of trees.

I looked up and saw the leaves and branches of the trees shaking, disturbed by the sudden rushing of air coming out from a strange object. I could hear the whistling sound the object made as it glided smoothly and nearer towards where I stood. I watched in awe, in great fear and deep curiosity . . . I wanted to run away but my feet were tied to the ground. I tried to scream in fright but my voice didn't come out. I even attempted to close my eyes to blot out this unreality, but they continued gazing straight at the object. Was I being controlled by some power behind it? Was I seeing visions that were only created by my imagination? Or were they real? Thank God! It was real.

The object, which was a flying saucer, landed on legs about 100 feet from me. A sliding door opened and a steel stairway protruded outwards then dropped to the ground. A figure came out, accompanied by two others. They were all dressed in a tight-fitting, black-brown suit which extended from their black, heavy-looking shoes up to their head. The only parts not covered by this suit were their faces and hands, which were covered by a thin, transparent, black material. Around their faces they wore a glass visor with two tube-like things starting under their chins, hanging on their shoulder blades and extending to their back, ending in a sort of oxygen tank, from where they took their supply to breathe. They had the faces of human beings and must have been about five feet five inches to six feet tall, with lean but strong-looking bodies. I could not judge the exact height because they were not all of the same size.

Someone said something which seemed like a question, but I couldn't understand a word. I remained quiet and watched for further developments. Hearing no reply from me, the leader pressed a button on a small box he was carrying and presently the box gave out a red light whose beams fell right before my eyes. I did not feel any effect at all, except that it was soothing to the eyes . . . Maybe it did have an effect on me after all, because after a few seconds the leader switched off the red light and asked me a question in English, though he had a slight accent, rather similar to that of Germans.

'Would you like to be one of us?' he said.

'No,' I answered unemotionally, in a dry and uninterested tone.

'Would you like to visit our place?' the leader asked, in a kind, gentle tone with no element of threat, harm or evil in it.

I still answered, 'No.' Maybe I replied automatically, without stopping to think, or maybe it was because I was frightened out of my wits, but I cannot describe the exact feeling I had at that moment. All I can say is that I was somewhere between reality and unreality.

The leader looked at me as I reached for something inside my coat pocket. 'Stop!' he shouted, and I did so immediately, out of fear. A beam of red light fell on my pocket. Shortly afterwards he said, 'You can take out your eyeglasses.' How could he know my glasses were in my pocket? Was the red light responsible for this strange information? I put on my glasses and observed their countenance. The leader's companions were all quiet, absorbed possibly in their own thoughts, sizing me up and the immediate surroundings.

'Finally, the leader ventured forth a message which I have put down in writing as best as I can remember,' Bobby continued. 'The message may be interpreted by people in different ways, according to their own liking and thinking; however, I believe it has great significance.'

We come to your planet not for a visit but to deliver a message which may well serve as a warning to mankind. We *cannot and must not* reveal from which planet, star or moon we come from, because of the imminent danger of your people contriving all possible means and resources to conquer space and eventually to try and conquer us, although we could fight back and wipe out your mean and selfish humanity.

Within your planet, there is continuous strife among nations for power and domination, and within every nation there is dissension and dissatisfaction among the masses. Within a family and family relations there is still enmity, intrigues and conspiracies, and within an individual's mind there is still a continuous struggle between good and bad, evil and purity, generosity and selfishness. Why? Because there is so much selfishness on your planet, meanness, cruelty, wickedness and evil; more than good, purity and generosity.

Observe carefully the great mass of humanity killing each other through centuries of war and strife. And for what purpose? For power and domination; the intense desire to dominate and subjugate . . . There are thousands of good people on your planet, but the mean and selfish humanity outnumbers the good by millions and millions.

Unless there is a radical change starting now, from an individual's mind, within a family and family relations, within a nation and between nations, your people will all be destroyed by their own selfishness for power and domination. Not until your humanity is completely wiped out will there ever be peace on Earth. The danger of atomic war between nations is imminent . . . It may not be now, it may be centuries from now, but the end will still come. There is no turning back for your people. Some day you will all be wiped out by your own greediness, and *if* a few good people live through, then they will propagate and breed an unselfish humanity and no longer will there be a continuous strife between nations, within a nation, within a family and family relations, and within an individual mind, and there will be peace on Earth at last.

There is a great and possible danger, too, that your humanity's intense desire to conquer, eventually seeking power and domination over the other planets, will mean only a complete massacre for Earthmen, because other planets will retaliate with terrifying power and force, only because of their fear of your selfish humanity coming to their planets and spreading greed and evil around. This is our message. Transmit it and let humanity beware!

'The leader looked at me after he had finished his message and smiled. Was it out of pity, friendliness or mockery? I am sure it was a smile of pity because of the inevitable end the Earth will come to some day.'

As the leader was on the point of re-entering the craft, he turned round to address Bobby with a final message:

Your seeing us will greatly affect your body as well as your life a great deal. There are three possible consequences for you: you might die in April 1963 because of shock; we may take you to visit our place; or you might not die in April 1963, in which case you will lead a normal life again but with occasional mental and visual contacts from us until such time as we decide not to bother you again.

'The leader gave me a last farewell smile and went inside. The steel stairs went up, and the saucer got ready to leave,' wrote Bobby. 'I stood in the dusk and watched it whisk away, leaving me in a predicament I felt that something had gone away from my life. I cannot describe how lonely and helpless I felt . . .'

Not technically-minded, Bobby provided few configurational details of the craft, though he did tell me that a series of what looked like rectangular windows surrounded the central section, where the door opened out. I made a sketch in 1964, with his guidance. Interestingly, it accords with that made by the British engineer 'H.M.' in South Africa in 1951 (see p. 92), whose account was not published until 1977. Also, his description of the box held by the leader – presumably a translating device – is reported in a number of other cases, such as the Leeds encounter in 1976 (see Chapter 17).

Assuming some substance to this story, one can only wonder at the lack of perspicacity displayed by these particular extraterrestrials. Why select an individual of a nervous disposition who would hardly dare to tell even his own family – much less humanity at large – about his experience? Even if he had broadcast the message to all and sundry, would it have made the slightest difference? Of course not. In any event, it is published here for the first time, nearly four decades after the event.

'I had to write down my strange experiences because only by writing could I attain peace of mind,' wrote Bobby. 'I seek no publicity of any kind because it will surely ruin me in my own country.'

Bobby was affected deeply by his experiences. Sometime in the spring of 1963 he did indeed suffer a mild heart attack in Vienna, as he had been warned might happen. Fortunately, he made a full recovery. Although no further physical contacts ensued, he told me that from the end of 1962 to July 1964, the extraterrestrials communicated with him by means of images, projected on to the mirror of his room in the house in Beckenham, southeast London, where he lived at the time with a number of other students, including John Bingham. The images seemed to be generated by, or projected from, a red light in the sky. From a small point on the window pane, a beam of red light fanned out until it struck the mirror on the wardrobe opposite. Within a circular image of about one foot in diameter, a face was seen and a voice heard. The discourses were mainly philosophical, Bobby told me, though he was reluctant to disclose any of the information he received in that way.

During the time I knew Bobby personally, I found him to be a gentle and honest soul, not lacking in a keen sense of humour. He is, incidentally, also a fine pianist who for many years was professor of piano at a certain university on the east coast of the United States. I find it difficult to believe that he fabricated the story, if only for the reason that he was invariably reluctant to discuss it – even with me.

I asked John Bingham, who had known Bobby for longer than I had, if he believed the story. 'I would say that I *know* he was telling the truth,' he responded, 'because of a certain extraordinary experience I had when he was having those communications, when I also saw an identical projected image. That proved to me he was not telling lies.'[12]

Bobby believed that most people who have had the 'rare privilege' of witnessing 'strange beings or visions from other planets' suffer from depression and other sequelae, owing to the futility of trying to prove their experience to others. Others can be driven to the point of madness or even suicide. 'A few tend to laugh off these strange happenings, only to fail in the end,' he wrote. 'Others who are strong-willed and full of initiative and drive are able to forget about these incidents by erasing them from their minds, and are then able to settle back to normal lives . . .'

NOTES

1 Picco, Héctor Antônio, 'Trip on board a UFO', *Crónica*, Buenos Aires, 14 December 1995, translated by Jane Thomas Guma.

2 Picco, H.A., 'UFO base in the Gulf of Mexico', *Crónica*, 15 December 1995.

3 *Concise Medical Dictionary*, Oxford University Press, fourth edition, London, 1994, p. 511.

4 Frohse, F., Brödel, M., and Schlossberg, L., *Atlas of Human Anatomy*, fifth edition, Barnes & Noble, New York, 1959, p. 150.

5 Picco, 'UFO base in the Gulf of Mexico'.

6 Picco, H.A., 'God, UFOs and the Absolute Energy', *Crónica*, 16 December 1995.

7 Creighton, Gordon, 'The Italian Scene – Part 4', *Flying Saucer Review*, vol. 9, no. 4, July–August 1963, pp. 10–11.

8 Ibid., pp. 11–12.

9 Bowen, Charles (ed.), 'Sindbad the Sailor', *Flying Saucer Review Case Histories*, supplement no. 14, April 1973, pp. 14–15, translated from *Lumières Dans La Nuit*, Contacts Lecteurs, series 3, no. 5, January 1971.

10 Silvano, Ceccarelli, 'Mario Zuccalà's Strange Encounter', *Flying Saucer Review*, vol. 8, no. 4, July–August 1962, pp. 5–6.

11 Mesnard, Joël, 'The French Abduction File', translated by Claudia Yapp, *MUFON UFO Journal*, no. 309, January 1994, pp. 7–9.

12 Interview with the author, 9 February 1997.

Chapter 12

Continuing Contacts

While some extraterrestrials in the 1950s and 1960s dropped by occasionally delivering doom-laden lectures to hapless contactees, most seemed determined to avoid any contact. Whatever their agenda, they continued to evince as much interest in rivers, lakes, seas and oceans (the 'hydrosphere') as they did in drier lands. This should not be surprising, given that the hydrosphere constitutes nearly three-quarters of planet Earth.

'Having been a pilot in the Fleet Air Arm, I always was a firm believer in UFOs because of radar,' Sir Mark Thomson, now a company chairman and private pilot, informed me in 1995. 'Even 30 years ago, military radars were sufficiently reliable and sophisticated as to be able to determine whether an object was a UFO, a weather balloon, an aircraft – or somebody's imagination!' Sir Mark went on to relate an incident that occurred *circa* 1963, when he was a Royal Navy lieutenant flying twin-jet Sea Vixens, in the aircraft carrier HMS *Victorious*:

> During one foul night in the Indian Ocean an 'object' approached the carrier task force at extremely high speed and executed some extraordinary manoeuvres physically impossible for any known man-made machine. Although I was not a witness to the incident, I did learn that the object was tracked for some time by numerous radars in several ships.[1] [2]

Other naval encounters occurred in 1963, some especially interesting ones involving unidentified submarines. According to former British naval intelligence officer and biologist the late Ivan Sanderson, sometime in that year the US Navy conducted a series of exercises off the coast of Puerto Rico to train personnel in the detection and tracking of submarines. More than five surface ships were involved, including the aircraft carrier USS *Wasp*, the command ship, as well as several submarines and aircraft.

Sanderson learned that a sonar operator on a destroyer reported that one of the submarines had broken formation in an apparent attempt to pursue an unknown underwater object. Similar reports came in from all the other ships and from the sonar-equipped aircraft. According to one of Sanderson's sources:

no less than thirteen aircraft (including submersibles and aircraft, one must suppose) noted in their official logs that their underwater tracking devices had latched on to [a] high-speed submersible. All of which is said to have immediately been reported to COMLANT [*sic*] in Norfolk, Virginia. At this point, all the reports become somewhat vague and obscure. Various numbers of people, in various numbers of ships, are alleged to have observed or heard the sonar blips caught by their own operators, and all to have concurred in the fact that this object was being driven *by a single [screw] at more than 150 knots* . . . Thus, the object recorded above beat anything that we can do at the present stage of our technological development, by nearly four times in speed.

Sanderson also learned that the unknown submarine was tracked for four days as it manoeuvred, including to depths of 27,000 feet.[3] (The greatest measured depth in the Atlantic Ocean is the Puerto Rico Trench, found just north of that island, at 30,246 feet, about which more later.)

Yet another incident said to have occurred at sea in 1963 is reported by Dr Jacques Vallée, the distinguished UFO researcher, who learned that a US Polaris nuclear-powered submarine interrupted its lengthy submerged mission in the Atlantic to surface while all personnel were ordered to remain below.

A few superior officers went up to the tower. They are said to have come back down with three humanoid bodies in clear plastic bags. The sub dived again and rallied to the East Coast at top speed. The [submarine] had accomplished none of its stated objectives, which included the test firing of several missiles. As for the beings, they looked like shaved monkeys. Perhaps they were indeed monkeys, recovered from a classified space experiment.[4]

Perhaps so. Unfortunately, Dr Vallée could not provide me with details that might help determine if that was or was not the case. Given the highly secure and single-purpose nature of the operations of ballistic-missile-armed nuclear-powered submarines of this period (and later), it seems odd indeed that such a mission would have been interrupted and, in effect, aborted in the manner described by Vallée. Only events quite exceptional would have led to such a change in the normal pattern of those operations.

We return now to cases involving contact with extraterrestrials where, in contrast, the witnesses have provided an abundance of details, the better to verify the purported encounters.

CONTACTS FROM COMA BERENICES
Among the most frequently published photographs of flying saucers are those taken in New Mexico in the 1960s by the contactee Paul Villa. Curiously, little has been published either about the details of the pictures or about Villa and his claimed encounters.

Apolinar (Paul) Alberto Villa Jr. was born in 1916, of Native American, Spanish, German and Scottish descent. He claimed to have been taught telepathically by extraterrestrial intelligences from the age of five, and though failing to complete the tenth grade of school, he seemed to have been well versed in subjects such as mathematics, electricity, physics and mechanics. He also had an unusual talent for detecting defects in engines, generators and other such machines, a talent that served him well in his profession as a mechanic, first in the Air Force and later as a private citizen.

FIRST CONTACT

Villa claimed that ten years prior to photographing his first series of saucers in 1963, he had been contacted by extraterrestrials while he was working for the Department of Water and Power in Los Angeles. At Long Beach one day in 1953, a strong inclination suddenly came over him to go down to the beach, a feeling he did not understand at the time. There, he said, he met a man about seven feet tall. Initially, Villa was afraid and wanted to run away, but the man called him by name and told him many personal things about himself. Villa realized that he was communicating with a 'very superior intelligence', and he then became aware that this being was a 'spaceman'.

> He knew everything I had in my mind and told me many things that had taken place in my life. He then told me to look out beyond the reef. I saw a metallic-looking, disc-shaped object that seemed to be floating on the water. Then the spaceman asked me if I would like to go aboard the craft and look around, and I went with him.

Villa reported that the saucer occupants were human-like in appearance, though more refined in face and body. Also, they had an advanced knowledge of science, as evidenced by their craft and from the information given by them.

Villa was informed that the galaxy to which our Earth belongs is as a grain of sand on a huge beach, in relation to the unfathomable number of inhabited galaxies in the entire universe. Because of the aliens' technological advancement, their spaceships could penetrate the Earth's radar detection systems, so that they were picked up on our radar screens only when they chose to call attention to their presence in our skies. Their craft were constantly active around our planet, and more and more sightings and landings would take place to increase public awareness of their existence. They said they were here on a friendly mission to help our people; that they had bases on the Moon; that Phobos, one of the two moons of Mars, was hollow and had been artificially constructed, and that a Superior Intelligence governed the universe and everything in it.[5]

SECOND CONTACT

On 16 June 1963, Villa's space contacts telepathically told him to drive alone in his pickup truck to a site near the town of Peralta, about 15 miles south of Albuquerque. There, at 14.00, he claimed to have seen a flying saucer which he estimated to be about 160–170 feet in diameter. (In an earlier estimate he gave it as 70 feet.) The ship 'posed' and hovered at low altitude and at various distances as Villa took photographs of it framed by the trees, and sometimes showing his truck in the foreground. He used a Japanese-made Apus folding camera with a Rokuoh-sha f4.6 75mm lens (which I have examined), with 120 format Kodacolor film.

Two of the photos show the ship as it flipped on edge with its lower part rotating, apparently, according to Villa, to indicate that the space people had created an artificial gravity-field within the craft; thus they remained completely comfortable no matter what attitude the craft assumed relative to a planet's surface. According to the late Coral and Jim Lorenzen of the Aerial Phenomena Research Organization (APRO), one of these images (see photo no. 3 in colour plates) fails to show the branches of a tree in front of the craft as it should do were it 170 (or even 70) feet in diameter as claimed, indicating that the 'craft' might be a small fake. It could be the case, though I have been unable to establish this to my own satisfaction, working from enlargements I personally reproduced. Villa himself claimed that the branches of the tree were swaying from a 'huge rush of wind' generated locally by the craft, which might have accounted for the apparent inconsistency. In other photos from this series, where nearby vegetation and also Villa's truck provide useful reference points, the saucer appears to be a large object at considerable distance from the camera, thus reducing the likelihood of fraud.

At one point, Villa said, the craft hovered about 300 feet above his truck and caused it to rise slowly into the air to three or four feet for a few minutes. Also, when the craft was about a quarter of a mile away, a flexible, controlled 'rod' could be seen, apparently probing the ground and trees at different angles and curving into different shapes. While this was going on, a small, shiny, remotely controlled sphere, six to nine feet in diameter, exited from the main craft and disappeared behind trees, then reappeared and shot off at terrific speed, while glowing a reddish colour.

When the craft hovered a few hundred yards away, according to Villa, between the tree tops, the bottom section was tinted 'amber red, like hot metal', but the colours changed from a 'shiny chrome to a dull aluminum' back to amber. At one point it became so bright that it shone almost unbearably. As it passed over Villa's head, he felt not only heat from it but also 'a prickly or tingling sensation all over my body'.

Although the upper, domed structure of the craft could be turned independently from the lower section, Villa learned, it appeared to remain stationary during flight, while the lower section rotated at different speeds. It made a whirring noise that sounded something like a 'giant electric motor or generator'. At other times it gave off either a buzzing, 'pulsating' noise, or it suddenly became totally silent as it moved about in different directions.

At about 14.30, the craft landed on tripod legs and nine 'beautiful' crew members – five men and four women – disembarked through a previously invisible door. These beings ranged in height from seven to nine feet, said Villa, and were well proportioned, immaculately groomed and dressed in tight-fitting one-piece uniforms. The colour of their hair ranged from 'fiery golden' to 'polished copper' to black. Villa was told that they came from the 'constellation of Coma Berenices, many light years distant'. (Coma Berenices is a constellation notable for the large number of galaxies it contains.) In addition to communicating telepathically, they were also able to speak many of our languages. During the 90-minute conversation with Villa, they spoke in both English and Spanish (Villa's native tongue), but when conversing among themselves, they spoke in their own tongue, which sounded like 'something akin to Hebrew and Indian'.

The craft operated as a mother ship for nine remotely controlled monitoring discs, manoeuvred from instrument panels in the mother ship, Villa was informed. These could pick up imagery and sound from areas to which they were directed, and relay them to TV-monitor panels in the mother craft, a remote-viewing technology first described by George Adamski.

The 'vents', clearly seen in some photos surrounding the central section of the craft, were openings possibly used for 'collecting and ionizing atmospheric gases'. These vents were not used nor left open outside a planet's atmosphere, where the magnetic lines of force are further apart, but, like the 'door' underneath the craft, were 'hermetically sealed', either manually or automatically, after leaving the atmosphere.

The hermetic sealing is accomplished by removing all foreign substances from the basic elements of the parent metals, and a device is used to charge both pieces that are to be sealed together, either positively or negatively, depending on how a certain metal is naturally charged. The hermetic sealing of two or more metals cannot be accomplished unless they are first neutralized and then all charged with the same polarity. The carbon elements, however, being amphoteric [chemically reacting as 'acidic' to strong 'bases' and as 'basic' towards strong acids] and combining equally well with positively or negatively charged elements, cannot be charged. 'Tubes' are used to achieve hermetic sealing, using this carbon principle.

Villa reported that some crew members carried a miniature version of these tubes that appeared made of an aluminium-like material, about eight inches long and one inch in diameter, tapering slightly from the centre outwards. These devices could be used to paralyse any animal life form, including man.

The 'Coma Berenicians' were peaceful and expressed a desire for Earthlings to rise above their aggressive, warlike instincts. Love, they said, is 'the most powerful force in all the universe' which, used correctly, could transform the hearts of men. 'When the law of love rules the minds of the men of Earth,' Villa learned, 'then the people of other worlds will come in great numbers and share with us their advanced sciences . . .'[6 7]

THIRD CONTACT

Paul Villa's second series of colour photographs was exposed at several locations in April 1965. The most interesting photographs from this series were those taken on Easter Sunday, 18 April, at about 16.00, in an area 20 miles south of Albuquerque, close to the bed of the Rio Grande river, one of several areas I visited together with Villa in 1976. At one point the craft, which he told me he estimated at about 150 feet in diameter, projected a beam of light that caused a small bush fire (a Biblical 'burning bush'!), then another beam shot out and extinguished it. Smoke from the fire is visible in trees just below the craft, and just above and to the left of the tailgate of Villa's truck (see photo 9). The craft also produced a 'miniature tornado', causing the lower branches of some trees to appear blurred. The turbulence was so high, Villa reported, that he thought he and his truck would be blown away. Suddenly, the wind ceased, as if it had been 'switched off', and the surrounding air became quite hot and there was dead silence.[8]

Based on the atmospheric hazing or 'thickening' effect, whereby an object becomes increasingly less well defined in imagery the greater its distance from a camera, these photographs show a large craft at considerable distance from Villa's camera.

The ship landed on telescopic tripod landing gear that can be seen protruding from the bottom of the craft in photos 7–8. The three crewmen had light-brown hair and tan skin and appeared to be about five feet eight inches tall. Villa claimed to have talked with them for nearly two hours about personal as well as general matters.[9]

Villa told me that three scientists from the University of New Mexico, Drs Klein, Ulrich and Lincoln LaPaz, whose earlier and official UFO investigations are described in *Beyond Top Secret*, surveyed the landing site with a Geiger-counter and a magnetron, with negative results.[10]

REMOTELY CONTROLLED PROBES

Among Villa's most remarkable colour photographs are those showing various remotely controlled discs and spheres, which had been launched from a manned disc, about 42 feet in diameter. This series was taken on 19 June 1966, three miles west of Algodones and 30 miles north of Albuquerque. Photo 10, for example, shows a disc on the ground, one of several said to be from three to six feet in diameter, complete with tripod landing struts. According to Villa, the struts did not telescope into the saucer as in several other cases, but instead seemed just to shoot in or out of the bottom. Allegedly, the contraption protruding from the top of the dome was an optical device incorporating a combination of prisms and lenses. The entire unit could be retracted into the dome and could also swivel from the bottom in a circular motion, or oscillate from side to side.

In photos 11 and 12, a similar craft is seen in the air and on the ground, accompanied by four or five spheres. The disc was said to be about three feet in diameter. On landing, it bounced, shot up in the air, then landed again in almost the same place. It made no noise while on the ground, but when it shot into the air, it gave off a sound 'like an electric motor under load; an unsteady sound which seemed to pulsate'. The small spheres, no more than three inches in diameter, rotated around the larger, six-inch sphere when away from the disc. When close to the disc, however, the larger sphere always remained near the top, and the other spheres whirled around it at different speeds and orbits. The following is Villa's additional, non-verbatim description of the appearance and behaviour of the spheres:

> They also manifest in a cascade of changing colors, from a shining aluminum to a gleaming chrome then to a bright red or the sparkling blue of a welder's torch. With the smaller spheres, this change in coloration did not take place rapidly but was a gradual modulation from one color to another in a pulsating rhythm. But with the larger sphere the manifestation was different: as the small spheres careened about the larger one, the latter would change instantly from its shining chrome-plated luminescence to red, blue, green and even yellow. At times, too, the spheres would get glowing hot, resembling the frequently seen and often reported 'fireballs'. Their speed and maneuverability were incredible, for they flitted about like butterflies or raced crazily at high velocity in an array of orbital patterns.

Some of the discs and spheres had 'flexible, probing antennas resembling the antennae of certain insects', though these are not visible in the photos.[11]

One of the many controversial aspects of Villa's claims is his assertion that he himself made one of the remotely controlled discs, which is shown in photographs 14 and 15. This 'experimental craft', a few feet in diameter – similar to the others though lacking landing struts – was made according to 'exact specifications' given him by his 'space friends'. The photos show the craft in flight, 'monitored' by one of the spheres. During this test flight, said Villa, the disc fell to the ground due to a 'slight error' on his part, though the fault was soon rectified.[12]

The problem here, of course, is that, apart from the absence of landing struts, it is practically impossible to differentiate between this disc and all the others, a corollary being that *all* the discs and spheres were manufactured by Villa. I do not believe that a useful argument is as simple as that. Even were this the case, one would have to ask how it was that Villa could have managed such an elaborate, impressive hoax, inasmuch as the discs and spheres evince no indication of being suspended or superimposed. William Sherwood, formerly an optical physicist for Eastman-Kodak, told me that, based on his studies of the prints and on his discussions with Villa, all three series of photographs are genuine.

What became of the experimental disc? Villa told me that it had been destroyed when his house had burned down some time after the 1966 contacts. He did show me a later experimental craft, also supposedly made with the guidance of his space friends, of a much deeper shape than the earlier one and 18–20 inches in diameter, which he had photographed during a test flight. Regrettably, perhaps suspiciously, he seemed reluctant to let me examine it at close quarters, though it no longer functioned.[13]

DISCUSSION

In personal discussions and in correspondence with Paul Villa, I pointed out the inaccuracy of certain prophecies he claimed to have been given by his space friends, such as that 17 nations would have the atomic bomb by 1966 (30 years later, however, one could make the case that 17 nations had some approach, at least, to an atom bomb), and that Ronald Reagan would be elected president in 1976 (though he was thus elected four years later). Furthermore, some of the information supposedly imparted to Villa is spurious. Mars, purportedly used as a base by his space friends, had 'canals' and even 'pumping stations' as well as 'cacti and other plants' in certain locations. Atmospheric pressure at ground level, he told me in 1976, was equivalent to that at 12,000 feet on Earth (it is actually less than one-hundredth of the Earth's), and there were some high stratus clouds (true, but this was already known).[14] Undaunted, Villa passed on the information he had acquired, as well as his photographs, to a certain Mr Martin at the Jet Propulsion Laboratory (JPL) in Pasadena, California, which runs NASA's unmanned space programmes, including the 1964 Mariner, the 1976 Viking and the 1997 Pathfinder probes of Mars.

Was it possible, I asked Villa, that his space people had lied to him? 'No,' he replied, 'the space people, as you write, are not lying to so-called contactees, they just don't divulge hardly anything about their plans. Why should they? People would just make money from that info; besides, how can humanity appreciate anything if it is beyond their capacity to understand?'[15]

Accusations that Villa made money with his photographs are without foundation: he lived in very modest circumstances, spending much of his time and money sending free copies of his colour photographs to all and sundry. 'We write to premiers, kings, governors, leaders all over the world,' he wrote to me.[16] In 1967, with Villa's permission, Ben Blazs of UFO International copyrighted and sold sets of the photos, but Villa himself saw very little of the money, he told me.

Villa claimed that the Walt Disney Studios, as well as the US Air Force, had studied the negatives of his photographs and could not fault them. Dr Edward Condon, who headed the University of Colorado's investigation team which, sponsored by the Air Force, studied the subject of UFOs from 1966 to 1968, reportedly said 'they were the best he had ever seen', Villa told me.[17] There is no mention of either Villa or his photos in the *Scientific Study of Unidentified Flying Objects*, edited by Dr Condon.[18] Villa reported to William Sherwood that after studying his camera and the negatives, Dr Robert Low, co-ordinator of the Colorado project, said that although the team knew that his pictures were 'good', they could not use them because the committee was committed to an essentially negative conclusion.[19]

Villa, who died of cancer in 1981, shunned publicity throughout his life, avoiding interviews with the media and rarely granting meetings, even to researchers. Perhaps this was due in part to some disturbing threats: he claimed, for instance, that he had been shot at once in his pickup truck (I saw the bullet hole in the side window), and that helicopters frequently hovered around him. Most researchers who did manage to spend time with him found him genuine. 'He certainly never tried to use his unusual personal experiences for monetary gain,' wrote Bill Sherwood. 'To me he seemed always humble and sincere, unimpressed by the attention he received from the Secretary-General of the United Nations, U Thant, who called him at his workshop to discuss his experiences with the extraterrestrials.'[20] During the 40-minute telephone call in 1970, U Thant reportedly also discussed the worldwide UFO situation.[21]

In my own correspondence with him, Villa sometimes waxed effusive about politics and religion, seldom decisively addressing questions I posed. Our discussions in person in 1976 proved more fruitful. Three different groups of extraterrestrial beings were coming to Earth, including 'certainly one that is good', he asserted. 'We are not under observation, since they are here all the time.' The space people had hundreds of bases within our solar system, including many on Earth, Mars and Venus. Some groups came here simply as tourists. Villa's group liaised with about 70 contactees in the United States and about 300 worldwide.

Their craft, when not completely silent, made sounds like a 'musical saw' or an 'enormous generator', or a 'clanking noise'. Unlike some other craft, those belonging to Villa's group did not ionize air as a by-product of propulsion. Water was as essential to them as it is to us, and it was first vaporized before being taken on board the craft. In spite of their 'phenomenal abilities', the space people were not superhuman, which brings forward an interesting point.

Villa drove Lou Zinsstag and me to those sites in the vicinity of Albuquerque where he had taken photographs of craft and conversed with the crew (who would not allow themselves to be photographed). Though some were dressed in the traditional one-piece suits, he said, most were dressed in more typical human clothing. At one of these sites, beside the Rio Grande near Algodones, I asked him what the other crew members were doing while he conversed briefly with a man he assumed to be the pilot. 'Oh, they were just bathing their feet in the river,' he replied, without hesitation.[22]

At the time, that reply, delivered without so much as the bat of an eyelid, astonished me. Eventually, though, it contributed to a growing conviction that Paul Villa's story contains essential elements of truth. As so often is the case, contactees' stories seem to mix truth and fiction, and something of a problem is presented, then, in sorting out the wheat from the chaff.

A COMPELLING CASE FOR CONTACT

Of all my case files involving extended contact with extraterrestrial, quasi-human beings, there are few which I have found to be completely convincing. The following one involves a witness whom I knew as a friend for 30 years; a compelling case for contact, and one in which I believe the extraterrestrials did not impart any false information. That is not to say that I believe every word, for a few inconsistencies in the witness's story, owing to a tendency to embellish at times, emerged over the years. Because she was reluctant even to have the story published after her death, which occurred in 1995, I will refer to her only as Joëlle.

Joëlle was born in St Petersburg, Russia, of French and Russian parents, in 1914. During the Second World War she became a passive member of the Maquis, the French resistance to the Nazi occupation of France. After the war, she worked in Paris for the Ministry of Armaments, then came with her family to live in London. By a strange coincidence, her two daughters, Frédérique and Isabelle, were form-mates and friends of mine at the Arts Educational School, where, in addition to the usual curriculum, we studied acting and dance (1952–53), and it was then that I first met Joëlle. We were not to meet again until 1967, in the company of her husband and Lou Zinsstag, from whom I learned of Joëlle's contact story.

It was in September 1963, when Joëlle was in the Sheffield area conducting a house-to-house field survey for a market research company, of which she was a senior partner, that her extraordinary adventures began. The survey included questions relating to domestic appliances, and at one house she was struck by the number of very modern-looking gadgets in the living room, none of which was on the market. Queried about these, the lady of the house (whom I shall call Rosamund) responded that her husband was a scientist who regularly tested the latest devices to assess their practicability.

Joëlle noticed a large radio transceiver, and was informed by Rosamund that her husband was an amateur radio ham who talked with people all over the world. To demonstrate, Rosamund turned the set on, then left the room temporarily. Hearing a very brief message in English, Joëlle wrote it down on the back of her survey notepad. When Rosamund returned, Joëlle said that a message had come through, but did not say that she had written it down. Looking suddenly shocked, Rosamund switched the set off, explaining that her husband would never forgive her if he knew she had turned it on without his permission.

Later, back at her hotel, Joëlle pondered the message. 'Will be at Blue John tomorrow, 4.30 p.m. – Mark', it read. It meant nothing to her at first, but later she became intrigued and made a few enquiries. 'Blue John' turned out to be the Blue John Caves, near Castleton, in Derbyshire's Peak District, the name deriving from the French *bleu-jaune*, given to the blue fluorspar mineral found in Derbyshire. Wondering if perhaps she had uncovered a spy-ring, Joëlle determined to find out what was going on.

THE BLUE JOHN ENCOUNTER

On the afternoon of Monday 16 September 1963, Joëlle set off by car to return to London, via the Blue John Caves. Arriving in the vicinity of the caves at around 14.30, she parked in a vantage spot overlooking a mildly sloping valley, ate her packed lunch, then waited to see what might happen.

Shortly before 16.30, Joëlle noticed a brilliant light in the sky, which she first took to be the Sun. It was moving, though, and when it came to rest, several hundred yards from her position, the brilliant glow ceased, and she could now see that it was a highly unusual disc-shaped aircraft, approximately 20 feet or so in diameter, supported on tripod landing legs with inverted mushroom-shaped pads on their earth-contacting ends. Beneath a cupola could be seen several circular windows. After a pause, a man – presumably 'Mark' – stepped out from the other side, dressed in a blue one-piece suit and a cloth helmet of some sort. Simultaneously, a man came out of a car parked some distance away and began walking down the slope towards the craft. Joëlle recognized the car as the one that had been parked outside Rosamund's house. After the two men had greeted each other warmly, Mark turned towards the craft and signalled briefly to the other (presumed) crew member(s), then both men headed towards the car and drove away. The craft began to glow and lift off the ground, retracted its landing gear, hovering momentarily before shooting off at a fantastic speed.

A SAFE HOUSE

At that time, Joëlle did not accept the fact of flying saucers. She assumed that this was a highly advanced aircraft, perhaps of Soviet origin, its occupant a spy liaising secretly with Rosamund's husband. So she decided to wait before driving to Rosamund's house, finding out what she could, then perhaps reporting the matter to the police.

Half an hour later, she knocked at Rosamund's door. The scientist (whom I shall call Jack) opened the door cautiously and asked what she wanted. Joëlle gave the excuse that she had interviewed his wife the day before and needed to double-check some questions. Jack made as if to close the door, but at that point Mark – now dressed in terrestrial clothes – interjected. 'That's alright, Jack,' he said, 'let her in.' Reluctantly, Jack opened the door and showed Joëlle into the living room.

'Why don't you tell us the real reason you're here?' began Mark.

'Because I need to check some questions with Rosamund for my market research survey.'

'That's not true, Mrs——.'

Joëlle swore to herself.

'Tut, tut,' said Mark, teasingly. 'You shouldn't swear like that.'

How had he known that she had sworn?

'You came here,' continued Mark, 'because you saw my craft and wanted to find out what was going on, didn't you?'

Reluctantly, Joëlle admitted the truth. And from then on, she was 'let in' on the alleged alien liaison. The discussion that night lasted well into the small hours. At first incredulous, she gradually accepted the sensational truth: that Mark was indeed a man from another world. For the next 15 months or so, Joëlle had a total of about eight and a half hours of meetings with Mark and another member of his race, a man who, because his deep voice sounded like that of the actor Valentine Dyall, was given the name of 'Val'. These meetings reportedly took place at several locations in England, including at least two in Joëlle's London flat, near Earls Court.

ALIEN BASES

Joëlle told me that, having no knowledge of the subject at the time, she began by asking some 'rather stupid' questions. Later, after reading a few books, she was able to make more sophisticated enquiries. Her first question, naturally, related to the origin of the visitors. This was one of a number of things that Mark and Val politely refused to discuss in precise terms: they responded merely that they came from a planet, similar in many ways to Earth, located in another solar system. They also stated that we are not alone in our solar system, and implied that they had bases on two (unspecified) moons of Jupiter. Interestingly, it was reported in 1997 that signs of life, in the form of molecules containing carbon and nitrogen, had been detected on two of Jupiter's largest moons, Ganymede and Callisto, based on data gathered by NASA's Galileo spacecraft. For many years, some astrophysicists have speculated that life might exist in the warm water lying beneath the frozen surface of Europa, the smallest of Jupiter's moons.[23] Thousands of years ago, said Mark and Val, their people had bases on Mars and on the Moon. They also revealed that they had a number of bases on Earth, located in South America, Australia, the Soviet Union and elsewhere (though not in the United Kingdom).

Although *Homo sapiens* originated on Earth, the visitors explained that, to speed up human evolution, they had on two occasions genetically 'interfered' with us. While similar in appearance, Earth humans and extraterrestrial humans evolved separately. Because of their advanced evolution, aliens live longer than Earth people. Val and Mark were extremely refined, fair-skinned, with perfect teeth and a not immediately noticeable peculiarity about the eyes. On one occasion, Joëlle says she saw, though did not meet, a dark-skinned man who was a member of the same group.

Mark and Val said they were liaising in great secrecy with a team of scientists from several nations, having initially established their English contacts through Jack eight years earlier. None of those names was ever revealed to me. In addition to Jack, Joëlle met two other such scientists, one of whom worked at the Woomera rocket range in Australia, set up jointly with Britain at the end of the Second World War. As to the purpose of the extraterrestrial missions, this was another question that they declined to answer precisely. 'We are not here for entirely philanthropic purposes,' was all they volunteered on one occasion. Whatever the mission, it demanded considerable dedication from the scientists, some of whom ostensibly worked with the aliens at the bases, or even (on rare occasions) travelled to their planet, necessitating their going 'missing'. Ideally, therefore, those without family responsibilities were involved.

In *Alien Liaison*, I discussed an alleged alien base located at, or in the vicinity of, Pine Gap, America's most secret facility in Australia, some 15 miles from Alice Springs. According to information supplied by Professor J.D. Frodsham in 1989, three hunters returning from an all-night trip witnessed a 'camouflaged door open up in the grounds of the base and a metallic circular disk ascend vertically and soundlessly into the air before disappearing at great speed'.[24]

Officially a 'Joint Defense Space Research Facility' sponsored by both the American and the Australian defence departments, Pine Gap serves principally as a downlink site for reconnaissance and surveillance satellites.[25] It was established by the Central Intelligence Agency in 1966 and is run jointly by the CIA and the National Security Agency (NSA). According to one American observer: 'The Australians have accorded the [Pine Gap] facility remarkable hospitality. People and cargo routinely fly in and out, entering and exiting without the burden of customs or immigration checks. The place enjoys almost extra-territorial status.'[26] According to one of my sources, formerly a CIA employee, Alice Springs is considered to be a 'reward' posting – which is *not* to say that an actual alien base exists, or did exist, at Pine Gap. Nevertheless, there are some intriguing early references to the alleged existence of such a base, located '1,400 miles from Sydney' (which could place it in the Alice Springs area), in letters written by George Adamski. In 1951, for example, he wrote to a correspondent as follows:

> Under very interesting circumstances I had previously been told of a big space laboratory 1,400 miles from Sydney [which] has been in operation for the past three years. I was made to understand that space ships could be landing there [and that] a communication system could be going on through this laboratory between earthmen and spacemen . . . It wasn't given to me as definite fact, but as a possibility from which I was to draw my own conclusions.[27]

If there is any truth to this rumour, the implication is that the laboratory was functioning in about 1948, years before Pine Gap was officially known as a satellite intelligence-gathering and relay base. In replies to questions from the same correspondent a few months later, Adamski explained that he had acquired the information in 1949 from a scientist attached to the Chilean government, a former commanding officer in the Chilean Air Force. 'A communication system is definitely going on,' wrote Adamski, 'not only there but in [the United States] as well.'[28][29]

Regarding the existence of alien bases in the United States, in January 1952, prior to his first contact in the Californian desert in November that year, Adamski spoke with a marine engineer from Alaska who claimed that spacecraft regularly landed in a certain area in that state. According to the unnamed engineer, the 'space people' he saw ranged in height from three to six and a half feet.[30]

I include the foregoing information from Adamski for three reasons: first, it pre-dates any publication relating to the existence of alien bases on our planet. Secondly, as discussed in Chapter 7, Joëlle claimed to have met a similar group of extraterrestrials to those who contacted Adamski in 1952 (and who regrettably were obliged to discredit him). Joëlle's contacts also informed her that they had a base in Australia, location not specified, where they liaised with a team of human scientists. Finally, one of my most reliable and well-connected sources has learned that a number of such bases exist worldwide, and that a limited liaison between extraterrestrials and our people was established in the late 1940s. Interestingly, the locations of two of these bases were given as somewhere in Alaska – and Pine Gap.

CONTINUING CONTACTS

On one occasion, Joëlle said she was invited to inspect a spacecraft at close quarters. This turned out to be the same craft as the one she saw at a distance in September 1963. On this occasion, in the vicinity of the Welsh border, one of the scientists was being taken to a base in South America. Joëlle told me she was poor at judging sizes, but estimated that the width of the landing legs was about three inches and that the inverted mushroom-shaped pads were possibly an inch or so wider. Apart from a series of round portholes, no further details of the craft could be discerned, as it was dark at the time. Also, the entry point was out of her view. She was not allowed to go aboard, though she did touch the hull, which later caused her to feel 'slightly ill'. Shortly afterwards, she and Rosamund drove up to the top of a nearby hill to watch the craft take off. With a sound as of a swarm of bees, it rose vertically, slowly at first, then shot off, illuminated, at an angle.

Joëlle said she helped the visitors in a number of ways. Once, they asked her to translate a certain Russian manuscript at the British Museum. Also, on more than one occasion, she cooked meals for them at her London flat. Both Mark and Val had 'perfect manners', enjoyed drinking wine with their food, and had a great sense of humour. They stressed a desire to be treated normally. 'We may be thousands of years in advance of your people,' they said once, 'but please don't look on us as angels.'

Mark and Val did not rely on telepathy to communicate between themselves; they also spoke their own language. When communicating at a distance with the scientists, they used a type of radio system with pre-arranged 'secure' frequencies, using tiny radios strapped to their wrists. More sophisticated methods of communication could be used, as Joëlle was to discover. Arriving back at her flat on one occasion, she was astonished to see Val standing in the living room.

'How on earth did you get in?' she asked, as she went to greet him.

'Don't come near me – don't touch me!' he said. 'Just calm down. *I'm not actually here.*'

Val went on to explain that what she saw before her was a projected image, effected mentally between minds as a means of enhancing communication from a distance. 'Maybe it was, as he said, just a picture in my mind,' Joëlle told me. After a short discussion and a farewell, the 'picture' simply faded out. This particular phenomenon has been reported in a number of contact cases, including that of Cynthia Appleton (see pp. 207–8).

THE HOME PLANET

On one occasion at her flat, by means of a certain technical device, Joëlle said that her friends projected for her some three-dimensional still images (similar to our holograms, though more realistic) of their home planet. Certain kinds of trees could be seen, as well as houses, mostly circular in shape though not all of identical design. Tubular-shaped vehicles, which travelled just above the ground, were shown. These could hold up to four people and were programmed to stop at certain points, unless otherwise desired. Animals shown included cows, similar to certain of our breeds, though smaller.

Joëlle learned that weather on the home planet was not as drastically contrasted, neither were the seasons the same, as on Earth. The aliens did not eat as much as we do, and consumed a great deal more fish than mammal meat, which was seldom eaten. Fruits were plentiful, and a fermented drink similar to wine was produced.

No separate countries or governments existed, as such, though from what Joëlle could gather, there was a type of 'council'. No social or racial divisions existed. Though there was no money, a system of 'credits' was used. One did not get something for nothing and everyone had to contribute to society in some way. Even those normally engaged in, for instance, scientific work, took their turn at performing more menial tasks. Couples restricted themselves to two children, who matured much earlier than do humans. There were no hospitals: injuries caused by accidents, for example, could be healed by sophisticated machines.

Music was enjoyed, though different from ours. Stringed (not bowed) instruments were mentioned. Val and Mark made a point of emphasizing how much they liked our music.

It was implied that travel between the visitors' solar system and ours was 'virtually instantaneous', Joëlle told me, though they declined to give her any details as to how this was effected. In any case, she felt that she probably would have been unable to understand the *modus operandi*. They did explain, however, that certain differences in their planetary environment made it difficult for them to live on Earth without periodic 're-conditioning', a process similarly reported by some other contactees, such as Howard Menger. From what little Joëlle was told about this, I infer that these difficulties related principally, perhaps not wholly, to atmospheric pressure and gravity. While Mark and Val were working here, it was necessary for them to undergo re-conditioning or 'decompression' about every four days, either in the spacecraft (including a giant carrier craft) or at their bases. They required no more than four hours' sleep a night.

The visitors pointed out to Joëlle that, were she to visit their planet, 'you may not see us'. This *could* imply that they existed in another dimension or 'frequency', though Joëlle was inclined to the view that our less well-developed physical senses, *vis-à-vis* our limited perception of the electromagnetic spectrum, would be responsible for this condition. She always emphasized to me that, in spite of their technical, mental and spiritual advancement, her extraterrestrial friends were physical beings, requiring physical sustenance and transport. She also had the impression that they were not necessarily dependent on planets, their carrier craft being completely self-sufficient.

THE HOME OFFICE

In 1967, three years after the last of her meetings with Mark and Val, Joëlle claimed to have received a visit from two representatives of the Home Office in London. The men began by asking questions relating to the 'disappearance' of Jack and Rosamund and some of the other scientists, who by this time were supposedly living 'elsewhere', perhaps at a base in South America. Joëlle presumed the men had located her from one of the missing scientists' address books they found at his home. In any event, they were knowledgeable about the story. Joëlle politely refused to answer certain questions. 'You don't really expect me to answer that, do you?' she would reply; a response which seemed to please the investigators.

A SPIRITUAL LINK

During the course of her meetings with Mark and Val, Joëlle learned a great deal. She did not tell me everything, and on a few occasions I noticed that when she might have been on the point of a keen revelation, she suddenly stuttered to a halt. She believed that somehow she had been hypnotized to prevent her disclosing any sensitive information; a hypnotic block effected without recourse to any conventional induction method.

If Mark and Val were reluctant to discuss their origin and technology and the actual purpose of their mission, they were sometimes more forthcoming on other subjects of discussion. Generally, they explained once, they preferred to exert their formidable powers of telepathy when influencing humanity, though on occasions they had interfered directly. They would do so in future, for instance, in the event that a nuclear catastrophe threatened to destroy our planet – perhaps with severe consequences extending beyond it which might impinge on them. Other extraterrestrial beings were coming here, they said, who were not so well disposed towards us, though no further information was made available.

In addition to the visitors being responsible for genetically 'upgrading' the human race on two occasions in our distant past, it was alleged that a few of our great spiritual leaders, including Jesus, were genetically 'engineered' by a type of artificial insemination, in an attempt to instil Earth people with spiritual concepts. The reluctance of this particular group of extraterrestrials to communicate with humanity at large was due mainly to the fact that we simply are not psychologically or spiritually ready for contact with a higher civilization, and it is necessary for us to evolve independently. Essentially, Joëlle was informed, we are spiritual beings, surviving beyond death.

For Joëlle, the experiences with her friends remained a treasured and vivid memory for the rest of her life.

Seldom do we hear nowadays of encounters with spiritually advanced extraterrestrials. Have they left our planet for ever? Are they alive and well but engaged in less ambitious projects? In any event, I often wonder if a principal reason behind their presence here on Earth is related to the very survival of our planet – as an alien base.

'What a beautiful planet,' they once remarked to Joëlle. 'Such a pity you're destroying it . . .'

NOTES

1 Letter to the author from Sir Mark Thomson, Bt., 31 January 1995.
2 Interview with the author, 14 February 1997.
3 Sanderson, Ivan T., *Invisible Residents: A Disquisition upon Certain Matters Maritime, and the Possibility of Intelligent Life under the Waters of the Earth*, The World Publishing Co., New York, 1970, pp. 20–2.
4 Vallée, Jacques, *Forbidden Science: Journals 1957–1969*, North Atlantic Books, Berkeley, California, 1992, p. 309.
5 Green, Gabriel, 'The Paul Villa Saucer Photos', *UFO International*, The Amalgamated Flying Saucer Clubs of America, Los Angeles, no. 21, 1964, p. 3.
6 Ibid.
7 Blazs, Ben, 'Villa Set No. 1' (fact sheet), UFO International, PO Box 552,

Detroit, Michigan 48232, 1967.

8 Blazs, Ben, 'Villa Set No. 2', UFO International, PO Box 552, Detroit, Michigan 48232, 1967.

9 Green, op. cit., p. 4.

10 Interviews with the author, Albuquerque, New Mexico, 1–2 September 1976.

11 Blazs, Ben, 'Villa Set No. 3', UFO International, PO Box 552, Detroit, Michigan 48232, 1967.

12 Ibid.

13 Interviews with the author, 1–2 September 1976.

14 Letter to the author from A.A. Villa, 1 June 1977.

15 Letter to the author from Villa, 25 June 1975.

16 Letter to the author from Villa, 17 July 1976.

17 Letter to the author from Villa, 24 August 1976.

18 Condon, Dr Edward U. (Ed.), *Scientific Study of Unidentified Flying Objects*, Bantam Books, New York, 1969.

19 Saunders, David R., and Harkins, R. Roger, *UFOs? Yes! Where the Condon Committee Went Wrong*, Signet Books, New York, 1968.

20 Letter to the author from William T. Sherwood, 1 June 1970.

21 Letter to the author from Sherwood, 28 January 1982.

22 Letter to the author from Sherwood, 1 June 1970.

23 Arthur, Charles, 'Vital ingredients of life are discovered in Jupiter's moon[s]', the *Independent*, London, 3 April 1997.

24 Letter to John Lear from J.D. Frodsham, then Foundation Professor of English and Comparative Literature, Murdoch University, Western Australia, 13 December 1989. The story was additionally confirmed to the author by Professor Frodsham.

25 Burrows, William E., *Deep Black: The Secrets of Space Espionage*, Bantam Press, London, 1988, p. 190.

26 Pilger, John, *A Secret Country*, Vintage, London, 1992, p. 199.

27 Letter from George Adamski to Emma Martinelli, 30 September 1951.

28 Letter from Adamski to Martinelli, 24 November 1951.

29 Letter from Adamski to Martinelli, 16 January 1952.

30 Ibid.

Chapter 13

Neither Rhyme nor Reason

So conditioned are we to the notion of alien omnipotence and the perfection of their technology that when something goes wrong, as reported in a number of cases described earlier, we are inclined to doubt, or to wonder if the event was staged as part of an elaborate deception. Why should this be so? The following case, reported by two witnesses, is unique inasmuch as it involves the observation for over four hours of humanoid operators carrying out repairs to a landed disc, followed by the arrival later of another disc whose operators rendered assistance to the disabled craft. The incident, investigated in depth by Ted Bloecher and Dr Berthold Schwarz, occurred in an area one mile north of New Berlin, New York.

Mary Merryweather (pseudonym), who majored in music at Ithaca College, was staying with her mother-in-law at the latter's home when the incidents occurred, on 25 November 1964. At 00.30, Mary stepped out on to the porch and noticed what she took to be a shooting star, followed by another, more unusual light, which came down and 'followed along the brook' a few hundred yards away from the house. 'It occurred to me that this was an unusually bright light, a brightness and intensity I had never seen before,' Mary told Bloecher. 'Not only was the visible part [of the sighting] strange, but there was a kind of low hum, like a drone-hum combination [that] never changed pitch.' She called her mother-in-law to join her on the porch.

During the next half-hour, three cars passed by: two slowed down, one stopped briefly, with the light hovering above or following them. The occupants of the cars, apparently frightened, accelerated away from the scene. Before the arrival of the third car, the lighted object had come to a stop several hundred feet directly across the road from the house. 'Then it kept going north north-west and went up on the side of this mountain about 3,800 feet away,' said Mary. 'Then it settled down just below the ridge of this hill.' Because of the cold night air, Mary returned to observe the lighted object from inside the house, together with her mother-in-law, as their English springer spaniel lay quivering at their feet.

THE HUMANOIDS
The two women took turns observing the object through five-power

binoculars. There seemed to be movement around the landed object. Mary reported that:

> the light seemed to be underneath it and it apparently was sitting on legs, because the bottom of the object was up from the ground, far enough so that these – I'll call them 'men' for lack of knowing who they were or what they were, because they were built like men – could get under this thing, if they got down on their hands and knees, or sitting down; they lay down under it like a man does working under a truck or a car . . .
>
> I could see them coming around this vehicle and they brought with them their boxes of tools, like tool chests or something, and one of these chests took two men to carry it . . . They appeared to be coming around something in a semi-circular movement, as if they were walking around a round vehicle . . . a light on the bottom of the object [was] so intensely bright I couldn't make out the form of the object . . .
>
> There were about five or six [men]. They seemed to be dressed in something like a skin diver's wetsuit. It was a dark colour, and their hands were visible apart or out from the wrist of the suit; their skin was lighter than the suit they were wearing . . . I could see the muscular build of them, their spinal column; they were standing on two legs like we do, and they worked with arms and hands that were like ours. The only difference was that they were slightly taller than we're accustomed to seeing people [an estimate based on the size of the bushes she could see in the lower part of the field on the hillside].
>
> The only ones I could see well were the ones up close to the vehicle where the light was shining on them, and most had their backs to me, or their sides . . . They seemed to have hair, like we do [which] seemed to be well-barbered, fairly close to their heads. The profile of their faces . . . was like the profile of a man's face.

The Repairs Begin

The men seemed to be working with tools 'like a man would use to work on a piece of machinery that had gone bad, or [on] a motor,' Mary continued. 'They took something out from underneath the center of their vehicle and let it down, gently, with their hands.' Sometime prior to this, *another* vehicle was seen to land on the crest of the ridge, just above the first one.

> Four or five more 'men' joined the ones who were working on the ground. It was just after [the first crew] had removed whatever they took out of the center of it, which seemed to be like a motor or a power supply . . . The four or five other 'men' joined them and they also began to work. I could see 'men' standing in the foreground, down the hill a little way. I could see them cutting long – what looked like – heavy cable, because it arced, or fell in a loop as they were holding it between them. They were cutting it in exact lengths and they worked quite hard at doing this . . . the cable appeared to be dark, and they used it in fixing this piece of machinery.

It was by now 01.15. The men left the motor or power source directly underneath where they took it from and set to work on it. 'And while I watched them work, and cut and struggle,' said Mary, 'they were walking around, were sitting or half-lying down, leaning on an elbow, and kneeling. There were about ten or twelve men in total – I couldn't be absolutely sure, because they were coming and going, and bringing things and taking things back to the vehicles. I couldn't see the figures without the binoculars . . .'

Frightened, Mary's mother-in-law wondered whether she should call the police or some government agency. 'Well, I hate to,' Mary responded. 'You know, if we call someone, they're going to come up here with guns and firearms and bother them, and they just want to get that thing fixed and get away.'

I'm sure they saw me after that car decided it would go away . . . My mother-in-law [also] felt that we were watched. She said, 'I am sure that they realized that we did not call the authorities, that we weren't going to, and wouldn't.'

At exactly four-thirty by our kitchen clock, the 'men' got down in a team and there were nine of them – there were some behind, a group of three, that were evenly spaced around this piece of machinery, and there was a line of six 'men' behind them; they seemed to be holding something, or seemed to be ready with something . . . Then, all together, they picked this thing up and moved it directly upwards and tried to fit it into the bottom of this vehicle. It went right up, maybe eight inches, and then it seemed to go off at an angle. You could see the bottom, like a plate, or like the bottom of a motor [which] was tilted, instead of being level. As they tried to get it in, they were turning it, too, like screwing a screw in; they turned it a little, and it went back a little bit, but it wouldn't go up there the way it should.

They got it up into the vehicle, I think, except for the last three, maybe four, inches of it, and it was just off, it wouldn't fit, it wouldn't go . . . so then they carefully retraced everything they had done and set the thing back down on the ground again. They worked on it another ten minutes, and then they tried it again, the same method, and it wouldn't go . . . They retraced their steps again and put it back down on the ground and worked on it another ten minutes. These 'men' that had been cutting cable, cut something else that was like cable, only it seemed to be a little lighter, and they cut shorter pieces. They worked, and they were hurrying . . .

This attempt also failed. Finally the men took out the piece of machinery, set it on the ground and worked on it yet again, for about three minutes. 'And they very carefully picked the thing up and it went back in.'

THE DEPARTURE
There was just enough light for Mary to see that the part of the vehicle

facing the witnesses was round, and that the bottom section tapered upwards. She estimated its diameter at 25 to 30 feet, and the length of the landing legs (which also tapered) at six to seven feet. 'Now, whether it tapered up to a cone-shape or was rounded on top, I don't know,' Mary added. 'Just before they got this thing into the centre – and it seemed to be cylindrical, I don't know what the top was like – this intense light came out from underneath the vehicle . . .'

It was a minute before five minutes off five. I could see them quickly pick up everything they could pick up and the 'men' from the vehicle above them on the hill ran with their material up there; these 'men' were running like a man running with something extremely heavy, two 'men' with the tool boxes – the one that required two to carry. There were at least two more tool boxes . . . because there were two other 'men' who were laboriously running. They ran around the side and I didn't see them after that. It looked like they were picking up cable pieces these other men had left just before that; they ran up the hill with them, and I didn't see them any more, either.

At five minutes off five, the vehicle on top of the hill left. It went straight up – I don't know how many feet – and it shot off, almost like an instantaneous disappearance, in the direction that it had come from, west south-west. A minute later the other vehicle rose straight up, went to the crest of the hill, rose a little further again, and shot off in the same direction that the other one had left in, at the same speed. And that was it. It had been a long night . . .

TRACES

The following afternoon, Mary and her mother-in-law went up to the site where the incident had taken place. Mary found three places where 'something cone-shaped and round at the bottom, very heavy and spaced in a triangle about 15 to 20 feet to a side, had set into the ground'.

They were at an angle like they were the legs of a tripod, [with] something on it that was very, very heavy, because one of them had set on a rock and broken it, and gone down a little ways into the ground where it was bedrock, or maybe shale. The impressions on the bare ground that didn't have any rock underneath were about 14 inches wide and up to 18 inches deep. The shallowest hole was about four inches deep. There were two sets of these, one at the top of the hill and one down the slope. They were set like an equilateral triangle – one hole wasn't any further from the other two.

On the Monday after the sighting, Mary went with her husband to search for traces of the cable, and about 50 or 60 feet below the lower set of holes he found a three-inch piece of peculiar wrapping.

It looked like a strip of something they had missed. The outer part of it looked like the wrapping, something like a brown paper towel, only it wasn't like our paper towels . . . And in the center of it – it had been cut out laterally – you could see the strip, maybe an inch long more or less, something that looked like very finely shredded aluminum strips [about $\frac{1}{16}$ of an inch thick] laid in there, and it was as long as the piece of paper and had been cut, and had the color and feel of aluminum, although it wasn't aluminum [and] didn't behave like aluminum. Aluminum will crumple and this didn't crumple. You couldn't crease it.

Though Mary's husband was a chemical engineer, he did not press for analysis of the material, and although the family searched for the sample when Bloecher interviewed Mary in 1973, unfortunately it was not found. Interestingly, a number of UFO landing sites elsewhere in the United States in the 1960s involved residue of aluminum-like strips. These strips were often found in bundles, and in some cases may relate to aluminum 'chaff' discharged from Air Force planes for radar counter-measures, though it is either coincidental or curious that some have been found near UFO landing sites (and, in one instance, inside the mouth of a cow believed to have been mutilated by aliens). Investigator Don Worley sent me samples from a 1968 landing site he investigated in Indiana. I arranged for analysis of a sample at the Oak Ridge National Laboratory in Tennessee. Unfortunately it was lost, so I then sent two more samples to a private laboratory in Kent, England. According to Dr Anthony Fish, the samples consisted of aluminum with trace impurities of iron and silicon, similar in composition to 99.5 per cent aluminum cooking foil, the major impurities of which are silicon and iron.

'Mary Merryweather has at no time sought publicity as a result of her unusual observation,' Ted Bloecher commented. 'To the contrary, she has gone out of her way to avoid it and has discussed the incident with no more than a dozen people, most of whom were family members or close friends.'[1][2][3]

OF ROBOTS AND HUMANOIDS

Encounters with human-type beings accompanied by apparent robots have been reported all over the world. One of the most intriguing such cases occurred in the Cisco Grove area of Placer County, near the Loch Laven lakes of Northern California, in September 1964. A thorough investigation was conducted by Ted Bloecher and Paul Cerny, whose report I summarize here.

During a hunting trip on 4 September, 28-year-old Donald Smythe became separated from his companions. For protection, he decided to spend the night in a tree on a mountain ridge. After about two hours, he noticed a light moving in a ziz-zag manner. Believing it might be a heli-copter, he climbed down and lit three fires on large rocks to attract atten-tion. The light made no noise. It made a sweeping half-circle around the witness, moving over a canyon on the south side of the ridge. In addition

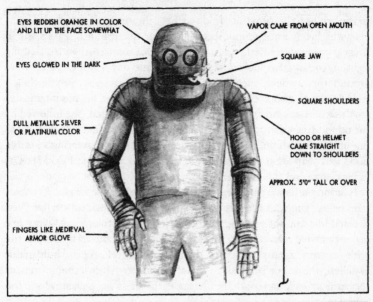

EYES REDDISH ORANGE IN COLOR AND LIT UP THE FACE SOMEWHAT

EYES GLOWED IN THE DARK

DULL METALLIC SILVER OR PLATINUM COLOR

FINGERS LIKE MEDIEVAL ARMOR GLOVE

VAPOR CAME FROM OPEN MOUTH

SQUARE JAW

SQUARE SHOULDERS

HOOD OR HELMET CAME STRAIGHT DOWN TO SHOULDERS

APPROX. 5'0" TALL OR OVER

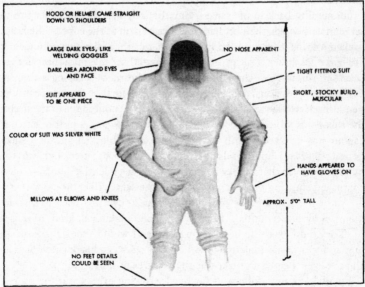

HOOD OR HELMET CAME STRAIGHT DOWN TO SHOULDERS

LARGE DARK EYES, LIKE WELDING GOGGLES

DARK AREA AROUND EYES AND FACE

SUIT APPEARED TO BE ONE PIECE

COLOR OF SUIT WAS SILVER WHITE

BELLOWS AT ELBOWS AND KNEES

NO FEET DETAILS COULD BE SEEN

NO NOSE APPARENT

TIGHT FITTING SUIT

SHORT, STOCKY BUILD, MUSCULAR

HANDS APPEARED TO HAVE GLOVES ON

APPROX. 5'0" TALL

Fig. 17. The robot and spaceman encountered in a terrifying ordeal by Donald Smythe in Cisco Grove, near the Loch Laven lakes of Northern California, in September 1964. (*International UFO Reporter*)

to the bright light, Smythe could see three illuminated rectangular 'panels', which remained motionless for four or five minutes. From one of these panels came a flash, and a dark object of some sort, with a flashing light on top of what looked like a dome, came down, appeared to move around the witness, then landed. Shortly afterwards, Smythe heard sounds of something 'crashing through the brush' on the mountainside.

A few minutes later, two 'humanoid' figures emerged, one followed by the other, dressed in 'some kind of light-coloured, silver or whitish-looking uniform, with puffs around the sleeves and joints', wearing a kind of hood and with what appeared to be large, dark eyes (see Fig. 17). The beings came and stood under Smythe's tree.

Not long afterwards, more noises were heard in the brush. A 'robot-like' being, about five feet in height, with glowing reddish-orange eyes, dressed in a kind of metallic uniform, approached the tree. Moving in a less articulated manner than the others, it scattered the embers of the fire with its arms, then returned to the foot of the tree. As the humanoids watched, the robot emitted a puff of odourless white vapour from its mechanical-looking mouth, rendering the witness unconscious for a few minutes.

On coming back to his senses, Smythe, alarmed, began to set fire to some match books, then his hat, and threw them at the robot. The latter backed off, together with the humanoids, but they all returned once the fire had died down. This process was repeated for several hours. In desperation, the witness shot three arrows at the robot, causing a flash of light, but no apparent damage, each time. For further security, Smythe tied himself to the top of the tree with his belt. At one stage of the night, the robot was joined by another, similar entity. Again, a puff of gaseous vapour was directed at Smythe, who once more blacked out for a short period. The two humanoids then tried to climb the tree, but Smythe managed to keep them at bay by shaking the tree violently. The first robot continued emitting puffs of vapour, with the same effect on the witness.

'And I tried all kind of goofy things, you know; just tried to distract them,' Smythe recounted. 'I tried yelling and making all kinds of noises . . . They didn't seem to hear, [though] when I would shout, these two in human form would look up . . . All I could see was a black patch of face, and the eyes. I couldn't make out any features of the face.'

Finally, at dawn, following further harrowing events and the sounds and temporary appearance of at least one more white-clad humanoid, both robots and humanoids departed in a 'cloud of fog'.

SHOCK

On Smythe's return home, his wife was shocked by his appearance and demeanour:

I knew something was wrong when I saw him. He was as white as a sheet [and] his eyes were dazed looking. He spoke to me in a very shaky voice. He had dark circles under his eyes. He looked terrible. His arms were covered in pitch, also his pants and T-shirt. He had small scratches all over his arms. He came in and didn't even say hi, hello or anything . . . He then proceeded to tell me about his Cisco Grove experience. His hands shook and his voice was subdued . . . he was on the verge of crying [and] was so badly shaken that he took a week off from work.

Suffering from chest pains and breathing difficulty, Smythe went to his doctor for a check-up, but nothing wrong was found. For a year and a half after the incident, Smythe suffered from terrible nightmares about the robots. Beginning in 1969, he began to experience occasional loud buzzing noises in his ears, which he believed were associated with the beings he had encountered. On one occasion, while camping with his wife, the buzzing sounds preceded a sighting of a 'big light' which moved swiftly over a ridge, followed a few seconds later by another, smaller light.

INVESTIGATIONS

Donald Smythe made out a report of the incident which initially was sent to Victor Killick, a retired professor of astronomy, who forwarded it to officials at Mather Air Force Base. On 25 September 1964, Smythe was interviewed at his home by a Captain McCloud and Sergeant R. Barnes of McClellan Air Force Base. Smythe offered his full co-operation and handed over one of his arrowheads for analysis. Nothing unusual was found. He also provided the Air Force investigators with a map of the area. Before leaving, the investigators suggested the following possible explanations for the incident: (a) a group of Japanese; (b) a group of teenagers playing a prank; or (c) a group of Air Force trainees. Suggestions of drinking and hallucination were also voiced by McCloud.

Two or three weekends later, Smythe returned to the scene of the incident, together with his brother, a friend, and one of the hunters. Apart from a few of his cigarette butts, none of the materials he had left there could be seen.

Civilian investigators were impressed by Smythe. 'Having just reviewed the case files on this fascinating and unusual encounter,' wrote Paul Cerny in 1995, 'there is absolutely no doubt in my mind that this incident is factual and authentic. I have spent considerable time and many visits with the main witness, and along with the testimony of the other witnesses, I can rule out any possibility of a hoax.'[4]

AERIAL ENCOUNTER OVER JAPAN
Captain Yoshiharu Inaba was flying a TOA Airlines Convair 240 from

Osaka to Hiroshima, Japan, on 18 March 1965, at 19.06, when a 'mysterious, elliptical, luminous object' appeared, just after the plane passed Himeji.

'I was flying at the time at an altitude of about 2,000 metres,' reported Captain Inaba. 'The object followed for a while, and then stopped for about three minutes, and then followed along my left wing across the Inland Sea for a distance of about 90 kilometres until we reached Matsuyama on Shikoku Island. It then disappeared.'

Initially fearing a collision, Inaba made a 60-degree turn to the right, but it was at that point that the object made an abrupt turn and positioned itself alongside the port wing. It emitted a greenish-coloured light and affected the automatic direction finder (ADF) as well as the radio. As the co-pilot, Tetsu Majima, tried unsuccessfully to contact the Matsuyama tower to report the observation, he heard frantic calls from the pilot of a Tokyo Airlines Piper Apache, Joji Negishi, who said he was being chased by a mysterious luminous object while flying along the northern edge of Matsuyama City.

Captain Inaba, a veteran pilot with over 8,600 flying hours, said that it was the first time he had encountered such an object. Weather conditions were good that evening, with a full moon. A test carried out two nights later by TOA Airlines ruled out the possibility that the pilots had seen the reflection of light from their plane.

At around 19.00 on the same evening, three workers from the Chokoku Electric Power Company in Fuchu, near Hiroshima, reported sighting a strange object over Yuki Town. 'It was shaped like a triangle whose top radiated a brilliant light,' said one of the witnesses.

According to a message relayed from the *New York Times* office in Tokyo to the TOA Airlines office, a group of 'flying saucer experts' from the US Defense Department, the Federal Aviation Administration and the Palomar Observatory was being sent to Japan to interview Inaba and Negishi. 'The American mission is believed to be interested in the case,' it was reported, 'because there have been several mysterious aviation accidents and flying saucers might have been involved.'[5][6][7]

ALIEN BASES RUMOURED IN ARGENTINA

In 1965, rumours of an alien base in Argentina flourished following numerous sightings of peculiar craft in the Valley of Loretani and a nearby ravine, located 60 kilometres southwest of Córdoba.

The first incident occurred at 20.00 on 15 July, when Rubén Busquets, the owner of a tree plantation, together with his wife Diana Loretani and daughter Marcela, noticed an unusual object as they were returning home from the Hotel de la Entrada. As Señor Busquets described it:

The object was big and very luminous, of a bluish colour, but varying from moment to moment, sometimes to orangeish-red. Its shape was that of a truncated cone, though we were unable to see clearly where the upper part of it ended owing to the beam of light directed upwards out of the object. The lower part was circular and convex. Taking the nearby hills as my gauge, I reckon that it was some 10 to 15 metres in diameter. It was motionless and made no sound. At one moment it shone a beam of light on to us . . .

After a long pause, it dropped down vertically, and the terrain hid it from us, but we could still see the glow from it. Then finally the glow went out. We went on up to the house, but before we reached it we met one of our *peones* who was lying on the ground, having been thrown by his horse . . . this peasant had seen it too, and his horse had been so terrified by it that it had thrown him.

Thereafter, sightings of UFOs flying over the valley and dropping down into the ravine became a regular occurrence for the Busquets family. A circular mark, some seven metres in diameter, was discovered at a spot where one of the craft seemed to have landed. On another occasion, Señora Busquets was able clearly to see windows on one of the craft. The family was not alone: by this time, labourers on the plantation had become quite accustomed to seeing the phenomena.

One day, as Señor Busquets sat in his car beneath some power-lines, watching one of the UFOs, he received a powerful electric shock. Concerned that the UFO might bring down the power-lines, he moved his car away from the area. The UFO immediately turned off its light, leaving only a reddish glow that eventually faded out. On 24 July 1965, the Busquets invited a group of people to observe the strange craft. All of them saw a 'shining cigar with a black band around it' which remained stationary for about 10 seconds before plunging down into the ravine. The witnesses included two lawyers, Dr Felix Cochero and Dr Fortunato Columba. On 5 September 1965, Captain Omar Pagani, who then headed the Argentine Navy's UFO investigations, observed a UFO through a telescope as it approached and dazzled witnesses (including two meteorologists), blocking out the entire field of view of the telescope as it did so.[8]

MISSING TIME IN MADAGASCAR
The witness in the following case, a certain Monsieur Wolf, was born in Germany but took up French nationality after serving in the French Foreign Legion. The remarkable story which follows is taken from an interview with Wolf by H. Julien, an investigator for the French journal *Lumières Dans La Nuit.*

'It was in May 1967,' began Wolf. 'I was in Madagascar, and serving in the Foreign Legion. We had just been on a reconnaissance exercise in bush terrain. We were in hourly radio contact with Central Head-quarters. We had halted at noon in a clearing about 100 metres wide and began to eat. The weather was fine.' Then it happened.

Suddenly we observed the arrival and descent of a machine of indefinable colouring. I am colour-blind myself, but I can state definitely that the thing shone very brightly, and was of the colour of a new coin shining in the sun-light. Around it there was an intense, dazzling glow. It came down with the motion of a falling leaf, and you would have said that there must have been some sort of accident – it was like a shining egg on the end of a piece of string. It came down very rapidly. And we felt a powerful ground shock when it landed. And then a piercing whistling sound. By now the craft was no longer luminous.

After that the whole thing was unbelievable. There were twenty-three of us Legionnaires, with one officer and four non-commissioned officers. And we were all paralysed. All of us saw the machine land and take off again, *but none of us perceived the lapse of time*. Let me explain: when the machine had departed, we all recovered the use of our limbs. We were all in exactly the same positions and the same places as we were when it had landed. But when we checked up on the time, we realized that it was now 3.15 p.m. *Two and three-quarter hours had passed without our perceiving it*. We had missed three radio rendezvous with Headquarters. Our officer got a fearful ticking off for it, for he was incapable of giving any effective explanation.

THE MACHINE

The machine was smooth, with no visible doors or windows. It appeared to bear no markings. No antenna. It was like a smooth egg, twice as high as it was wide. I can't say what its exact size was, given the amount of vegetation in the clearing. But, comparing it with the height of the trees as it was taking off, you could reckon that it was between seven and eight metres high. In its base it had several openings, of which we were able to get a good view as it took off. There were flames coming from them – not normal flames . . . One could have taken them for flames, but they must surely have been some-thing else. Something like what you see when you use a welding machine to cut metal. Each of the openings emitted a 'flame'; the whole thing produc-ing one big thick short flame about one-twentieth of the length of the machine itself.

It had legs. I did not see them, because of the vegetation, but on the ground there were three marks, set in a triangle, where it had stood. In the middle of the triangle there was a charred crater three metres deep – a crater which widened out towards it base. At the bottom of the crater there were some crystals of all colours, like bits of broken glass. The bottom of the crater was full of them, especially in the corners. It was like a vitrified ring.

When the craft departed, it rose up slowly until it was above the trees. Then it vanished at a fantastic speed, as though sucked up into the sky. It left a sort of trail behind it.

SWORN TO SECRECY

> Headquarters ordered us not to approach the landing site and not to discuss the matter among ourselves. Some specialists arrived by plane from Paris to interrogate us. We were made to swear on oath that we would keep it secret. We were visited by the doctors and we were made to undergo tests. For two days after the event we all had violent headaches, with a buzzing in the ears and a powerful beating in the area of the temples. We were not told the results of the tests made on us.[9]

Not the least surprising aspect of this important case is the missing time experienced by the Legionnaires, with not one of them being aware of the fact until afterwards; a classic side-effect reported in abduction cases. But were any of the men actually abducted into the craft? Was there any interaction at all? We may never know. The 'buzzing in the ears' brings to mind the case of Donald Smythe, described earlier, who suffered similarly following his harrowing experience in 1964.

This case is also important in that it lends weight in several respects to the theories of UFO propulsion propounded by the British aeronautical engineer Leonard Cramp, in his remarkable book, *Piece for a Jig-Saw*, published a year earlier (and now republished in the United States).[10]

MORE MYSTERIOUS SUBMARINES

At about 18.15 on 30 July 1967, the Argentine steamer *Naviero* was some 120 miles off the coast of Brazil, opposite Cape Santa Marte Grande (Lat. 28 48 S, Long. 46 43 W) in the State of Santa Catarina. As the officers and crew were taking their evening meal, Captain Julián Lucas Ardanza was notified on the intercom by one of the officers that something strange was near the ship.

Arriving on deck, Captain Ardanza could see a shining object in the sea, no more than 50 feet off the starboard side. Cigar-shaped, with an estimated length of 105 to 110 feet, it emitted a powerful blue and white glow. No noise could be heard and no wake could be seen in the water, neither was there any sign of a periscope, railing, conning tower or any other superstructure, such as would be expected of conventional submarines.

The mystery craft paced *Naviero* for 15 minutes at a speed estimated by Captain Ardanza at 25 knots, as against the 17 knots of his own vessel. Suddenly, the unidentified submarine dived, passed right underneath *Naviero* and vanished rapidly in the depths, glowing brightly as it went.

Because the *Naviero* was carrying explosives, and in case the crew might panic about being 'pursued' on account of this cargo, Captain Ardanza and his officers assembled the crew to tell them what had happened. In subsequent interviews with the Argentine press, Ardanza said that he had seen nothing like it during his 20 years at sea. Chief Officer Carlos Lasca described the object as 'a submergible UFO with its own illumination'. The Argentine maritime authorities officially classified it as an 'Unidentified Submarine Object'.[11]

For three weeks in February 1960, it was reported that the Argentine Navy, aided by United States experts, depth-bombed and called for the surrender of two mysterious submarines. The submarines lay in the bottom of Golfo Nuevo, a bay north of Rawson, separated from the ocean by a narrow entrance. They were chased by the Argentines all over the bay and each time they were trapped they managed to slip away mysteriously. Finally, Argentine Navy Secretary Gaston Clemente told newsmen that the patrol would be called off. Intrigued by these events, contactee George Adamski later asked his space friends for an explanation. 'The answer,' he asserted, 'was that they were spacecraft.'

> They were studying the bottom of the ocean to learn about conditions on our planet that are not yet indicated on dry land areas. A number of such craft are making a thorough study of underwater lands and naval ships of many nations have encountered them. For the most part, confidential, official reports of such encounters have been described as 'fantastic'. But here again, although our friends would like to surface and make themselves, and what they are doing, known, our fears keep us in a state of hostility that prevents their doing so. Instead, they pass their findings on to others of their people who are working among our scientists and in other important places throughout the world. In turn, and in time, this information will be given to the people as findings from IGY [International Geophysical Year] research.[12]

A TRAGIC ENCOUNTER

It was 16.00 on 13 August 1967. Inácio de Souza, manager of a large *fazenda* (plantation) between Crixas and Pilar de Goias, about 100 miles northwest of Brasília, was returning to the house with his wife, Luiza, when they saw three strange-looking figures playing about like children on the fazenda landing strip. At first, Inácio thought the people were naked, but his wife had the impression that they were dressed in skintight pale-yellow clothes. They appeared to have no hair.

On spotting Inácio and his wife, the strangers began running towards them. It was then that Inácio saw a peculiar 'aircraft', shaped like an inverted wash-basin, at the end of the landing strip. It appeared to be over 100 feet in width and was touching, or almost touching, the ground. Frightened, Inácio sent his wife into the house, reached for his 0.44 calibre carbine, took aim at the nearest figure and fired. Almost instantly, the 'aircraft' emitted a beam of green light which struck Inácio on the head and shoulder. He fell to the ground unconscious. As his wife came rushing out of the house to help him, the three strangers ran back to their 'aircraft', which took off vertically, making a sound like the swarming of bees.

The owner of the fazenda, a wealthy and well-known man (who asked that his name not be revealed), flew to the property three days later, having been informed about the incident in São Paulo, where he lived. He learned that for the previous two days, Inácio had complained of numbness and tingling of the body, as well as headaches and nausea. On the third day, in addition to these symptoms, Inácio began to suffer from continuous tremors of the hands and head. The owner took the sick man to a doctor in Goiâna, about 120 miles to the south. Burn marks, in the form of a perfect circle 15 centimetres in diameter, were found on de Souza's torso and head. Blood tests revealed that he was suffering from 'malignant alterations of the blood', i.e. leukaemia. The doctor warned the owner that the patient had about 60 days to live.

Inácio's weight began to decrease. He suffered great pain, and yellow-ish-white spots, the size of a fingernail, appeared all over his body, just underneath the skin. He died on 11 October 1967, aged 41. In accordance with her husband's wishes, Luiza burned his bed, mattress, bedclothes and clothes, as he was afraid that whatever had caused his terminal illness might be transmitted to his family.

'So far as I am concerned it was just another case of cancer,' stated the sceptical doctor at Goiâna, in response to questions from investigators. 'I advised the *fazendeiro* to "forget" what his employee said had happened, since he (the fazendeiro) had not been an eye-witness.' Unable to accept that Inácio's condition was the result of an encounter with space beings, the doctor attributed the story to a hallucination brought on by leukaemia.[13][14]

There seems little likelihood that this disturbing encounter was in fact the product of a hallucination. A hoax seems equally far-fetched. Inácio de Souza was described as a simple, honest and trustworthy man, with no motive for such a hoax. Would he and his wife have compromised his life and their livelihood by making up such a story?

The figures initially seen 'playing about like children' bring to mind a number of similar cases, such as that of José Higgins, whose 1947 encounter with aliens in Brazil is described in Chapter 3. Had Inácio not taken the drastic action of shooting at one of the figures, perhaps the out-come of this encounter might not have been so tragic.

A POLICEMAN ABDUCTED

One of the most fascinating and, for me, compelling alien contact reports is that of Nebraska police patrolman Herbert Schirmer, who encountered a landed craft and its crew in Ashland, Nebraska, in the small hours of 3

December 1967. Initially, Schirmer recalled having seen a 'flying saucer' on the edge of the highway at 02.30, reporting it as such in the police log-book. Further details emerged six months later, during time-regression hypnosis with a professional hypnotist, Loring Williams.

As Schirmer got out of his patrol car to investigate, quasi-human beings approached, paralysed him temporarily with a 'greenish gas', then took him on board the football-shaped craft, which rested on tripod land-ing legs. A small exterior catwalk surrounded the centre of the craft. He found himself in a room about 26 feet by 20 feet and about six feet high, with portholes, 'computer-like' screens, and two triangular-backed chairs. The four crew members were nearly five feet tall, dressed in tight-fitting silver-grey uniforms that enclosed their long, thin heads, with a short aerial-type antenna coming from one side, belts with a flashlight-like 'gas-gun' in a holster, and gloves and boots. They had an emblem on their large chests that looked similar to that of the ancient winged serpent (akin to the caduceus reported by Dan Fry – see p. 68). 'The skin on their faces was sort of grey-white, a pasty dough colour,' Schirmer told the journalist Warren Smith. They had thin, slanting eyebrows above large, oriental-type eyes. 'These eyes were not actually like oriental eyes,' he explained, 'they looked more like cat's eyes.' The nose was longer, flatter and more prominent than those of humans. 'There wasn't much lip to them. They were more of a slit in the face . . .'[15]

One, who appeared to be the crew leader, communicated to Schirmer that they were extracting electricity from a nearby power-line, an opera-tion he observed through a porthole. Later, the electricity taken was replaced. 'When they land, an invisible [electromagnetic] force field is thrown around the ship in a circular pattern [as] a defence mechanism,' said Schirmer under hypnosis. 'In some way which I do not understand, they draw a type of power from water. This is why we see them over rivers, lakes and large bodies of water.'[16]

The craft, just over 100 feet in diameter, was propelled by 'reversible electro-magnetic energy', creating inertia-less, gravity-free flight. A crystal-like rotor in the centre of the craft was linked to two large columns. 'He said those were the reactors,' Schirmer recounted under hypnosis. 'Reversing magnetic and electrical energy allows them to con-trol matter and overcome the forces of gravity.' The craft supposedly was made from pure magnesium.

Schirmer was shown a small, saucer-shaped device that could be launched from the larger craft. As also described identically by George Adamski, the device could transmit real-time audiovisual data. 'They send the little saucer down to check out an area before they bring the big ship back in,' Schirmer told Smith. 'The pictures from the little baby saucer show up on the vision screen inside the ship. The best way I can describe [it] is to compare it with a baby moon hub–cap, which the kids buy to dress up their cars.'

Allegedly, the aliens came from a nearby galaxy, but had bases in our solar system, including on Earth (see Chapter 19). 'I'm not certain they are from the places they said,' explained Schirmer. 'This might be something to throw us off guard.' The crew leader said that, to protect themselves, the aliens deliberately confused us; that contacts were selected on a random basis for that very reason. 'He said they left things to pure chance. If there isn't any rhyme or reason to something, it is bound to puzzle the governments of the world and UFO investigators . . . I was very impressed by their security.'

After about 20 minutes, Schirmer was escorted off the craft, and he recalls watching it take off. Accompanied by a high–pitched whine, a red-dish–orange glow appeared on the bottom of the ship, the tripod legs retracted and the craft shot into the sky.

'I had a bad headache that morning,' Schirmer told Warren Smith. 'There was a weird sort of buzzing sound in my head. If I started to fall off to sleep, the buzzing noise got louder. I also had a red welt running down the nerve cord on my neck, right below my left ear.' For a long time after the encounter, Schirmer suffered from severe headaches.

Schirmer's case was brought before the University of Colorado UFO study group. He was flown to Colorado for standard psychological tests, including a hypnosis session, attended by Dr Edward Condon himself, head of the study group. Staff members told Schirmer that a negative report on his experience would be given. 'They said their work was being checked by the Air Force and other government agencies before it was published. I was told that the whole Condon Committee was a cover-up designed to get the Air Force off the hook following so many sightings in 1966.' Schirmer continues:

> Some of those guys felt that the Central Intelligence Agency was messing around. One staff member had done a lot of the field investigations. He was picked up on a narcotics charge. The police went right to his house and directly to a cache of marijuana. I know police work. You have to have a tip-off to know where to find something like that. Several people believed that this fellow was probably set up, framed, because of something he had found out during his field work.[17]

MORE LONG-HAIRED INTERLOPERS
Another landing of humanoids at a fazenda in Brazil took place in the early part of January 1968, at the Lagôa Negra Fazenda, near Lagôa dos Patos, in the Brazilian state of Rio Grande do Sul. This time, there were five witnesses: the owner of the fazenda, his wife, son and daughter, and the fazenda manager.

The incident occurred sometime between 20.00 and 22.00, on a clear night. Initially, a disc-shaped object, about 10 metres in diameter and three metres high, was seen 'floating' two metres above the ground, beside a grove of eucalyptus trees, observed at a distance of less than 400 metres from inside the fazenda house. With a cupola on top and a sort of protuberance underneath, the craft had a metallic gleam and emitted a powerful, cold, reddish light that penetrated through the chinks in the windows and the doors of the house, causing the witnesses' eyes to burn.

The first two figures to appear beside the disc seemed about two metres tall, dressed in white overalls with a broad white band at the waist, and high, dark-coloured collars. The faces of the beings were described as full, white, with long hair hanging down to their shoulders. The beings also had large bare feet, long hands, and a rather rigid manner of walking, without bending their legs.

Afterwards, three more beings appeared. These were no more than 1.4 metres in height, wearing chestnut-brown overalls with a similar-coloured band around the waist, hair as long as the previous two, and shod in small boots. They walked rapidly, though not leaving the area beneath the disc. Momentarily, the two taller beings moved away from the disc and went towards a wire fence, getting as far as a ditch that ran beside it. After following this until they were at a point halfway between the disc and the gate, they retraced their steps. They then walked away from the disc a second time, went up to the gate via a different route, halted in front of a small wooden bridge over the ditch, and returned once again to the disc. For a third time, the taller beings left the area of the disc, following the route they had first taken. This time they crossed the bridge, came to the gate and opened it, entered, closed the gate, then headed towards the house.

By this time, the fazenda-owner and his manager had left the house and taken up a position lying under two palm trees, enabling them to observe the disc and its occupants without themselves being seen. The wife and children stayed inside the house. Frightened by the red light that had penetrated the house, the son lay in bed and covered himself with the bedclothes. The five household dogs, normally fierce towards strangers, at no time seemed to be disturbed, even when the interlopers approached the house. The manager, who was armed, decided to challenge the intruders, but the owner ordered him to keep quiet.

'When the beings had got to about 60 metres from the house,' reported investigator Jader Pereira, 'the daughter was able to see their features clearly because the whole area around about was completely lit up by the light from the disc, and she exclaimed: "Mother, they look like Saints!"' This exclamation frightened the mother, who decided to call her husband to come back inside the house. As she opened the door and called

Fig. 18. The long-haired interlopers who landed in full view of five witnesses at a *fazenda* near Lagôa dos Patos, in the Brazilian state of Rio Grande do Sul, in January 1968. (*Terence Collins/FSR Publications*)

him, the two beings halted. 'They did this several times,' said Pereira, 'until finally they turned round and went back to the disc along the same route by which they had come. Then all five of them entered the object, which rose up vertically, apparently with a slight rotary movement.' The entire incident lasted for about twenty minutes.

The following day, the witnesses discovered two kinds of footprints: one large, as though from bare feet, with very long toes and angular heels; the other prints small, with a smooth heel, while the forepart of the sole showed 'a sort of five-pointed star' in the centre. Unfortunately, no plaster-casts were taken.[18]

The humanoids described in this case compare interestingly with earlier descriptions of similar beings, such as George Adamski's 'Orthon', with his long hair and 'chocolate-brown' overalls. According to the investigators: 'The family are people who are held in high regard locally and in the Municipality of Viamão. None of them had ever hitherto had any interest in the subject of flying saucers.'[19]

RUMOURS OF ALIEN BASES IN PERU
During the summer of 1968, the rural populace in the Peruvian Andes became so alarmed by numerous 'flying saucer apparitions' in their area

that the district authorities of Huaraz dispatched a commission, accompanied by armed police, to investigate reports that the 'saucers' were using a particular area close to Lake Yanacocha, in central Peru, as a base. Many farmers claimed to have seen luminous objects 'shooting downwards at great speed', as though to land, on an extensive plain lying between Lake Yanacocha and Lake Pumacocha. Twice weekly, it was alleged, the 'saucers' appeared in broad daylight, then vanished swiftly, leaving landing traces on the ground.[20]

According to UFO investigator and engineer Antonio Ponce de Léon, a 'saucerdrome' (*platillodromo*), had been discovered at Chumo, in the Sicuani area of southern Peru. De Léon reported that he had learned about the 'saucerdrome' from the local Indians.[21] Rumours proliferated in 1968 that an alien base existed in the depths of Lake Titicaca. Many local witnesses reported sightings of 'saucers' heading towards the lake and vanishing therein.[22]

In 1993, many inhabitants of Huaraz, capital of the department of Ancash, north of Lima, claimed that flying saucers had bases in the highest peaks of the Cordillera Blanca, a stretch of the Andes. The highest such peak, El Huascarán, is 7,000 metres; others vary from 5,000 to 6,000 metres. According to witnesses, flying saucers were regularly seen to emerge from the lakes in this area. On 28 October 1993, for example, numerous farmers, shepherds and others reported sighting a UFO emerge from the lake on Mount Carhuac. According to Osterling Obregón, a teacher, Mount Carhuac was lit up with different colours when the UFO rose up from the 5,000-metre peak.

'For four minutes I watched the UFO rise slowly above the Cordillera Blanca,' stated Obregón. By the time the teacher returned with his camera, the object had disappeared. Ten minutes later, what appeared to be the same object was seen by another teacher, Juan Gómez. The object emitted brilliant lights and described a series of aerobatic manoeuvres.[23]

CLAMP-DOWN

It was Sunday 9 September 1968. Professor Wilton Ribeiro was walking on the beach at Itaipu, near Niterói, to the east of Rio de Janeiro, when a strange object, emitting beams of orange light, suddenly descended silently about 200 metres away. The object came closer and hovered about 10 metres above the sea, spinning on its vertical axis while making a humming sound. Other witnesses included João Abud, juridical assessor of the Secretariat of Justice of the State of Rio de Janeiro, and Professor Sohail Saud, a teacher of business studies. Professor Saud said that the object, which he described as a large disc, made a number of low-level passes over the beach before landing briefly. Occupants could be seen inside the craft, he claimed, though he was unable to describe them

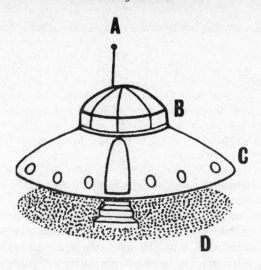

Fig. 19. The landed craft near Macédo, State of Sâo Paulo, Brazil: A. Antenna with red light. B. Dome, in segments. C. Skirt, seemed to be spinning clockwise. D. Patch of violet light beneath. (*Terence Collins/FSR Publications*)

in detail, other than that they were wearing 'helmets'. Several witnesses were reported as having suffered from shock.[24]

On 21 November 1968, a young Brazilian woman claimed to have witnessed a landed craft and its occupants at a bus stop near Macédo, in the State of Sâo Paulo. The bus had halted while the driver took a customary break. On waste ground about 40 metres away, the witness noticed a shining, metallic object, standing or hovering close to the ground. It was of a similar shape to that reported by George Adamski and others, with several exceptions: the cupola appeared to be divided into four segments, with an antenna at the top; a row of circular, ever-changing lights surrounded the rim or 'skirt'; and there was a set of three steps beneath a large (stationary) entrance in the rim (see Fig. 19). Standing in front of the craft were three 'men', about two metres in height, wearing skintight, shining black suits, leaving only the faces bare. One of the entities held 'a sort of tube' under one arm, about 60 centimetres in length and seven centimetres in diameter, surrounded by a spiral coil. Two thin protuberances came out of one end of the tube.

Between the witness and the entities, about 20 metres from the bus and with their backs towards her, a crowd of about twenty people, grouped behind three armed policemen, confronted the entities. On the side of the road opposite the bus were parked two police radio-patrol cars. As the two parties continued facing each other, a brilliant, silver-coloured beam of light suddenly shot out from the tube. As investigator Nigel Rimes reported:

> The beam was directed at the party of Brazilian police and bystanders, the front ranks of whom (including the policemen) immediately ceased all movement and were 'paralysed'. She noticed however that a number of the others who were not in the forefront were also affected, and she saw several fall as though in a faint. She also noticed that the entity did not swing the tube itself, but swung his whole body round, still holding the tube in position under his arm.

The entities walked calmly and slowly back to their craft, which then took off and climbed away rapidly.

Though no corroboration of this extraordinary report came forth, to my knowledge, both the investigators – one of whom was Willi Wirz, managing director of the *Brazil Herald* – were impressed by the witness. 'Our young lady seemed to be an entirely sincere and truthful person,' reported Nigel Rimes. 'She made it clear that she desired no publicity and that she had only come forward with this information because she felt that it was her duty to do so.

'Our conclusion so far is that this case certainly seems to be genuine, and that the military authorities have clamped down on other witnesses . . .'[25]

NOTES

1 Bloecher, Ted, 'UFO Landing and Repair by Crew: Part I', *Flying Saucer Review*, vol. 20, no. 2, March–April 1974, pp. 21–6.

2 Bloecher, Ted, 'UFO Landing and Repair by Crew: Part II', *Flying Saucer Review*, vol. 20, no. 3 (published December 1974), pp. 24–7.

3 Schwarz, Berthold Eric, MD, 'New Berlin UFO Landing and Repair by Crew', *Flying Saucer Review*, vol. 21, nos. 3/4, 1975, pp. 22–8.

4 Bloecher, Ted, and Cerny, Paul, 'The Cisco Grove Bow and Arrow Case of 1964', *International UFO Reporter*, vol. 20, no. 5, winter 1995, pp. 16–22, 32.

5 *Japan Times*, Tokyo, 21 March 1965.

6 *Mainichi Daily News*, Tokyo, 22 March 1965.

7 *Mainichi Daily News*, Tokyo, 23 March 1965.

8 Creighton, Gordon, 'Further Reports of UFO Bases', *Flying Saucer Review*, vol. 15, no. 2, March–April 1969, pp. 15–16. The original account appeared in *Asi*, Buenos Aires, 7 September 1965.

9 Julien, H., 'A 1967 Landing in Madagascar', *Flying Saucer Review*, vol. 23, no. 1, January–February 1977, pp. 29–30, translated by Gordon Creighton.

The original article appeared in *Lumières Dans La Nuit*, no. 160, December 1976.

10 Cramp, Leonard, G., *UFOs and Anti-Gravity: Piece for a Jig-Saw*, Adventures Unlimited Press, One Adventure Place, Kempton, Illinois 60646, 1996.

11 Galíndez, Oscar A., 'Crew of Argentine Ship See *Submarine* UFO', *Flying Saucer Review*, vol. 14, no. 2, March–April 1968, p. 22.

12 Adamski, George, *Flying Saucers Farewell*, Abelard-Schuman, London, 1961, pp. 44–5.

13 Bowen, Charles, 'A Fatal Encounter', *Flying Saucer Review*, vol. 15, no. 2, March–April 1969, pp. 13–14.

14 Brazil in Throes of Big Flap', *The APRO Bulletin*, March–April 1969, pp. 1, 5.

15 Norman, Eric, *Gods, Demons and UFOs*, Lancer Books, New York, 1970, pp. 169–93.

16 Blum, Ralph, with Blum, Judy, *Beyond Earth: Man's Contact with UFOs*, Corgi Books, London, 1974, p. 117.

17 Norman, op. cit.

18 Pereira, Jader U., 'The Remarkable Landing at Lagôa Negra', translated by Gordon Creighton, *Flying Saucer Review Case Histories*, supplement no. 5, 1971, pp. 3–4.

19 Ibid., p. 3.

20 Creighton, op. cit., p. 16, translated from an EFE report, Urcos, Peru, 31 July 1968, as published in *Sur*, Malaga, Spain, 1 August 1968.

21 Ibid., translated from *Expreso*, Lima, Peru, 27/29 September 1968.

22 Ibid., pp. 16, 19, translated from an *Agence France-Presse* despatch, Buenos Aires, 11 November 1968.

23 'UFO Base in Highest Mountains in Peru', EFE, Lima, 31 October 1993.

24 'Brazil: Landing on a Beach near Rio de Janeiro', *Flying Saucer Review*, vol. 17, no. 2, March–April 1971, p. 30, translated by Gordon Creighton from *O Dia*, Rio de Janeiro, 10, 11 September 1968.

25 Rimes, Nigel, 'Baleia Entities Seen Again?', *Flying Saucer Review*, vol. 15, no. 2, March–April 1969, pp. 6–8.

Chapter 14

Contrasting Encounters

Carroll Wayne Watts, a 39-year-old cotton farmer living in Loco, near Wellington, Texas, saw his first 'spaceship' on the morning of 8 February 1967. Together with another sighting six weeks later, it was to be a prelude to one of the most bizarre, fascinating encounters ever reported; an encounter that led to nationwide press coverage, though with a sinister sequel of events for Watts and his family. Perhaps owing to the seemingly ludicrous nature of the encounter, and Watts's subsequent 'confession' to having perpetrated a hoax, it is seldom cited in the literature. Here follows the story, based largely on an unpublished report he wrote together with his wife, Rosemary.

At about 11.00, Watts saw what he first took to be a fast-moving aircraft, flying from the northwest on a southeast heading, at an estimated altitude of about 1,300 feet. 'I thought it to be a jet at first until I noticed that it didn't have any wings and didn't leave exhaust streams,' said Watts. 'It looked like a long cylinder shape at that height.'

On 21 March, at about 15.00, Watts was measuring cotton land on one of his farms when he noticed another peculiar aerial object. This time it was much closer. 'It was about 200 feet off the ground, traveling about 50 miles per hour,' he recalled.

> The front of it was raised at about a 30-degree angle. When it passed me at the closest point of about 200 yards, I could see an opening in the front which looked to be a window. There was also a window on the side which was oblong. The ship looked to be shaped like a cylinder about 100 feet long and eight to ten feet across. The color was a light dull gray which did not reflect the sunlight. It traveled about three-fourths of a mile northwest and turned northeast and set about a 70-degree angle and completely vanished in about 30 seconds. It didn't make any noise nor did it leave a vapor trail.

A CLOSER ENCOUNTER

On the evening of 31 March, Watts's wife had gone with their children to a church meeting. At around 22.30, returning home after visiting his parents, Watts spotted a light in a field by an old house on his farmland. 'There was some equipment stored in the house so I thought that I should go up and take a look around,' he said. 'The light at first looked

like a car with four headlights on bright. As I approached, I noticed the
lights were much brighter.'

> I stopped within 200 feet of the vehicle when I noticed it was something other
> than a car. It was about 100 feet long, eight to ten feet high and ten feet across.
> It was shaped something like a .38 bullet, long and round on one end and the
> front end came to a point where the light was and sloped a little downward.
> It was between a dull gray and aluminum in color. It didn't reflect light.

Watts pulled up about 50 feet from the craft and got out of his truck to
take a closer look. He walked around behind it, then for about 20 feet
along its left side. Here he noticed a slight ridge where he assumed there
might be an opening.[1] In an interview with reporter Tony Kimery, Watts
described what followed:

> I thought it must be some new aircraft the Air Force had developed and that
> it must have made an emergency landing, or something. I also thought that
> there might be injured crewmen aboard, and I wondered how to find out,
> since there weren't any windows or doors. I scrounged around and found an
> old rotting fence post and pulled it out of the mud, and started banging and
> sounding out the machine by hitting it with the post.[2]

Suddenly, a door he had not seen slid open. 'The door opened from
the top and came down to make steps,' Watts explained.[3]

> Inside, there were no crew or anything, just machinery and all kinds of meters
> and dials, lit up by this strange bluish light. Then there was a loud crackling,
> like the beginning of a Victrola record, and then a voice, sounding like it came
> from a machine or was recorded, began talking to me. It knew my name and
> everything and it told me that it wanted to give me a physical examination.[4]

'I asked them what was the reason for the physical and they said [that]
a man had to pass a rigid physical before he could stand the flight. They
told me to stand in front of this machine if I would take the physical. The
machine reached almost from the floor to the ceiling. They also told me
that they had a machine that could go within 300 yards of a building or
house and could tell how many were in the house and also what age [the
occupants] were.

'After they told me this, they asked me again if I wanted to step inside,
take the physical, and experience some of these things. I told them "no",
that I didn't think that I wanted to. The only other things that I could see
in this compartment were some gauges on the ceiling, which I couldn't
read, and some maps on the wall.'

> The maps were about three feet square but only about 12 inches from the
> floor. The complete map was a light gray color. It looked like they were large
> scale maps of land. There were seven or eight crooked lines running diagon-
> ally that looked like rivers and also moon-shaped markings that could have

represented mountains. It had latitude and longitude lines on it.

'After I was asked the third time to take a physical, I got a little jumpy and decided that I had better leave,' continued Watts. 'After I left, it raised about three feet and turned south down in a pasture. There was no noise at all coming from it except for the time when the door opened. When it was standing still, it never did touch the ground. When it took off, its lights changed from a fluorescent light to an amber or soft red light. It had only this one large light on the front . . . about 20 inches across.'[5]

Without bothering to turn his truck around in the thick mud, Watts ran back the half-mile to his home. After hearing her husband rather incoherently explaining what had just happened to him, Rosemary Watts called the police in Wellington, 11 miles to the northeast of their farm. Chief of Police Donald Nunnelly, a relative of Watts, together with Collingsworth County Sheriff John Rainey, arrived and accompanied Watts to the scene of the incident. Nothing could be found, except Watts's truck, stuck in the mud.

Sheriff Rainey informed the Air Force, and a lieutenant interviewed Watts. The lieutenant was unable to reassure Watts or offer any helpful advice in the event that another, similar incident should occur.[6]

ON BOARD

On 11 April 1967, at about 20.30, Watts saw a light, similar to the one he had seen on 31 March, and in about the same location. An hour later, he left his house to see if he could find the 'spaceship' again. 'Since I couldn't get any information from the Air Force or any other source as to what the other ship was,' he explained, 'I wanted to find out more about it myself.' After driving around slowly for about a quarter of an hour in a three-mile radius of where he had seen the light, he turned into a road leading to an old vacant house to turn around.

Just as he started to back up, Watts noticed that the area around him was getting lighter. At first he paid little heed because there was a lot of lightning that night. The light became brighter; simultaneously, the engine of his pickup 'made three or four vibrations and died'. He tried unsuccessfully to start the engine. 'About that time, I looked in my rear-view mirror and saw a light reflected in it. I stepped out of the pick-up and turned and faced the back of the pick-up.' There he saw an object, different from the others he had seen.

About 10 feet from the rear of the pick-up was an object shaped similar to an egg. However, the bottom wasn't as round as the top. It was about 30 feet in length and about 15 feet in height in the center. (At times, these smaller crafts jerk about 10 feet from side to side. They have two antenna-type rods extending down about three feet. When they landed and picked me up, these rods were about six inches from the ground. This made the ship about three and one-half feet from the ground.)

There were three lights attached to it; one on each end and one in the center close to the top. These lights rotated from red to yellow to green, but each time that they changed there was only one on red, one on yellow, and one on green, and then they would start over with each one a different color at the same time.

This small craft had a clear bubble on the top, about three feet high and four feet in diameter. The door opened down from the top and formed steps. There were two men standing on the steps and two men standing in the doorway.

One of the men said, 'Let's go and let you take your physical examination.' I hesitated a minute before I answered. I asked them if they would harm me in any way, and they said, 'No.' I also asked them if they would bring me back, and they said that they would bring me directly back. Since I had the choice of either running across the field or getting aboard with them, I chose to go with them.

I had to stoop down a little to get through the doorway. I'm five feet eleven inches, and the doorway [was] about five and one-half feet in height and about 30 inches wide. The room into which I walked was about 12 by 10 feet. There was a door in each end. I didn't notice a handle on either door. I didn't get to look into these other two rooms. The floor was flat.

As I stepped through the door into the spaceship, the two men who were already in the room motioned for me to sit down in a reclining chair that was in the center of the room. There were also four chairs on the opposite side of the room. There were two in each corner, made similar to an automobile bucket seat. When the four men sat down in these chairs, the chairs tilted back so that their heads were about one foot from the wall.[7]

The main door or hatch slammed shut, making a sound similar to that of a heavy car door being closed.[8]

TAKE-OFF

When the 'men' sat down, one of them pushed a button on a small control-panel and a light came on underneath the button.[9] The lighting dimmed to near-darkness, and there was a slight jolt.[10] Immediately after the ship had taken off, a high-pitched whine could be heard, which by Watts's estimation lasted for about one and a half minutes.

A few seconds after this button was pushed, the ship moved forward, and since I didn't sit down in the chair as I was told to do, I was thrown backward against a partition. One of the men pushed a button on the panel again, and the ship slowed down. They then got up and helped me get into the chair and strapped it across my chest and above my knees.[11]

The metallic-looking chair turned out to be extremely flexible and comfortable, adjusting itself to Watts's body contours.[12]

> We had been traveling about five minutes and it got real cold. They didn't seem to react to the cold as I did. It got down around freezing. He pushed another button on the panel and in a few minutes it started getting a little warmer. It took about 10 minutes for it to get warm enough for me to quit shaking from the sudden drop in temperature.
>
> We traveled about 20 minutes from the time of take-off when the ship started slowing down and stopped. I heard four metallic clicks, and the wall opposite me between the two chairs slid back to reveal a round-shaped tunnel . . . about five feet high. When I first heard the clicks, the men got up and unbuckled the belts. The nearest man to me said, 'Let's go over here', and led the way through about an eight-foot tunnel into another ship.

WEAPONS?

'As we stepped from the tunnel into the first room, in the left end of it was something that resembled shelves or racks of some type [which] reached from the ceiling to the floor. There were four shelves which had eight to ten "weapons" on each shelf.'

> The weapons were made similar to a rifle or shotgun; however, the overall length was about 28 or 30 inches long. The 'stock' was about 12 inches in length. Where our guns have a receiver and trigger, these had neither, but a hand-grip on the stock. In place of a receiver, there were three oval-type discs about one-half inch thick about one and one-half inches apart and five inches around. The barrel started there and was about three inches round. On the opposite wall from these weapons was some kind of cabinet that had pigeon-holes in it [in which] were stacks of these round, slotted discs.

THE 'PHYSICAL'

'We turned right and walked through an opening which I didn't see as a door,' Watts's report continues. 'This man stepped to the side of the door and remained on the outside of the room. I was met a short distance inside the room by two more men who explained to me that they were conducting a survey and would appreciate it if I would co-operate and submit myself to a physical. This room looked like the same room that I looked into on March 31 when I saw the larger ship.'

> In this room, there was a strong odor that smelled like sulphur. It burned my eyes and I could feel it going down my lungs a little. I could see it every once in a while coming in from a vent around the floor. I had noticed this odor a little before in the small craft and some in the first room that I walked into on this large craft.
>
> They had me walk on to the center of the room where the machine was located, to undress and take the physical. This machine was about six feet tall and 36 inches wide, and cupped in front. They put me in a position in front

of the machine, about one and one-half inches from it. There was a large throw-switch about a foot from the top of the machine. They pulled this switch down and a light came on on each side of this machine, about five inches from the top. There was a narrow band of glass between each light, about three-eighths of an inch wide. There was also a round control-knob on the left side, about three feet from the floor, which they turned until these two lights came together.

There were four gauges over this machine that were molded into the ceiling. I couldn't read any of them. Each one had a needle on it, but I couldn't read the writing. While they were adjusting the machine, there were round, flexible needles coming out, one about every inch. The needles were about the size of a hypodermic needle. Instead of piercing the skin, they came out and just laid down against the skin. I stood and faced the machine about five minutes and then they turned me around. They put me in position again and re-adjusted it until the lights came together again.

During this time, the machine transferred its information into another machine, starting at the side and extending out about five feet. Out of the other end came a paper about 18 inches wide. One of the two men who seemed to be in control of my taking the physical walked back and forth from me to this paper, about every two minutes. I stood there about five more minutes, and they turned the switch off and said that it was complete. They walked down and got the paper and came back by me, and I asked them how it turned out. He said, 'All I can tell now is that you are about eight pounds over-weight.'

An Attempt at Verification

Watts was told to get dressed. The beings walked into another room where the four others were, gathered in a close group and examined the paper. After Watts had re-dressed, he looked around and noticed a desk and chair on the right side of the room. On the desk lay a stack of maps, about three inches thick and a yard square. One map lay directly in front of the chair, as if the beings had been studying it. 'On the other corner of the desk,' claimed Watts, 'was a stack of papers and pencils and also a paperweight.'

Like the famous Brazilian abductee Antônio Villas Boas, who tried to take a 'clock-like instrument' to furnish proof of his experience 10 years earlier, Watts felt he ought to purloin the paperweight.

The paperweight was about two by three inches, with squared corners and about three-eighths of an inch thick. It looked like metal to me, but wasn't heavy like steel. I looked to see if all the men were in the other room and picked up the paperweight and put it in my pocket underneath my billfold. I stood there a short length of time when one of the men who took me up there walked into the room and came directly over to me. He lifted the paperweight out of my pocket without saying a thing. I caught his arm with my left hand,

and he hit me across the back of the head. I was unconscious for about 15 to 20 minutes.

THE RETURN

On recovering consciousness, Watts found himself back in the small craft again, strapped in the chair. 'And we were moving,' he added. He asked the man closest to him what the purpose of the physical was. 'We are conducting a survey,' came the blunt reply. Asked to explain the purpose of the survey, the man merely responded that 'they would make maps of the complete planet before they were through', because, said Watts, 'the people on Earth would in a short time reach the Moon and eventually build better and bigger equipment to go out farther, and that's when they would step in.

'They said they didn't have wars where they came from; that they put all their energy to work for scientific matters. I asked him if they were more intelligent than we, and he said, "Yes, I am 162 years old. I have already lived three times longer than you will live." He said that they were from the planet that we call Mars.

'About that time the ship slowed down, and they unstrapped me and opened the door for me to leave. The last thing they said to me was, "If we need you, we'll contact you." They brought me back to the same spot where they picked me up where my pick-up was. I was on the two ships about one and a half hours in all.'[13]

Meanwhile, back at the farmhouse, Rosemary worried about her husband. She had seen the light of the 'spaceship' in the distance, and telephoned the Wellington Police Department. Donald Nunnelly and John Rainey came to her house and began looking for Watts, who eventually arrived back home, too disturbed to go into details about his latest experience that night. Nunnelly noticed that Watts had an unusual twitch in his face and that he was nervously massaging the back of his neck.[14]

THE BEINGS

In Watts's report, the entities are described in great detail:

The four men in the small ship that took me up were dressed in white coveralls. Their shoes were of material or soft leather that resembled a high moccasin, with soft soles that didn't make any noise when they walked, like mine did [a peculiarity reported in other cases].

The two men who gave me the physical were dressed in white, two-piece suits. The coats had a flap down the front so that I couldn't tell what kind of closure was on them. They also had a pocket on each side of the flap, just below the waistline on the coat. The one who seemed to be in charge had [an] insignia on his collar.

These men were about four feet high [or so]. They varied two or three inches at the most and weighed from 115 to 130 pounds. They were a little thicker than ordinary through the chest. The color of the skin was between a white and a very light gray. They had no hair on their heads nor eyebrows.

Their facial features differ from ours. They had only a small slit about one-half inch long for an ear. Their eyes started in the center, about where ours do, but go around the side of their head to about where our hairline starts. Their eyes are about one inch up and down at the thickest point, which is the center, and tapered to each end. They can see to the side as well as to the front. The pupil in their eyes is oblong and about one inch long. You could see about three-fourths of an inch of white on each end of their eyes. Three of the men had brown eyes, one had green, one blue, and one, who seemed to be in charge of the ship, had blood-red eyes. He also had something on a little white belt around his waist which I took to be some type of weapon. It was about four or five inches long and two inches wide, with a round handle, but I didn't see a trigger or barrel like we have on our guns.

Their nose didn't protrude out from their face like ours. It curved over some type of bone structure which had two slits below it, about one-half inch long with a small hole about the size of a kitchen match beneath each.

Their mouth was more of a straight line with very thin lips. The only expression that they ever made with their mouth was to smile. I never did see one with his mouth open. Their conversation was done through mental telepathy, and they can read your thoughts. However, I talked to them naturally, and their thoughts just came to me someway. I never could hear the conversation between them, only when one was sending thoughts directly to me.[15]

FURTHER ENCOUNTERS

At about 22.00 on 21 May 1967, Carroll Watts looked out of his front door and saw the light and outline of the same, or a similar, cigar-shaped craft, less than a mile away. 'My wife and I watched it for about one and a half hours that night,' said Watts. 'It would dart sideways a few minutes and go back each time to the same spot where we first noticed it. The light was bright enough that we could see the outline of the ship, but it wasn't a glaring light. It was still there when we retired that night. We debated about going down there, but decided against it since our children were already asleep, and it was too late to take them to my parents' house to stay.'

The following night, Watts spotted the 'spaceship' above a field about 200 yards away, so he decided to take some 8mm movie film of it. 'Two weeks later we saw them across the road from our house in a pasture,' said Watts, 'and I went out and took a picture of the ship.'

On 7 June 1967 at 11.00, while repairing a fence on a farm he owned, about a mile northwest of his house in Loco, Watts saw the large ship approaching from the north. He ran to his car, grabbed his Polaroid camera and managed to take seven black-and-white pictures. 'The ship was about 300 feet in the air when it came over me, made a circle and left,' said Watts. Of the seven photographs, four developed clearly.

At about 15.00 on 11 June 1967, Watts spotted the ship again, on the farm where he had encountered the 'Martians' on 11 April. He was planting cotton at the time. As before, the ship approached from the north and headed south. Later, Watts claims, it returned and landed on a ridge in a pasture. One of the beings got out of the ship for a short time and walked around. 'During this time I was down in a low place west of this ridge. I was able to get one picture of the man before he ran and got back into the ship and left, going north.'

Watts took seven Polaroid colour photos of the ship, most of which came out satisfactorily. The eighth is less so. 'Since it was a very bright sunny day,' explained Watts, 'the man appears to have hair, but it was a shadow on the back of his head. He was facing toward the sun which accounts for the shadow. I was about 60 yards from him when this picture was taken. He is rather broad-shouldered for his height, and their neck is a little thick compared to ours.'[16]

My enlargement of this photo, supplied by Rosemary Watts, shows what can be construed as the head and top half of the trunk of a being of indeterminate type: unfortunately, it is out of focus or blurred. 'We would certainly have liked to have had a full view of him and also one that was more clear,' wrote Mrs Watts to Henry Johnson, the then husband of Madeleine Rodeffer, 'but it was the last of the film in the [pack] and he didn't have time to reload the camera to take another one.'[17] Nonetheless, the photo is interesting. Another photo, one of those showing the large cigar-shaped ship, is excellent (see plates).

THE DISTURBING AFTERMATH

Learning of the Air Force-sponsored UFO study at the University of Colorado, Watts decided to inform the study team about his experiences, and to show them his photographs. Reportedly, the two investigators who came to interview him were the late Robert Loftin, author of a pro-UFO book,[18] and William Courter, a private detective.[19] According to Rosemary Watts, while these two men were interviewing Watts at his home, on the morning of 26 February 1968, 'their motel room in Wellington was entered and a lot of material pertaining to this case was stolen.'[20]

On the night of 26 February, the story of Watts's experiences hit the Associated Press (AP) newsrooms. Watts had decided to make a press statement. The following morning came the headlines: 'Farmer Talks With UFO Occupants'; 'Farmer Who Rode with Spacemen Wants Story Tested'; 'Everyone Loco in Loco? Citizens Vouch for Man Who Rode in UFO' – and so on.

The AP story reported that a set of some of Watts's Polaroid prints had been given to the astronomer Dr J. Allen Hynek, at the time an adviser to the Air Force's Project Blue Book. Hynek told AP that preliminary examination showed no signs of fraud. 'If this is a hoax, it is a very clever one,' he said. 'There's no question the story is preposterous,' he added. 'The question is, is it true?'[21]

In March 1968, Watts's original Polaroid colour prints were sent by registered mail to the University of Colorado. Subsequent attempts to have the photos returned were unsuccessful. The photos, it seemed, had been lost. As Tony Kimery elaborated:

> The only available prints were a set of black-and-white copies that were made by Dr Condon [head of the University of Colorado UFO study] and sent to Mr Loftin. These copies were in his possession and, presumably, lost during the motel break-in. Loftin had made copies of three of the photographs, which he distributed to the news media. These three photos are the only remaining from the original series of seven [and] are of poor quality due to the fact that they are second generation prints.

Soon after the story appeared across the United States, Watts issued an oral statement to the press to the effect that the entire story had been an elaborate fabrication. He gave the reasons to Kimery:

> I had full intentions of trying to prove my story to be true before anything was released, and had been working with the Associated Press for several days when the UPI heard of the case and butted in. They released a short rundown of the story that same night, and as soon as the story hit the wires, I received my first threat to forget about trying to prove it. Therefore, I decided to call it a hoax and forget the whole thing. There were several serious threats made to me and my family, and I found out that I had stumbled on to something far more serious than I had expected.

It was Dr Allen Hynek who suggested to Watts that he undergo a polygraph examination. Watts readily agreed. The test was administered by L.R. Wynne, owner of the Amarillo Security Control Company in Amarillo, Texas. According to several delayed news reports, Watts failed the test. In spite of the examination results, however, Watts stated that Wynne believed his story to be true. After Watts's 'confession', Wynne told reporters that he considered the Watts story to be a fabrication.

The night before Watts went to Amarillo, an unidentified caller advised him not to make the trip. Watts ignored the advice. The following afternoon, as he was driving to Amarillo, he noticed a blue-green Plymouth parked beside the road with the hood up. An attractive young blonde lady was trying to wave him down. Assuming she needed assistance with her car, Watts pulled over. As he stepped from the cab of his pickup, he was stunned by a blow to the back of his neck. He fell to the ground, dazed, and rolled over, though he managed to focus on his aggressors: two men, both about six feet tall, weighing 190 pounds, and in their early thirties. Wearing turtleneck sweaters underneath rather expensive-looking suits, they were very tanned, clean-shaven, with a rather 'Greek' appearance. Both were holding exotic-looking rifles of an unknown make. They spoke with precise voices. The young lady, evidently the decoy, smiled as the two aggressors threatened Watts: he was to fail the polygraph examination or face 'possible consequences' on his return home.

That night, a large black car, with no lights, passed slowly by the Watts residence, drove up the road a short distance, then returned, opening fire with a shotgun. The shot did not hit the house. According to a report by investigator Steve McNallen, based on an interview with Police Chief Donald Nunnelly, Rosemary Watts called the police, who arrived on the scene later. Watts had apparently rushed out of the house and returned the fire with his M-1 carbine. It is not known whether any of the shots hit the car or its passengers.

Watts and his family were not the only residents of Loco to be threatened. Investigators learned of other cases involving UFO witnesses who had been approached by unidentified authorities and told to keep quiet about anything they might have seen, 'for reasons of national security'.[22]

SUPPORT

In interviews with the Associated Press, Police Chief Donald Nunnelly and other law enforcement officers in Collingsworth County described Carroll Wayne Watts as 'a stable family man, a church-goer, out of debt, and with no motive for fabricating his story'. Furthermore, Nunnelly said he knew of as many as 50 people who had seen similar strange lights in the area, and four or five who had glimpsed a cigar-shaped object. 'This is the sort of thing they will talk about, but only over coffee with their very closest friends,' added Nunnelly. 'They don't want to wind up in an institution. I've known of some law officers who got mixed up in things like this and lost their jobs and everything.'[23]

At 04.00 on 11 April 1967, the day Watts was taken aboard the craft for his 'physical', Mickey Kendricks was awakened by what he thought was a truck driving into the freight depot behind his home. When he looked out of the window, however, he saw a cylindrical-shaped object, estimated to be 20 to 30 feet long, with red and yellow lights revolving around it.

On 3 November 1967, Hazel McKinney and two female companions were driving to work at Childress, Texas, when they noticed a huge bright light in a field beside the road. Suddenly, the object, now appearing silver-grey in colour, 'angled-up' and disappeared. 'It was big enough to drive a car in,' said Mrs McKinney. 'It was shaped like a cigar – one end was round.'

Sometime after Watts's story hit the headlines, Childress Police Chief Alvin Maddox had been laughing about it with two colleagues. Shortly afterwards, driving ten miles north of Childress and eight miles south of Loco, Maddox observed a large bright light hovering in the sky at about 500 to 1,000 feet. 'I took after it in my car, driving about 105 miles per hour,' said Maddox. 'I followed it about 14 miles, but it left me.'[24]

So concerned was Watts about his experiences that he asked for a Congressional investigation. Nothing came of it, of course, following his confession.

There are many parallels in Watts's story, not just with published cases but also with reports which remained unpublished until many years later. Watts had little or no interest in the phenomenon, and by all accounts had not read any books on the subject prior to 1967. I have not met him, though I did speak on the telephone to his wife a few years after the incident. I found her to be sincere, and genuinely concerned about the whole affair. Surprisingly, both Carroll and Rosemary Watts temporarily retained a positive attitude towards what must rank as one of the most bizarre encounters with extraterrestrials. As the couple concluded in their report:

> What you have just read is a true story, and we have tried in every way to tell it as near to what happened as possible . . . We feel like if the public can hear about these things and know about what to expect, before they come face to face with them as we did, it will make it easier for them to understand and accept. We have written this because of the nightmare we lived when we had our first experiences with these UFOs.
>
> Since we have found out that these people mean no harm, we are not afraid of them anymore; instead, [we] accept them as friends unless they prove themselves otherwise. Only time will tell the outcome of this, but we pray that God will have His way in seeing that things turn out for the best . . .[25]

Things did not turn out for the best. Some years after these events, Watts began to have behavioural problems. His brother told investigator Don Worley that Carroll had become 'influenced' by a stranger. Paranoia set in, and Watts began to believe that certain people were out to 'get him'. Finally, out of mistaken fear, he pulled his gun on a law officer, for which he received a prison sentence. From his cell in Texas State Prison, Watts wrote to Worley:

'The incident cost me my wife, my children, $285,000, my freedom, my health [heart trouble] and nearly my life. Simply because something happened to me that I didn't understand, and that I talked about, and I think now that I would have been better off myself if I had died in the incident.'[26]

Sinister incidents such as these should serve as a salutary reminder of the dangers that can be involved in encounters with extraterrestrials –

particularly in their aftermath. But now, by way of contrast, let us turn to another, quite different kind of contact.

CATMAN AND ROBOT

During the Second World War, Walter Marino Rizzi served as an aircraft mechanic and as an interpreter for the Italian and German air forces. Later, he worked as a sales representative for a car firm in Bolzano, Italy, his territory including the Dolomite Mountains in the South Tyrol. At midnight one Saturday in July 1968, having spent the evening with a Dutch girlfriend in San Cassiano, he set off by car via the Gardena and Sella Passes to Campitello, where his aunt managed the Sport Hotel. Here he planned to spend the night.

'The weather was not very good, and I seldom saw a star in the sky,' Rizzi begins in his own report, an English version of which he gave to Lou Zinsstag in 1980. 'There were always dense fog-banks, and more than once I had to stop because I couldn't see the road. So I decided to stop at the first possible place and sleep in the car.' After driving through the Grödner Pass, he found a suitable spot – a sand-dump beside the road – and settled down in his car for the night.

At around 01.00, he awoke with a start. There was a strong smell of burning. Thinking that his Fiat 600 was about to catch fire, owing probably to a short-circuit, he checked the engine but found everything in order. While walking around the car, however, he noticed below him on the opposite side of the road, about 500 metres away, a powerful light shining through a gap in the fog.

'It looked like an illuminated hotel terrace,' Rizzi continued, 'but there are no hotels around there – nothing at all. I knew the place very well and had passed that way hundreds of times. Then, in a moment of clear visibility, I saw an enormous object bathed in a strange white light.

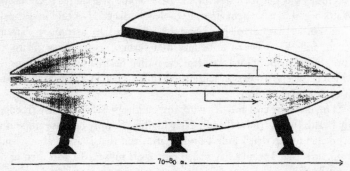

Fig. 20. A drawing by Walter Rizzi of the craft he encountered in the Dolomite Mountains of Italy in July 1968. (*Walter Rizzi*)

A light-enhanced frame from the 8mm movie film taken by George Adamski, in the presence of Madeline Rodeffer and others, at her home in Silver Spring, Maryland, on 26 February 1965. The craft was estimated by an optical physicist to be 27 feet in diameter. *(© Madeline Rodeffer)*

The huge Schmidt crater, situated on lunar upland near the Sea of Tranquillity, photographed in May 1969 by the astronauts on board the Apollo X lunar module, commanded by Lt.-Gen. Thomas P. Stafford. A Hasselblad 500EL/70mm camera and Zeiss Sonnar 250mm telephoto lens were used. *(NASA)*

The valley at the Blue John Caves, Derbyshire, where the contactee 'Joëlle' witnessed the landing of an extraterrestrial craft in September 1963.
(©*Timothy Good*)

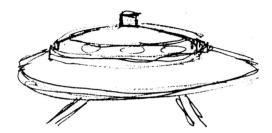

A sketch by 'Joëlle' of the craft. Note the similarity to some of the craft photographed by Paul Villa. The portholes were round, not oval as depicted.

Apolinar (Paul) Villa, the publicity-shy contactee who in the 1960s took a remarkable series of colour photos of what he claimed were various types of extraterrestrial craft. He is shown here during an interview with the author in 1976. *(© Timothy Good)*

The Apus folding camera, with a Rokuoh-sha f4.5 75mm focal length lens, with which Paul Villa took all his pictures, using 120 format Kodacolor film. *(©Timothy Good)*

Photos selected from the series taken by Paul Villa, 1963-6 (Copyright claimed in 1967 by UFO International, PO Box 552, Detroit Michigan 48232)

Photos 1-6 taken on 16 June 1963 near the town of Peralta, about 15 miles south of Albequerque, New Mexico.

1. Craft above Villa's truck.

2. According to Villa, the craft flipped on its edge with the lower part rotating.

3. Because no branch is shown in front of the disc, some believe it must have been a model suspended in front of the tree. Villa maintained however that many of the branches on the tree were vibrating, due to some effect from the craft; therefore, perhaps, it did not register sharply on the picture.

4. While some believe that the craft shown here is also a model suspended from the trees, evidence suggests that it is further away from the camera.

5. The craft at this point is purportedly about a quarter of a mile away from the camera. Note the out-of-focus foliage in the foreground, suggesting that the craft is not a small model.

6. Villa's most well-known photo, arguably the best of the series taken in 1963. The craft was a few hundred yards away. Although the upper, domed structure could be turned independently from the lower section, it appeared to remain stationary during flight and the lower section alone rotated at different speeds.

Photos 7-9 taken on 18 April 1965, about 20 miles south of Albequerque, close to the bed of the Rio Grande.

7, 8. In photo above, the lower branches of the tree at right were vibrating violently.

9. Just above and to the left of the tailgate of Villa's truck can be seen a small cloud of smoke. According to Villa, as a demonstration of some of their powers, the extraterrestrials shot a brilliant ray of light at the ground, causing the brush and trees to ignite. Examination of this photo indicates that the craft is a large object at a considerable distance from the camera.

10. This craft was one of several remotely controlled discs, varying in diameter from three to six feet. The tripod landing gear did not telescope into the disc but seemed to just shoot in or out of the bottom. Villa was informed by his extraterrestrial contacts that the rod protruding from the dome was an optical device which could be retracted inside the dome.

Photos 10-16 taken on 19 June 1966, three miles west of Algodones and 30 miles north of Albequerque, New Mexico.

11, 12. The disc shown here is about three feet in diameter, accompanied by smaller spheres. When it landed it bounced, shot straight up, then landed again in the same place. The smaller spheres, about three inches in diameter, seemed to rotate around the larger six-inch sphere when away from the disc. When close up to it, however, the large spheres always remained on top or close to it.

13. Photographed at about 11.00 am, this disc was also about three feet in diameter.

14 and 15 (OPPOSITE ABOVE). Paul Villa claimed that these two photographs show an experimental disc, three feet in diameter, which he made according to specifications given to him by his extraterrestrial contacts. He did not manufacture the smaller sphere, which he said was monitoring the not entirely successful test flight.

15.

16. Villa asserted that this photo, one of two taken between 12.45 to 1.00 pm, show a manned craft about 40 feet in diameter, in which were stored the smaller, remotely controlled disc-probes. The problem here is that it is difficult, if not impossible, to differentiate between these particular photos and the others in this series.

A portrait of 'Xiti', the extraterrestrial lady initially encountered by Ludwig Pallmann in India in 1964, painted by the Polish painter, Vera Waleska, and commissioned by Pallmann. Xiti had unusually large eyes, peculiar fingertips, and sometimes wore a delicate blue veil, depicted here. She is shown surrounded by some of the plants that allegedly were hybridized by the extraterrestrials, and (top right), the three disc-shaped spacecraft based at the plantation. Unable to reproduce on canvas Xiti's extremely unusual clothes—as described by Pallmann—Waleska painted her in the seventeenth-century French style.
(Vera Waleska/Ludwig Pallmann)

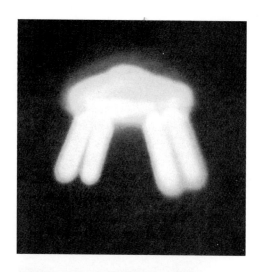

An aerial craft projecting beams of light, photographed by an anonymous doctor on 23 March 1974, near Tavernes, Var, during a wave of sightings in France.

No details are available about this enlargement of an interesting and probably genuine photo, published by the Californian researcher Gabriel Green in the late 1960s.

The Laguna Cartagena, part of the National Wildlife Refuge in Cabo Rojo, Puerto Rico, and (below) the El Yunque National Forest, where many islanders believe alien bases are located. *(© Timothy Good)*

The small creature killed by 'Chino' Zayas in a cave behind the Puerto Rico National Guard Camp at Santiago, Puerto Rico, in 1979 or 1980. For these photos, taken by Rafael Baerga a year or so after the incident, the creature was taken out of the jar in which it had been preserved in formaldehyde. Below is a photo of the creature's skull, clearly showing where it had been struck a fatal blow.
(Rafael Baerga)

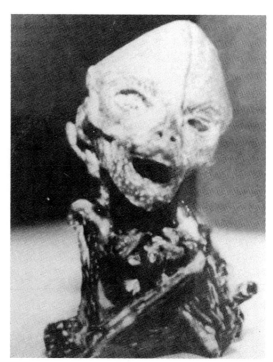

(RIGHT) The notorious 'chupacabras', reported in the mid-1990s by numerous witnesses in Puerto Rico. This painting by Jorge Martin is regarded as the most accurate representation of these creatures, which reportedly were responsible for the deaths of hundreds of animals. *(Jorge Martin)*

(BELOW) One of several Polaroid photos by Filiberto Caponi of a creature he encountered on several occasions outside his home in Ascoli Piceno, Abruzzo, Italy, 1993.
(© Filiberto Caponi)

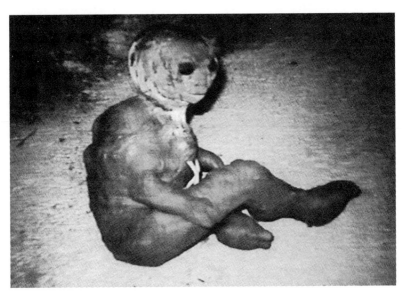

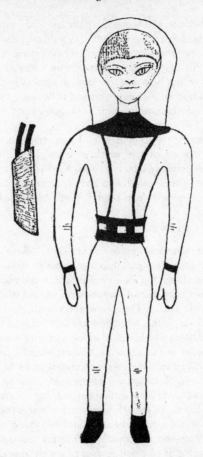

Fig. 21. 'Catman'. A drawing by Walter Rizzi of the alien being with whom he claims to have communicated. At left is the apparatus worn on its back.
(*Walter Rizzi*)

'A slope went down from the edge of the road, and, taking the torch, I descended and made my way towards the plateau where this huge object was. As I approached, I could see it more and more clearly, the fog becoming less thick. My heart was thumping painfully. I was not afraid (I have never been afraid of anything), but terribly excited.'

The 'saucer' was a beautiful silver colour, about 70 to 80 metres in diameter, resting on three legs about two metres from the ground. The legs were about two metres in diameter at the base. Everything was bathed in a fleecy white light and there was an intense smell of burning. As soon as I got within three metres of the disc I felt suddenly blocked, as if my body weighed 1,000 kilos; I was unable to move and had great difficulty in breathing.

The glass dome on top was brightly lit and I saw two 'beings' looking down. On the right side of the disc was a [cylindrical] robot, about 2.5 metres tall, with three 'legs' and four 'arms', holding the outside of the disc and turning it round. A circular 'trap-door' opened in the underside of the disc, giving out a violet and orange light, and someone emerged, dressed in a close-fitting suit and a glass helmet. The figure seemed to be about five feet four inches tall. 'He' approached to within a little more than a metre from me and raised his right hand.

I cannot describe my feelings when I looked at that creature with [its] beautiful eyes: it was a strange, gentle sensation; I felt as free and light as a feather, and was perfectly calm . . . As his head and neck were free beneath the glass helmet, which was a round globe reaching the shoulders (with two flat tubes leading upwards from behind), I was able to see him very clearly.

PHYSIOGNOMY AND ORIGIN

His hair was very short and light brown in colour, looking almost like fur. He had beautiful eyes, wider apart than ours, slightly oblique, like a cat's, the whites being light hazel, the irises a greenish blue, and pupils that contracted continually (when they dilated they were round, and when they contracted they were oval, like a cat's). The nose was very small, also like a cat's. His lips were small and very tight, reminding me somewhat of Greta Garbo, and when he smiled he showed his teeth, which were very white and regular. His skin was light olive and as smooth as rubber; when he turned his head (he could turn it right round), there were no creases on his neck.

He had very wide shoulders and narrow hips. His arms and legs were the first parts of him to hold my attention because they were a bit different from ours: from the hip to the knee they were much longer than below the knee, and the ankle seemed to me to be articulated like that of a horse. The arms, too, were very long – the upper part much longer than the lower – but I was unable to see the hands very well because he was wearing gloves, although he appeared to have very long fingers.

I asked him in Italian where he came from, and no sooner had I framed my question than the answer was already in my mind, as if I had always known it. The planet he came from was very far from our galaxy and about 10 times the size of ours, with two suns, one big, the other small. Their day is longer than ours and one third less light, while their night is very short. Their vegetation is quite similar to ours, but they have enormous trees and very high mountains. They have two frozen poles as we have, stony deserts, and animals similar to ours, though with some differences in structure and size.

I thought of asking him how they live and on what, and immediately I received my answer: he moved his mouth a little, but I did not hear his voice. The answer appeared in my mind – I suppose, by mental telepathy. He told me they do not work; everything is automatic, they are all equal and everybody has what he wants. There are creatures like apes, which do menial jobs such as cultivating fruit and vegetables, harvesting, and so on . . . He explained that they don't need their teeth as much as we do, as they don't eat meat, adding that, compared with them, we are like animals. They nourish themselves on fruit, cereals and seeds, and have, in addition, apparatuses that generate energy. Disease does not exist.

I 'asked' him why he had an olive-green skin, and he told me that the colour I saw was not the real one, because our system of magnetic colour receptivity was not like theirs (I did not understand exactly what he meant). [The colours we see represent slightly differing frequencies in the electromagnetic spectrum.]

He told me that their organism is much simpler than ours: they have only one digestive tract without all the paraphernalia [entrails] that we have. Their heart and lungs, however, are well developed because they need a large intake of air to feed the brain and purify the liquid that runs in their veins, which is of a different composition from our blood. Moreover, they have very powerful muscles in order to withstand the strong atmospheric pressure on their planet. In fact, when he emerged from the disc and came towards me, he had a leaping or loping gait, like the astronauts walking on the Moon.

I was continuously struck by the beauty of his eyes and wanted to ask 'him' if 'he' was a man or a woman. His eyes twinkled for a moment and he explained, with a smile, that he was neither; that their system of propagation was not through mating, like us.

Since I was examining him closely to see how he was built, he gave me to understand that his build was in conformity with the exigencies of life on his planet. The top of his head was bigger than ours because his brain was double the size of ours. Simply by means of thought and the emanation of [thought] waves they could do things that we cannot even imagine . . .

SPACE TECHNOLOGY

As I was only a metre or so from him, I tried a couple of times to reach out and touch him, and each time I was impeded. Meanwhile, on the right of the disc, the robot was working on the outside of the disc, where there was a kind of 'rotor' or 'ring' . . . projecting two metres [from the craft].[27]

'It looked like a multi-coloured, ring-like rainbow,' Rizzi told me in 1997.[28]

Every now and then it tilted down and the centre sharpened to a point, one half turning in one direction and the other in the opposite direction. I asked him whether the sharp point was to destroy meteorites . . . He smiled again, and explained that they do not destroy meteorites; they avoid them . . . In any case, the function of the external ring is merely to enable them to enter the atmosphere of the planets they are visiting . . .

In space, they travel by the main [carrier] disc, which is outside any magnetic field, using another source of energy: it has the same shape, but is very big, with a diameter of, I think, five kilometres. It is fitted with ledges for smaller discs such as the one I was privileged to see, or a number of very small discs without equipment, which are used for exploration purposes, since they have a system of magnetic propulsion with no speed limit in any atmosphere, and [which is] immune from temperature and other conditions. In the main disc, hundreds of their people can live as on their planet; they have an inexhaustible supply of energy and everything else that they need. Furthermore, they travel in 'neutral channels' through space, so that there is no danger of their being attracted by other planets, and also to avoid [asteroids] and defunct planets.

I asked him what defensive weapons they had, to which he replied that they could disintegrate anything, even at enormous distances. He indicated that I should pick up a stone about two metres away and throw it at the dome of the disc. I picked it up – it was [shaped] like a large potato and it weighed about a kilo – stepped back a bit to get a better angle, and hurled it. As soon as the stone approached the disc, a ray of violet light shot out and the stone exploded with a thud, disintegrating completely.

A Disturbing Message

I asked him why they didn't give us the benefit of their technical knowledge and stay with us for a while on our planet, and how long it would take to acquire their technical proficiency. He replied that, in the first place, it was impossible for them to interfere with the evolution of any other planet; that spending any time in our solar system would age them prematurely, and finally that we would never reach their standard of evolution, because the crust of our planet is too different, and in the near future there would be a shifting of the poles. This would result in a vast opening in the Earth's crust, bringing about the cataclysmic destruction of 80 per cent of the population, and leaving only a narrow strip of land still inhabitable for the few survivors.

God and Death

At this point I asked him if he believed in God. He looked a little abashed at first, then told me that God is everywhere; in us, plants, stones, grass, and nature – everything that exists.

I asked him how they die and how long they live. They die, he said, of exhaustion of cosmic energy, and live about a hundred times as long as we do, according to our concept of time.

Time to Go

Meanwhile, the robot had finished its job; it became smaller, the cylinder became narrower, and it moved to the centre of the disc, whereupon an orange light went on, and it seemed to float up and board the disc. It was clear that they were leaving.

In the dome, the other individual waved to me from time to time. I could not see 'him' very well, but he was like the one who was standing in front of me. All the while, the disc had been bathed in that fleecy white light, which did not cast shadows or hurt the eyes. I asked if I might have something of theirs, but he refused, saying it would be harmful to me.

I was so fascinated by that individual that I asked him to take me with them; that it didn't matter to me if I never returned. I was so upset at the thought that I should never see him again that I started to cry: I even went down on my knees and begged him to take me. Every time I made to embrace him I felt 'blocked'. Then he indicated to me to get up, his eyes shining in a strange way that made me feel warm all over, and he communicated to me as follows:

'You are very courageous, and you have been doubly lucky; firstly because if you had approached a metre nearer, right under the disc, you would have disintegrated; because the rotor was being checked, the magnetic field did not extend beyond the diameter of the disc; secondly, because you have had the chance to see us close up, and speak to us. However, neither you nor anyone else from this planet could ever remain with us – much less travel in our space ships.'

He then raised his hand as before. I felt myself pushed away from the disc by some unknown force, then he arrived beneath the disc and disappeared into the illuminated circular opening. In the dome there was still the other one who waved to me with those long arms. The white light began to dim, as I stopped – or, rather, I was allowed to stop – about 300 metres from the disc, the dome gave off a violet light, and also the outside of the disc became suffused with violet and orange.

At this point, there was a noise like a circular saw, the disc rose two or three metres, the feet folded in, then the light changed to light violet, becoming whiter and whiter. For a moment there was a whistle that I thought would shatter my eardrums; the disc began rocking, as though bidding me farewell, rose slowly to about 300 metres, then suddenly shot away at a terrifying speed.

'BACK TO EARTH'

I remained there, shaken, the tears streaming down my face, feeling totally desolate. I became aware that I was soaking with sweat, and the air seemed warm. I touched the ground, which also felt warm. The fog had disappeared, the sky was starry, and it was pitch-dark. I tried to switch on my torch, but it didn't work, so I groped my way back to the car. I tried to pull myself together after what had occurred. I pricked myself with a pin to see if I was awake, then relieved myself. Finally, I drove off towards the Sella Pass to get to my aunt's in the Fassa Valley.

The following morning in the hotel, Rizzi began making notes and sketches. He tried to tell some family members and friends what had happened to him, but they ridiculed him.

Three weeks later, Rizzi returned to the site of the encounter to take photographs of the marks of the three legs which had sunk into the ground under the weight of the disc. 'Much to my surprise,' he reported, 'in the area covered by the light that had been shining from inside the disc, the grass had grown to three times the height of the surrounding grass. I took a screwdriver from the car and dug up one of the longest blades of grass, together with roots and earth, and put it in a plastic bag to have it analysed in America.'

Two days later, Rizzi left for California, where his daughter lived with her American husband, a director of Pan American World Airways. When he passed through Customs at San Francisco Airport, an official asked him if the plant was marijuana. 'I told him it was a chrysanthemum shoot I intended to plant on my father-in-law's grave. He said okay, and let me go without analysing the plant.'

Rizzi and his daughter sent out a pile of letters describing his experience to all the addresses they had found in UFO magazines. 'Nobody answered, however, so I decided to keep my adventure to myself. I took a deeper interest in UFOs and from time to time read articles on the subject, often laughing at the rubbish people wrote.'[29]

'I should mention that my watch was losing as much as two hours a day,' added Rizzi. 'I took it to the watch-maker to put right, but it was no use, so I had to buy a new one. For about a month, I felt very tired the whole time, and I lost a lot of my hair, [but] before two months were up I was in good shape again and my hair had begun to grow back.'[30] Interviewed in 1997 at his flat in Bolzano, looking out on the spectacular Dolomites, Rizzi – a youthful-looking 77 years old – told me that it was as if his whole body had developed an allergy. Moreover, he became impotent for five months.[31]

DISCUSSION

Here we have another incredible yet intriguing and important encounter, one that begs all sorts of questions, of which the first must be: How truthful is the witness?

To begin with, owing to the ridicule incurred by Rizzi when discussing his experience with friends and relatives, he did not make his story public until 1979, in an interview for Radio Nord Bolzano. The story was first published in 1980 in *Flying Saucer Review*, based on a report written in less than accurate German by Rizzi (who is of Austrian-Italian parentage, Italian being his mother-tongue) and given to the investigators Hans and Daphne Markert, who subsequently passed it to Gordon Creighton. (With a few exceptions, the report I have used is the English version which Rizzi gave to Lou Zinsstag.) At their first meeting with Rizzi, in Germany, the Markerts were left with a very strong impression that he was telling the truth.[32] Lou Zinsstag was equally impressed, she told me.

Responding to the article in *Flying Saucer Review*, an incredulous Willy Smith, of the Center for UFO Studies, objected:

We [are told] that the [being's] planet is 'ten times the size of our Earth', although no indication is given whether this ten-fold increase refers to mass, volume or diameter. At any rate, it would certainly imply a much stronger gravitational pull, requiring a corresponding increase in cross section of the bone structure in order to cope with the additional weight. Yet, the creature is described as 'just like us', with a height of 1.60 metres, and not particularly sturdy. The same reasoning applies to the 'immensely tall trees', which in such a heavy gravitational field would tend to have a wide base and limited height.[33]

Rizzi told me that the home planet was larger in terms of volume, and that the aliens inhabited only the central, equatorial region.[34] In terms of Smith's objection, it is not necessarily axiomatic that a larger volume requires a greater mass, thus a greater gravity.

Among other objections, Smith notes that the landing site is described as completely uninhabited, whereas a photograph of Rizzi and others taken there shows a house in the upper left corner. I asked Rizzi about this. 'That's not a house,' he replied, laughing. 'It may look like a house, but in fact it's one of several huts used to store hay. Nobody lives there!'[35]

Some of what Rizzi claims to have been telepathically informed by the androgynous alien is inconsistent with the same kind of information in several other contactee accounts. I find it impossible to accept, for example, that the aliens' life-span is 100 times that of ours and that their galaxy is 'millions of light years away', as Rizzi told me. And although an estimated half of the stars in our galaxy are binary (where two suns move in an elliptical orbit around a common centre of mass), we have no proof so far for the existence of component planets in such systems. Interestingly, however, in 1998 NASA announced that the Hubble Space Telescope had captured an image of what may well be a very large planet (named TMR-1C) of a binary star system in the constellation of Taurus, some 450 light years from Earth.

These stumbling blocks do not devalue Rizzi's story for me. In the face of ridicule, he insists that this was the information conveyed to him. His story contains much that is new – and it has not changed over the years. In company with many other contactees, though, the encounter changed him. 'The experience changed my character greatly and had a profound effect on my attitude to all questions of religion or politics,' he declared.[36] 'It was the greatest experience of my life . . .'[37]

NOTES

1 Typewritten draft by Carroll and Rosemary Watts (undated).
2 Kimery, Tony L., 'Carroll Wayne Watts – Contactee, Hoaxer or Innocent Bystander?', *Official UFO*, vol. 1, no. 11, October 1976, Countrywide Publications, New York, p. 36.

3 Watts and Watts, op. cit.

4 Kimery, op. cit., p. 36.

5 Watts and Watts, op. cit.

6 Kimery, op. cit., p. 36.

7 Watts and Watts, op. cit.

8 Kimery, op. cit., p. 36.

9 Watts and Watts, op. cit.

10 Kimery, op. cit., p. 36.

11 Watts and Watts, op. cit.

12 Kimery, op. cit., p. 36.

13 Watts and Watts, op. cit.

14 Kimery, op. cit., p. 37.

15 Watts and Watts, op. cit.

16 Ibid.

17 Letter to Henry Johnson from Rosemary Watts (undated – early 1970s).

18 Loftin, Robert, *Identified Flying Saucers*, David McKay Company, New York, 1968.

19 Kimery, op. cit., p. 37.

20 Letter to Henry Johnson from Rosemary Watts (undated).

21 Associated Press, Loco, Texas, 26 February 1968.

22 Kimery, op. cit., pp. 37, 64.

23 Associated Press, 26 February 1968.

24 Kimery, op. cit., pp. 33, 36.

25 Watts and Watts, op. cit.

26 Worley, Don, 'Some Denizens of the "Black Nether-World" and their Abductee Victims', *Flying Saucer Review*, vol. 42, no. 2, summer 1997, p. 10.

27 Rizzi, Walter Marino, 'Flying Saucer Seen in the Dolomites', July 1968 (manuscript).

28 Interview with the author, Bolzano, 3 May 1997.

29 Rizzi, op. cit.

30 Rizzi, Walter, 'Close Encounter in the Dolomites', translated from the German by Gordon Creighton, *Flying Saucer Review*, vol. 26, no. 3, May–June 1980, p. 27.

31 Interview with the author, 3 May 1997.

32 Creighton, Gordon, 'Introductory Comments on the Rizzi Case', *Flying Saucer Review*, vol. 26, no. 3, 1980, pp. 21–2.

33 Letter to the editor (Charles Bowen) from Willy Smith, *Flying Saucer Review*, vol. 26, no. 6, November–December 1980, p. 31.

34 Interview with the author, 3 May 1997.

35 Ibid.

36 Rizzi, 'Flying Saucer Seen in the Dolomites', July 1968.

37 Interview with the author, 3 May 1997.

Chapter 15

The Plantation

The two passengers settled themselves down in their compartment for the train journey from Bombay to Madras on the night express. It was October 1964. Ludwig F. Pallmann, a German businessman dealing with the sale and installation of heavy machinery used in the large-scale production of food, eyed his travelling companion. The man was obviously a sahib left over from the British occupation, thought Pallmann. 'He was very well dressed, and had the unmistakable stamp of authority in his bearing.' About five feet ten inches or so in height, the stranger had a slim build and extremely long legs.

Pallmann struck up a conversation in English with the stranger and offered him a glass of whisky. 'My companion and I were the only two travelling in this compartment,' explained Pallmann, 'and after the introductory drinks we settled ourselves down in opposite corners and watched the landscape go by.'

It was not long before Pallmann noticed some unusual features about his companion, not least, the eyes, which were exceptionally expressive. In addition, the man had unusually long, slender fingers, the fingertips covered by some kind of protection, the like of which Pallmann had never seen before. 'For the rest, he was off-white in colour . . . a very light brown indeed; even his hair was light brown. I doubt if he weighed 100 pounds fully dressed, but for all his slim build he did not give the appearance of being physically weak. Indeed, on the contrary.' Something else attracted Pallmann's attention.

> I could not help observing that every time he inhaled, so he contracted his hands. Indeed, it seemed that he had to have recourse to deep inhalation with every breath, rather as though he suffered from respiratory trouble. Not that there was any external evidence of this, for, as I have said, he looked a very healthy specimen indeed.

When the stranger spoke, Pallmann was astonished to note that his voice did not appear to come from his mouth, but rather from a small 'speaking device' clipped to his chest. His speech, too, was peculiar. Although his English was impeccable, there was always a slight hesitation before he spoke. 'It was as if he was making the mental effort to say the

first word, after which all the others would come spontaneously,'
Pallmann elaborated. 'But that first word was laboured. It was a trick of
speech that was not at all unpleasant . . .'

The stranger introduced himself as 'Satu Ra'.

At nightfall, the train pulled into a large station. On the platform were
throngs of hungry-looking and miserable people, including an emaciated,
ragged woman and her fretful baby, who seemed to be suffering from
chronic malnutrition. Without a word, Satu left the compartment, made
towards the pair, and pressed a coin into the woman's hand and a tablet
into the baby's mouth. He then returned to the compartment. 'Please
come with me,' he said quietly to Pallmann. 'There is much to be done.'

Satu led the way to the last (third-class) compartment in the train,
which was swarming with passengers. Arriving at a group of young children, Satu began dispensing his tablets, at the same time speaking to
them in faultless Hindi. Then, seeing an old man squatting on the floor,
Satu spoke gently to him and also handed him one of his tablets. And so
it went on. Whatever the ingredients in the tablets, they seemed to have
a miraculously beneficial effect on the recipients.[1]

On arrival in Madras, Pallmann asked Satu where he came from,
having been unable to place his origin. 'There's no real mystery about
me, my friend,' came the response. 'I come from Cotosoti.'

'I've never heard of that,' said Pallmann. 'Where is it? In Central
America?'

'No, no, my friend. It is on Itibi Ra II.'

'Now I know you're joking.'

'I assure you I'm not.'

'I've travelled the world extensively,' said Pallmann, 'and I've never
heard of such a place. Whereabouts is it?'

Satu Ra merely pointed towards the eastern sky.[2]

'A SERVANT OF GOD'

'What he inferred,' continued Pallmann, 'was that he came from another
world, from another planet away out there in distant space. All my life I
had dealt with concrete things, concrete facts . . . Anything beyond that,
and I would be the first to admit that I was getting out of my depth. Yet
undoubtedly I had spent the last twenty-four hours in the company of a
flesh and blood person, albeit one with many remarkable attributes.'

Later, at his hotel, Pallmann received an invitation from a distin-
guished-looking Indian to visit a certain address in Madras, adding the
name of 'Mr Satu Ra'. So, the following morning, Pallman took a taxi and
found himself deposited at a palatial mansion, which turned out to be a
museum-cum-art gallery, where Satu was staying.

After greeting Pallmann, Satu pointed to an image of Lord Vishnu, one of the principal Hindu deities, with images of strange aerial craft painted on to the sacred cloth.[3] 'This is proof that earlier generations have observed the effigy of our out-of-space crafts having made an earlier landing,' he said to the bemused Pallmann, who was yet to be convinced of Satu's alleged origin. Perhaps 'Itibi Ra' might be an equivalent of the mythical Shangri-La, he wondered. Was Satu deluded? Or, more likely, was he trying to sell something?

'At the back of my mind was the thought that there must be a commercial reason for this invitation, and I expected some sales-talk from my host,' Pallmann continued. 'However, nothing of the sort happened.' The two spent a pleasant afternoon together, during which Pallmann was able to observe his host in a better light than hitherto.

He had the light brown skin of a Eurasian, huge dark eyes, a rather small mouth and an unusual chin line. The lower part of the jaw looked slightly deformed. Then, there were those finger-tip gloves which he seemed to wear at all times, even though the weather was extremely hot. Above all, there was this peculiarity of speech, this complete reliance upon an electronic gadget to reproduce his voice.'

Before the meeting concluded, Pallmann informed Satu of his travel plans, which included Kashmir, Calcutta and Benares. Benares, said Satu, would be the place where the two would meet again. But how? Satu handed Pallmann a curious ring, apparently made of solid gold, in the middle of which was what appeared to be a small piece of metal, sparkling like a diamond (see plates). 'Maybe this will indicate our presence in Benares,' said Satu cryptically, adding that the ages-old ring was 'the symbol of a human religion very close to ours'.

After thanking Satu profusely for the unexpected gift, Pallmann ventured to ask what profession his host was occupied in; a question that had been on his mind since the first meeting.

'I am a servant,' replied Satu, smiling. 'A servant of God, as are all my people . . .'[4]

XITI

On the evening of his second day in Benares, Pallmann was enjoying the night air in the garden of his hotel when the metal piece in the middle of the ring began to glow. Convinced it was a trick of the light, he tried moving it this way and that, but to no avail. If anything, the glow intensified. Suddenly, a mental image of Satu Ra came into his mind. 'The whole thing was ridiculous,' said Pallmann. 'It just could not happen. But there it was . . .'

Shortly after 21.30, the unmistakable figure of Satu approached Pallmann and greeted him warmly. The two engaged in a lengthy con-

versation covering a wide range of topics. Satu had a keen sense of
humour, and evinced a perfect understanding of Spanish when Pallmann
lapsed into that language from time to time. Although Satu's voice itself
was not peculiar, Pallmann noted, there was a quality about it 'that made
the hearer feel the inner meaning of words'.

> If he spoke of pain, then you almost winced at the word. If he spoke of love,
> then you were blanketed in the sensation of love. It is a difficult thing to
> explain. The gadget gave the voice a new dimension, a subtlety such as I had
> not heard in any other human voice. Again, this set me pondering . . .

Pallmann showed Satu some photographs of various temples he had
visited in Kashmir; different effigies to the numerous gods, pictures of
priests and worshippers, and so on. 'You have been there?' Pallmann
enquired.

'Yes,' replied Satu. 'But you know, my friend,' he added sadly, 'religion is blind, just like love.'

The two men sat down on a bench near the main entrance to the hotel
which afforded them a clear view of the bystanders. A group of uniformed hotel staff was loitering around the entrance, watching the girls
go by. When a particularly attractive girl, an airline hostess, walked down
the stairs, the reaction of the girl-watchers drew a smile from Satu. 'What
men won't do for the sight of a pretty face and figure,' he remarked. 'It is
the same in all the Universe . . .'

Asked by Satu if he could invite his sister, Xiti, along for the evening,
Pallmann readily agreed to the idea. 'Then I will summon her,' said Satu.
Naturally assuming that Satu would go to the reception area to telephone
Xiti, Pallmann was shocked to see him go into what looked like a trance.
'The expression on his face changed. It seemed as if that different chin
was suddenly possessed by lock-jaw . . . The eyes, too, were widely
dilated. Curiously, the light seemed to leave his eyes, as though someone
had turned off a switch at the back of the retina.' Just as suddenly, Satu
came back to normal, as if nothing unusual had happened.

'I for my part felt in need of a drink,' commented Pallmann, 'so I called
a servant across and ordered Scotch and soda for both of us.' As the servant departed, an attractive lady advanced upon the two men, whom
Pallmann assumed to be Xiti. As she approached, he thought it peculiar
that he had not noticed her before. It was almost as if she had 'material-
ized' in front of the men. 'My subconscious choice of mental words staggered me,' he continued. 'I had thought in terms of materialization. She
had come from a well-lit area. I could see everything and everybody in
the vicinity of the hotel entrance. Yet I had not seen her until she neared
the bench on which Satu Ra and I were sitting.'

Xiti, however, proved to be very much a material girl, and from the start, Pallmann found himself irresistibly attracted to her. She walked with a 'gliding, undulating movement, a movement in which body and arms moved rhythmically in a way that I had only noticed before with her brother'.

> There could be little doubt that this was Satu Ra's sister. There was that same different chin formation, those same compelling eyes, that same air of charm and of authority. And when we were introduced, she looked me straight in the eyes in a way that few women do. But there was no pert boldness in that look, merely fearlessness and utter frankness.
>
> Her every movement was a study in gracefulness. She was dressed in a glittering evening gown, as though she had just left a very formal reception. But although the ensemble was exotic in the extreme, there was no hint of the oriental about it, except that her tiny feet were enhanced by golden sandals. An orange half-veil accentuated rather than hid her matchless beauty.

Although Xiti spoke normally, it was evident that she employed the same technique for communicating as did her brother. On a small be-jewelled brooch around her neck was presumably an electronic gadget of some type. 'Her voice came from the heart of this fine, small brooch,' Pallmann elaborated, 'yet the sound synchronized with her lip move-ments. This was one of the refinements of the gadget. Never were lip movements out of phase with the sound.'

During the subsequent conversation, mostly in English, Pallmann decided to try an experiment. While Satu's and Xiti's understanding and use of Spanish were excellent, was it possible that a colloquial accent might confuse them? 'I persisted with my experiment, continuing to speak in Spanish, but ringing the changes, so that at one time I spoke as though I were a native of Spain, the next of Peru.'

> Their facial expressions changed as I altered my intonation. I could see their puzzlement reflected in their eyes. They looked at each other intently, as though they were listening to strange, unknown sounds . . . They seemed to be caught in some mental activity induced by imagined sensory impressions that were causing them some tension and ill-feeling. Immediately, they switched back into English. Thus I knew that they were not truly polyglot but were relying on some mechanical device . . . the gadget worn by Satu Ra and the brooch by Xiti.[5]

THE UNTOUCHABLES

The following morning, Pallmann was wandering along the banks of the sacred Ganges River in Benares when he was suddenly joined by Satu Ra. How he had known precisely where to find Pallmann, among the throngs of people, was a mystery. Satu led the way in the direction of the Ramakrishna Monastery. 'Because of certain primitive elements and castes which cling to Indian society,' said Pallmann, 'I was surprised to find my friend amongst these, the poorest and most miserable creatures: the untouchables.' Soon, they were mixing with other castes in the middle of the *mahabhinishkamana* ('the way to ultimate resignation'), 'the vast dumping ground where people, young and old, men and women, who are at death's door, are brought to await the end'. Satu began to go about his ministrations.

'I have never seen anybody, man or woman, professional or amateur welfare worker, act with such compassion and gentleness as I saw Satu Ra carry out his works of mercy,' continued Pallmann. Satu headed directly to a dirty, crying child, huddled over the body of her mother who had just died. With the utmost care, Satu washed the child as best he could and spoke comfortingly to her.

Shortly afterwards, Xiti, dressed in a green sari, appeared on the scene. She, too, was ministering to those children in need. Seeing one little girl covered with open sores, Xiti took out a yellow paste of some sort and covered the sores. 'The effect was little short of miraculous,' said Pallmann. The girl stopped crying and even managed a faint smile. 'The ointment seemed to be as much a panacea as the tablets that Satu Ra had dispensed on the train.'[6]

THE MOMENT OF TRUTH

That evening, Xiti having left the scene, Satu Ra hailed a rickshaw and informed Pallmann that they were going to visit the 'Children of God', the name given to a pitiful, primitive crematorium on the banks of the Ganges. 'I counted almost forty funeral fires, some so close that they were almost contiguous,' remarked Pallmann. 'We could see small, tormented limbs dangling outside the immediate orbit of flames until the consuming fire broke them off. The attendants snatched up the freed arms and legs and threw them back into the flames as if they were feeding a garden fire more twigs. It was a sickening experience . . .

'Life to these children had meant nothing but suffering and despair. What else could be expected in a country afflicted by poverty and starvation, set in a world riddled by fear, hate and war? These are reasons why nameless children are burnt at night beside a majestic Indian river.'

In a sad and reflective mood, the two returned to the hotel. It was there that Pallmann came to accept Satu for what he claimed to be.

Until that moment of truth, there were certain things that he had said that I had taken with the proverbial grain of salt. After the moment of truth, I was prepared to accept everything he said as gospel. Henceforth, as far as I was concerned, he, and Xiti for that matter, was always a witness of the truth. I don't know what alchemy it was that brought us to the moment of truth . . . Whatever the cause, on that memorable night I accepted that Satu Ra had come from another planet named Itibi Ra II, and that his people had discovered Earth in much the same way that Columbus had discovered the New World: by a deliberate voyage of exploration.[7]

FURTHER DIFFERENCES

There was something about Satu Ra's smile which bothered Pallmann: his teeth never showed. Perhaps this had something to do with the rather different jaw formation and the fact that his long, thin and sensitive lips always seemed to cover the teeth completely. Pallmann's curiosity had not gone unnoticed. Satu explained that, for many thousands of years, men and women on his planet had lived without teeth, gradually finding them unnecessary. However, on Earth, they did use a type of artificial support that kept the shape of their mouths similar to those of humans. On closer inspection, Pallmann noticed that Satu and Xiti had rather small tongues.

Pallmann's wish to inspect his friends' fingers was also granted graciously. When the protective covers were removed, the differences were immediately apparent.

In contrast to the feminine hand, the male finger-tips are flat and round, like little discs. Extremely sensitive they must be as there are no nails whatever, with the very rosy, fine and soft flesh extending to the very end of each finger. Xiti's hands were a true masterwork of nature: pointed and extremely thin, very long, entirely different from her brother's.

Both Satu and Xiti appeared to be very amused at Pallmann's mystification. 'But because of their kindness and frankness they came so much closer to my heart,' he added. 'They spoke to me like real friends, telling me also the reason for these differences.

'It seems that they are able to analyse sound, and perhaps are even able to "hear" through the sensitive nerves of their finger-tips. Also, at later times, I became sure of the fact that they were using their fingers as we would use our tongues for tasting and exploring, specially when doing biological research work . . .'[8]

THE YAVARI RIVER STORY

Ludwig Pallmann did not see Satu or Xiti in India again.

In Zurich, he took the peculiar ring to a jeweller, who commented that the pre-Columbian design and gold-work of the 'God' on its surface were unlike anything he had ever seen, and recommended that the ring should be shown to a specialist. 'The diagnosis of this expert was that he believed this to be a masterwork of great value, belonging to one of the earliest pre-Columbian dynasties,' claimed Pallmann. 'What intrigued them to the point of utmost curiosity was the metal insert, which I believe to be of extra-terrestrial origin.'[9]

A few years went by. Pallmann was busy installing milling and pulverizing plants in Argentina, Mexico, Colombia and Peru. While in Iquitos, Peru, he heard an interesting story from an Austrian tour guide, a rugged individual who had spent much time in remote jungle areas in that part of the country. 'On the other side of the Yavari River, I saw several white explorers turned native,' related the guide, describing an incident that had occurred while he was suffering from a fever. 'The funniest people I ever met; with hands so strange, that I thought them to be from a different world.' Pallmann pricked up his ears.

'They might have talked a lot of nonsense,' the guide continued, 'but they were such fine engineers. They even fixed a broken out-board propeller blade for me so that I could get back to the Yavari. For all their craziness, they were good at doctoring as well.' Pallmann bought the guide a drink and pumped him for more information. The Austrian opened up somewhat.

'This fellow with the funny mouth, a legacy of some fever, I suppose, gave me a tablet to swallow. I felt better almost immediately. Then he gave me some fruit juice. That was the best fruit juice I've ever tasted. Yes, they were white folk turned native all right. I told them to give it up, and come back to civilization, but they refused . . . I told the missionaries on the Brazil side of the border what I'd seen out there on the Yavari River. They wouldn't believe a word I said. Made out it was the fever. Said no white man would dare to go into cannibal country . . .'[10]

THE HOSPITAL VISIT

In early 1967, Pallmann was sent to the Maison Français Hospital in Lima for an operation on his right kidney. Fortunately, he was allocated a pleasant room with a private bath on the ground floor of the hospital, with an inner door that led to an antechamber and thence out to a patio-garden. He resigned himself to a three-day wait in what was then a heat-wave.

On the second night, racked with pain, Pallmann reached for the bell-push. It was almost three in the morning. 'My groping fingers failed to find the bell-push that would summon help and relief. But I did find something else: a hand that came from the pain-racked night and clasped my own. Tormented by pain as I was, I still felt a shock when I found my hand grasped by another slim, warm one.' It was Xiti.

Without a word, Xiti smiled, took the ring off Pallmann's finger, and gave him one of her healing tablets. Because of his pain, he had failed to notice that the metal inset was glowing. A faint light reflected from Xiti's talking device. Still without speaking, she ran her fingers over his fevered brow. The pain and fever immediately subsided, and he embraced her in gratitude. She stayed for the remainder of the night.

As they talked, Xiti told him about a nurse at the hospital, Maria Navidad, whom she had wanted to see, but had hesitated to do so for several reasons. It so happened that this young lady had been rescued as a baby by the *mestizos* (racially mixed people) and the Austrian tour guide, near the town of Pucallpa on the Ucayali River, and raised by the Catholic sisters who ran the hospital. Satu and Xiti knew Maria's mother, whom they had rescued after she had been terribly beaten in an area near their first landing site. Having been healed by her rescuers, Xiti explained, she had been taken to their home planet. Perhaps because she had been unable to adjust to the different planet, she had died soon afterwards.

Xiti predicted that Pallmann would be free of pain for six months. And so he was. 'I became the miracle patient of the famous Maison Français Hospital,' Pallmann declared. 'When the doctors came around to put me on the operating table, I had already eaten a very heavy breakfast, a thing I had not done for almost three weeks. I had gotten out of bed, a new man in need of a hot and cold bath. Feeling perfectly well, I had ventured outside and eaten in one of the little Chinese coffee-shops, the *chifas*, as they are called . . .'

> You should have seen the raised eyebrows, the looks of disbelief among the medical people when, instead of being wheeled into the operating theatre, I told them I was going to discharge myself as cured. They could tell by just looking at me that I was infinitely better, and although they agreed to postpone the operation, they insisted that I should remain in the hospital at least until the next day in order to make another series of exhaustive tests, and also to ensure themselves that I did not have a relapse.

Pallmann readily consented to the proposal. For the rest of the day he was submitted to a battery of tests. All proved negative.

Later that day, Pallmann asked Sister Marta at the hospital if he could meet Maria Navidad. The meeting was brief, and rather poignant. When Pallmann mentioned that he had heard about her rescue as a baby, and her mother, the nurse looked stunned, then tears ran down her cheeks. She spoke not a word, and Pallmann felt ashamed for having asked her for more information.[11]

AN ALIEN LIAISON?
The following morning, Pallmann left the hospital and checked in at the Savoy Hotel in Lima, having been unable to find accommodation at the Hotel Crillon, where he had a pre-arranged meeting with Xiti at the Sky Room that evening. Her entrance created quite a stir. 'Immediately, and because of the minute blue veil she wore,' said Pallmann, 'people noticed the subtle difference between her and "ourselves" – our people, from our

planet. By this, I mean not just cultured Peruvians, or the many Europeans and North Americans staying at this famous first-class hotel, but even the less instructed bell-boys and lift operators, stared at Xiti. But instead of finding her embarrassed or shy, she looked at me and everybody else with the greatest of ease.' Pallmann ordered drinks and the couple spent the rest of the night together.

During the next few days, Pallmann began to learn more about his friends from 'Itibi Ra II'. Xiti's feeling of security, for example, was apparently related to the advanced spiritual and mental perceptions practised by these people – what Xiti supposedly referred to in their language as 'amat mayna', or 'science of soul'. 'They are able to read our very thoughts,' Pallmann averred, 'and may be able to influence our thoughts should this be necessary because of security reasons.'

Xiti's interest in and enjoyment of music were immense. When passing a record store in Lima, for instance, she showed delight in the rhythm of the Colombian cumbia. 'Seldom have I seen a happier look on someone's face as this strange woman passed the record store,' remarked Pallmann.

To obtain the local currency, Xiti gave Pallmann several gold ingots, which he exchanged at a commercial house in Union Street. Even though quite a few 'adventurers' and certain Indian natives in Bolivia, Ecuador and Peru still traded in gold at that time, Pallmann claimed that 'the beautifully melted and carved ingots surprised the specialists'.[12]

INSIDE THE FLYING SAUCER

Pallmann gladly accepted an invitation from Xiti to meet her brother in Huancayo, a town high in the Andes about 130 miles east of Lima. On arriving at the station, on 17 February 1967, Pallmann did not at first recognize Satu Ra. 'He was dressed very much like the natives, with heavy woollen gear. There wasn't much difference to be noticed between his looks and the taxi driver whom he had charged with helping to unload all our baggage.' (Xiti, Pallmann remarked, had brought with her 'suitcases full of books, records, seeds, and God knows what else'.)

Some distance outside Huancayo, the taxi driver was paid and the three were left by themselves. It was while they were watching the sun go down from a peaceful lakeside that Pallmann claims he saw his first 'flying saucer'. It was an awe-inspiring experience.

'So much has been written and talked about on the subject of unidentified flying objects and a great deal of money has been spent by various military and private research investigators,' Pallmann explained, 'but despite all this, when you actually see a flying saucer for the first time, I believe that not one in a million scientific investigators would be able to explain the fantastic feeling that I experienced.'

Pallmann's subsequent description of the craft reads as if it were science fiction. In many ways more fantastic than descriptions given by other contactees, it is, nonetheless, fascinating, and should come as a challenge to critics who often find such descriptions suspicious.

> There was a soft but painful noise, or rather reverberation, as the saucer glided towards the edge of the lake, right to the spot where we were waiting . . . As the noise [reduced] a few decibels from painful to 'bearable', the saucer hovered, and opened up underneath its circular surface. Like a giant crooking his little finger, an embarkation device, soft and gripping at the same time, scooped us up and deposited us in some kind of 'antiseptic reception quarter'.
>
> Immediately, I became aware of the biological, vegetational, cellular structure – similar to soft polyethylene – embellished with exquisite designs and symbols. Only the flooring was a little harder, and I suppose the reason for this must be its mirror-like quality. Through this floor [one could see what looked like] a billion nerves and bloodvessels . . .[13]
>
> Inside the craft there was a discreet hum; the rhythm like the sound associated with low-voltage waves, or with turbines, as I thought then. Evidently, the reverberations I had heard and felt when first observing the flying saucer settle were either linked with particular manoeuvres, or were merely externalized noise.

Pallmann says he was 'stripped to the buff' to take a bath. During the bath, he fell asleep, then woke to find himself in a very comfortable, soft sleeping device, suspended like a hammock, but 'attached to many hundreds of fine and multi-coloured "veins" and "vessels". This, I later was told, is part of a "medical computer system" (health analysis during sleep forming only a small part of life-preserving treatments).'

Xiti, who had either risen earlier or not slept at all, brought Pallmann a kimono-like garment to wear. 'Breakfast' was not to his taste.

> I found the gelatinous-looking plants from their planet impossible to eat, and I tried the complicated arrangement of small containers from which I was supposed to sip. I was curious about the contents of these gadgets, so I started to sip at random. They all had a wonderful time just laughing like children about my lousy behaviour. The wife of one of the astronauts showed me how to do it. Nevertheless, I left practically with an empty stomach.

THE EYE

One of the most remarkable discoveries for Pallmann was the absence of doors, locks, keys or rooms, such as we know them (though private quarters are alluded to later). No mention is made of toilet arrangements. Everything, even the 'commanding cell', called 'yano' or 'the eye', formed part of the biological structure of the craft.

There was not one straight line, so to speak, in the whole space-craft, nor were the circular forms 'exactly' circular. At all times, the 'eye' of the craft is part of this body. This centre unit of the craft was so geared with other instruments that its power of involvement was complete. In other words, the 'eye centre unit' is some kind of an activated memory, a transmitting and receiving centre, similar to our brains . . . I was able to experience later how this individual brain of the 'saucer' became part of the giant system of cosmic generator-brains and, in particular, how this great individual unit had to be considered a minute part of the great memory computer on the home planet itself.

THE HOME PLANET

At all times, Pallmann was encouraged by his hosts to ask questions and to give them his impressions. From an 'observation post', he was shown images of the home planet (in a solar system located towards the centre of our galaxy, he learnt later), such as methods of transportation, food-processing installations, 'biological machinery' and various instruments. 'I even listened to a concert, and invaded some of their homes,' he claimed. 'I use the word "invaded" deliberately, because the involvement brought about by the "eye" makes one feel as if one is actually going to these places; going to the concert, for example, or visiting friends in their own home.'

Everybody looked happy in this Utopian society. 'Everybody seemed to be smiling, young and old. There didn't appear to be many unhealthy people about, but the "generator" was trained on to what they call "health centres", and I saw that even patients there were smiling.

'Linked with the all-pervading air of happiness on Itibi Ra II was an atmosphere of calmness and serenity. No one seemed out of patience. Nobody appeared to be in a hurry . . .'

Pallmann liked what he saw of the aliens' architecture. 'Most of their homes were built along river banks and the sides of lakes and other waterways. Their architecture was unlike any I had seen on Earth, except in futuristic exhibitions. They delighted in dominant colours . . .'

THE FACTORY PLANET

What intrigued Pallmann most about Itibi Ra II was that it was fused with two very small satellite planets. One of these smaller planets functioned as a giant biological artificial 'heart', pumping 'power' into the fused planet, while the other acted as a 'factory'. The Itibi Rayans, we are told, separate all artificial and mechanical working machinery from their normal, domestic surroundings.

Through the eye-generator I was able to peer right into the bowels of this factory-planet. To me, it looked for all the world like an opened-up octopus. The vast number of tentacles were, I suppose, the channels and cables tapping the power sources . . . It was from The Factory that my friends had boundless energy for any purpose they desired: for cosmic transportation, production, research, and climatic conditioning. Because of these vast and endless supplies of energy taken from Nature itself, they were no longer bothered by earthquakes, floods, hurricanes . . . and other similar disasters.[14]

THE AMAZON PLANTATION

According to Satu, extraterrestrial visitors discovered Earth many thousands of years ago. The 'Itibi Rayans' themselves, however, did not come here until 1946, when supposedly they established several plantations in South America for hybridization and research purposes. The plantation to which Pallmann claims to have been taken lay southwest of Iquitos, close to the Peruvian border with Brazil, between the Mirim River and the larger Yavari River into which the Mirim flows.

'The day after my arrival,' he wrote, 'I first became able to observe the surrounding area from my position near the giant "eye-generator". Because of certain quarantine restrictions, I was not allowed, during the morning, to step outside the craft and meet other members of the crew.

'I was wondering why the Itibi Rayans, of all the lovely places on Earth, had found this God-forsaken Green Hell, perhaps the worst spot in the entire world, to do research. "Surely," I said to Satu Ra, "you are not so afraid of the human race that you choose the worst possible spot on our planet?"'

'The climate here is perfect,' replied Satu, laughing.

Part of the research programme carried out by the Itibi Rayans involved the collection, study and hybridization of various plants and fruits required for their diet. 'Itibi Rayans, I soon found out, do suffer physically if they do not get enough new flavours,' Pallmann explained.

Of course, there are other reasons why these people from a different world go on these tremendous voyages of exploration. They are motivated to do so to keep fit, to keep on the move, to be superior . . . What has motivated Earth men to land on the Moon, and what shall motivate Earth men to step on to Mars and even Venus?

In order to grow new fruits and exciting new plant flavours, our friends have brought with them the most interesting specimens of vegetation, and they are real biological wonders. Some of these are cross-fertilized and transplanted with plants entirely different to theirs. Plants from our planet, and others, are used to produce new taste sensations . . .

During many hundreds of thousands of years, the people of my friends have evolved from meat eaters to vegetarians. I was informed that the evolutionary process in their food habits also brought in its wake a character problem . . . 'love of over-tasting'.

When the Itibi Rayans first found the tropical chirimoya fruit, for example, they became almost gluttonous, Pallmann related.

We shall never be able to experience what these conditions really mean, as our palates are not as sensitive as theirs. Having stopped eating solid food, having forgotten how to kill an animal and eat it, they also have entirely different nerves. No more war, on human beings or animals, except in self-defence. They have become planters, scientists, explorers, teachers, religious philosophers (cosmophilosophers), biologists, etc.

TELEVISUAL COMMUNICATION AND SPORT

During Pallmann's third day on board the spacecraft at the plantation, he observed Xiti talking to her parents on a two-way communication system similar to a television. On touching it, however, he discovered that, just like the spacecraft itself, it seemed to be of a biophysical nature. The conversation was in the Itibi Rayans' own language, which Pallmann described as 'a rather high-pitched melodic whispering, very charming and . . . with humorous undertones'. He was also able to see Xiti's parents' home and others in the neighbourhood. 'At one time, Xiti also spoke to a neighbour of her parents and the "eye-generator" had "gone" directly into the inside of her house, also very charmingly decorated with symbols similar to those I saw within the space-craft.'

To Pallmann's surprise, Xiti then 'switched channels', as it were, to a sporting event. 'I must say it was one of the hardest games I ever saw in my entire life.'

I understood that the huge amount of players, perhaps 7,000 young men, were engaged in a giant-size ball game, where a final selection of the most able and strongest also ended with the victory of the most intelligent team. To play this substitute for war, the young Itibi Rayans used several hundred computerized, electronically controlled gadgets, very similar to multi-coloured footballs. I understood that not the referee, but the different balls (perhaps also interconnected with each other) have to decide the game. To me, the whole thing looked like football on a giant chess board, played rapidly over a playing field of about five [square] miles. Decisions were made by those having reached a higher commanding status because of bravery and intelligent behaviour.

The hardness and often brutal behaviour of many lower-rank players really surprised me. It seemed to be absolutely contrary of what I had thought about the Itibi Rayans so far. I asked Xiti about it . . . she explained that these events have prevented war and bloodshed for many thousands of years. And yet, these games, she said, made it possible to keep the inborn and instinctive fighting condition of mankind intact. There is also the genetic reason to keep fit, to be healthy through hardship and sporting bravery.[15]

SPACE-TIME TECHNOLOGY

There were three spacecraft at the Mirim base, of which only one carried a crew. The other two were uninhabited supply craft. 'I must emphasize,'

Pallmann pointed out, 'that only science-fiction calls space-ships "flying saucers". That is a solecism of fantasy. I doubt if space-ships actually fly in the accepted sense of the word. They are propelled by cosmic waves.'

> A minimum fleet of twenty-seven to thirty ships are needed for operating within our solar system. The power units, or the carriers, are at all times above the control and supply ships. It is the carriers that arrange for the power to be switched on or off. The three-dimensional fusion of the carriers accords with the cosmic condition of the third dimension itself, and this makes it possible to [reach] a target at a very high speed, much faster, indeed, than the speed of light.

Several hundred thousand years ago, Pallmann was informed, the Itibi Rayans had been obliged to evacuate their dehydrating planet of origin (Itibi Ra), an evacuation involving several trips to and from the old planet to move people, animals, insects, plants, biological machines, recording devices, musical instruments, and so on. 'Indeed,' wrote Pallmann, 'only the necessity to survive had forced the Itibi Rayan scientists to think about travelling on to another planet and to create the necessary means of transportation.' Pallmann's elaboration of these 'necessary means' is hard to follow – and harder still to swallow.

> Only because of their highly advanced understanding of all life-creating ways of nature were they able to create and test a series of dimensional filtering and prismatic-type 'life-receiving' space-batteries, reacting to the inter-cosmic forces of colour, light, temperature, time, and other cosmic waves. The Itibi Rayans . . . created a new, fascinating 'interconnection' of cosmic batteries, reaching the dimensional scientific 'switch' from 'receiving' to 'sending' cosmic forces. In other words, instead of waves being received, activated and returned, they were able to move with the activated 'returned' waves themselves.

Assuming Pallmann's story to be neither the product of a deluded mind nor an outright hoax, such a vague elaboration might well be due to his own failure to grasp what was told to him. 'The biological structure of the space-craft makes it impossible – even for a technically trained man – to draw a blueprint,' he noted in his diary at the time. '*What* makes our brains, our nerves, transmit orders to our bodies to move heavy weights . . . Yes, they tried to explain! But I do not even know how a television circuit works, much less shall I ever understand this . . .'[16]

A PARADOX

It was implied that 'ordinary' human beings were as yet not conditioned to accept or mix freely with the Itibi Rayans. Yet, paradoxically, the local Indian peasants not only mixed freely with, but were employed by the extraterrestrials to work on their plantations. As Pallmann explained:

These Indians were employed on very humdrum tasks, keeping the area free of insects, because, despite the protective covers, insects did manage to find their way into the seedlings and saplings. The Indians looked upon their employers as light-skinned foreigners from another part of the world. I doubt if they gave a second thought to the rather unusual chin formation of the Itibi Rayans. In any case, the simple Amazon Indians would not have believed that people could come from other planets. They would have rejected the story in exactly the same way as most of us would reject the idea of several men having been landed on the Moon, if we had not seen it on TV.

'At first,' Satu Ra told Pallmann, 'the local Indians looked upon us with some caution. But then Xiti and I began to heal their wounds, and cure their sick. They soon came to accept us.'[17] On one occasion, Pallmann claims to have witnessed a group of Indians alighting from a spacecraft.

Out stepped the most audacious group of wild-looking but smiling savages followed by a bunch of serious Itibi Rayan explorers. There was plenty of excitement, but what really made me shake my head was this: these, perhaps the most feared man-eaters of the endless forests . . . were laughing and giggling like little girls. What an excursion it must have been . . .

The Itibi Rayans had known these men already since their first landing near Pucallpa. The amazing thing I discovered was the age of these Indians: all over 50 years, looking as healthy and young as those Indians being only 20 or 25. Another controversy! Had they been used as guinea-pigs by the Itibi Rayans?

LINISLAN

Though reluctant to talk about the possible use of the Indians as guinea-pigs, Xiti was more forthcoming when discussing certain discoveries her people had made regarding ancient South American civilizations. These discoveries were allegedly made during advanced forms of excavation in certain areas, while the Indians stood guard near strategic waterways and swamp-passages. The Itibi Rayans had located the remains of a huge long-lost city, given the name of 'Linislan', buried beneath a layer of seven feet of tropical growth. There, inside a temple, they discovered a huge pre-Columbian symbol, which they said proved that many thousands of years ago another extraterrestrial civilization had first landed on Earth. Xiti showed Pallmann a similar symbol on one of the control panels of the spacecraft.[18]

A FRUITFUL TRIP

By this time, Ludwig Pallmann was becoming increasingly concerned about his business affairs: a backlog of work awaited him in Lima. Yet so fascinating was the time he spent with the Itibi Rayans that when invited by them for a trip to Colombia, on 20 February 1967, he accepted immediately.

Another thing that bothered him was how Satu and Xiti, and others of their race, managed to travel around various countries without some sort of passports. What would happen if one of their men or women was arrested? 'Little did I know,' wrote Pallmann, 'besides the fact that Xiti used a perfectly imitated Argentine passport, that all Itibi Rayans know exactly what to do and what not to do. For instance, in many countries, it is useless to show a passport if that passport does not show an entry stamp from the airport police.

'On the trip to Colombia, Satu Ra decided not to use passports at all, but to proceed at night and only stay a very limited time and at a place where the chances of detection were absolutely out of the question.' Shortly after 22.00, the craft departed for Colombia. As Pallmann described the trip:

> The extremely short criss-cross over great altitude and distance was a disappointment. Exactly like on the first flight near Huancayo to the Mirim River base, I did not notice, see, hear or feel anything at all. But I did observe, and with the utmost interest, the immediate and very clever control-craft protection carried out in the darkness of what I was able to understand to be a huge delta swamp of the Magdalena River south of Barranquilla, [on the Caribbean Sea coastline of] Colombia.
>
> Within seconds the space-craft had covered itself with a special liquid coming out of a million pores which, besides being a perfect element of camouflage and natural colouring, also served as a bacteria and insect repelling agent. This only lasted about five to ten minutes.
>
> When finished, we immediately embarked in two very comfortable and very flat speed-boats [which] on both sides, and on the bottom, were propelled by tiny and silent generators. There was no motor at all but a great number of air-jets, working in absolute silence. I figured the speed [at] about 30 to 35 m.p.h., and the trip itself lasted well over an hour . . . I was only able to speak to Mr Satu Ra, as Xiti had not received clearance to join the party and all the other crew members did not carry language computers.

The group reached Barranquilla and found an isolated spot on the embankment. Most of the Itibi Rayans wanted to rest and observe the neighbourhood, but Satu invited Pallmann to see the night life of Barranquilla, Colombia's largest coastal town. Naturally, it was the fruit above all else which attracted Satu. 'Satu Ra displayed a naïvety that was astounding for one so astute as himself,' wrote Pallmann, who had given his friend some Peruvian money to purchase samples. 'He inspected the fruit, turning and prodding, but he did not buy anything. Instead, he offered a stall-seller money merely for the privilege of inspecting the stock, smiled politely, then moved on to the next stall. Each stallholder accepted the money with alacrity . . . I suppose they looked on the money as a tip given to them by an eccentric foreigner.'

Pallmann, meanwhile, having been starved of 'real' food for several days, devoured a grilled half-chicken, upsetting Satu in the process. 'I knew what he was thinking: that it was a crime to kill a bird just for a human being to eat it. At that moment, I must confess, I was out of sympathy with Itibi Rayan philosophy . . .' Meanwhile, another member of the crew, 'Mr Hua', second-in-command of Satu's spacecraft, appeared on the scene, and the trio set off for another market.

Examining a guayaba closely, Satu asked Pallmann for a detailed description of this, to him, unknown fruit. Having satisfied himself that the fruit could be cultivated, some was bought. An hour was spent looking for a specimen of the guayaba plant, but to no avail, so the following morning Pallmann went back to Barranquilla, where he was directed to the town of Santa Marta, across the river. Here, his search eventually bore fruit when he located some cuttings, which were handed to Satu at the rendezvous point the following night.

During the trip, Pallmann had bought himself a new camera, with a view to taking some photographs of the Itibi Rayans, their plantation and their craft. But it was not to be. 'As I feared, Satu Ra took a special interest in the camera,' wrote Pallmann.

> He told me about 'Amat Mayna', the science of soul based on ancient beliefs. Not that my friends believe in reincarnation, but, definitely, they do not care for photos and pictures because of certain implications. I understood that, besides certain security restrictions, they simply do not care about 'their looks'. They are devoid of all vanity, pride or feeling of superiority . . . During all the time, and particularly where Xiti was concerned, I never saw them use a mirror.

Satu confiscated the camera until Pallmann returned to Lima a week later.[19]

BACK AT THE PLANTATION

At the Mirim River plantation, Pallmann was show how the Itibi Rayan botanists went about their research and cultivation work. The plantation itself was laid out under huge green protective sheets.

> Air filters and humidifiers had been installed at strategic points so that, no matter what the weather, the plant biologists could always have controlled weather conditions inside the 'flavour station'. The main path through the plantation complex separated the station into two sections, each of which was made self-contained by means of coloured dividing sheets that were rigged tree-high.
>
> In front of the actual biology research laboratory was a wing consisting of several large tents [where] many vegetable 'guinea-pigs', which had been brought from Itibi Ra II, had been transplanted, and had then been used as

required for grafting on to samples of Earth vegetation [in order] to obtain as fine a strain of individual plant life as it was possible to get by uniting the best of Earth types with the best of the Itibi Ra types.

The biology research laboratory . . . was a series of interconnecting marquees, stretching for some 350 feet, and was some 60 feet wide. It was divided off into experimental bays, rather like the operating rooms of hospitals. In these bays . . . the finest instruments were used to dissect the cells of plants: the veins and stems were put under close scrutiny. X-ray pictures were taken, not the normal plate-type X-rays but a continuous record, rather like a roll of film. The plant 'surgeons' . . . could watch on separate left- and right-hand panels let into the wall. On these panels, the eye-computer projected a continuous report of the dissection as it proceeded. These television-type panels were studied throughout the entire process by special observation 'officers', who indicated their opinions to a chief scientific officer who controlled the actual work itself . . . The biologists sat at their work in the Oriental manner.

RELAXATION

Pallmann was invited to visit the bathing tents on the plantation. The Itibi Rayans, he learned, bathed at least twice a day: before going to work and when work was finished in the late afternoon. 'Their bathing habits are a combination of the Finnish and Japanese,' claimed Pallmann. 'The normal bath is like the Finnish sauna unit, and they have both wet and dry bath units. Because of their lack of inhibitions about nakedness, men and women bathe together.

'I noticed that Xiti, who is meticulous about her personal hygiene, was scrubbing furiously as if she had done filthy work in the laboratories. I remarked upon this, and she frowned a little. "Can you not smell?" she demanded . . . then she told me to my face that I had eaten meat. I laughed like an idiot. On two occasions I had eaten chicken in Barranquilla. Xiti grinned and pulled a face at me . . .'

The domestic arrangements on board the spacecraft also drew Pallmann's admiration.

The dining quarter and 'health' lounges were bright with decorations. Lovely, soft divans and deep cushions, gay with floral patterned covers, invited relaxation after the day's work. I admired the way in which the women, some being the wives of the astronauts, who shared similar jobs to the men all day, could shed their technical role during off-duty hours and revert to an essential feminism such as one experiences in Japan. They even took it upon themselves to see to the domestic side of the expedition.[20]

RELIGION AND SOCIETAL DEVELOPMENT

Asked about their religion, or 'cosmophilosophy' as Pallmann called it, Xiti and Satu said that their people make no distinction between 'God' and 'Nature', referring to them (in English at least) as 'God-Nature'.

'Disregarding the laws of Nature,' said Satu, 'is disregarding the laws of God, because God is Nature and Nature is God.'[21] The value of religion, Pallmann was told, should depend on the 'active role it is able to play in civilization's progressive and futuristic pattern'.

Regarding the future of our society, Satu predicted that a new social and political structure would be brought about within 100 years. As Pallmann described it:

> I was very much surprised when Satu Ra told me about a great feeling of friendship which shall come about over many nations on our planet because of a unique political situation I never believed possible. He mentioned [that] the whole planet Earth, within one hundred years from now, will benefit from the friendship he predicted between the United States and Russia.[22]

Satu Ra's prediction about the superpowers has come to pass. Let us hope for the predicted era of friendship on a wider scale.

HEALTH AND LONGEVITY

According to the Itibi Rayans, Earth is one of a number of planets, referred to as 'cancer planets', which are particularly prone to cancer. In addition to the known or suspected causes, they laid the blame on our modern, artificial and materialistic life-style (citing the lack of cancer among the Amazon Indians), as well as on other, sometimes inherited factors, such as fear, stress and sexual repression. Added Pallmann:

> They also know that we suffer as a result of many mental disorders, besides our many physical disorders, like blood and respiratory disorders. They have seen for themselves that our stomachs, hearts and glands are not working like theirs, that 80 per cent of us are suffering from some kind of constant tension and of what they know as . . . unnatural irregularity, leading to cancer.[23]

Satu Ra claimed to be 250 years old, measured in our terms; a modest age when compared with Walter Rizzi's alien, who communicated that his race lived up to one hundred times that of Earth people. Compared with ordinary human beings, Satu would have been in his early forties. He expected to die sometime between the years 2210 and 2220.[24]

Overpopulation was given as one of the main causes of misery on planet Earth. Satu and Xiti emphasized the need for both political and religious leaders to impose the strictest regulations to control our present growth of population.[25]

LOVE AND MARRIAGE

On Itibi Ra II, couples fall in love, marry and have children as we do, but, Pallmann was informed, marriages normally break up soon after the children enter an 'educational centre' when they are about six years old. (The 'seven-year itch', it seems, is not restricted to Earthlings.) At first shocked by this custom, Pallmann came to appreciate the fact that, if monogamy were practised on Itibi Ra II, some marriages would have to last for 400 years or more! Satu pointed out that, although incompatibility between marriage partners was normal after the seven-year period, compatibility, resulting in long-term unions, did occur. Such unions did not require the sanction of the equivalent of a registry office ceremony, or even a formal exchange of vows. Satu's current union (with a woman from another planet), he told Pallmann, had lasted for 90 years.[26]

Having also learned from Satu about the erotic behaviour of the Itibi Rayans, Pallmann had initially considered including some details in his book, but the publisher had advised against it. 'Actually, all I wanted to do was to describe the very healthy and natural behaviour of another civilization,' he explained.[27] Yet Pallmann, too, imposed censorship. 'There are matters on which I have had to maintain my privacy,' he wrote earlier, seemingly contradicting his later remark. 'As far as I am concerned, and especially as far as the sexual habits of the Itibi Rayans are concerned, I have tried to reveal exactly nothing, and I believe I do have the right to do this, simply because our own sexual habits are far from free.'[28]

On 26 February 1967, Satu sadly informed Pallmann that his people had received orders to evacuate their plantations in South America. The following day, he was taken back by spacecraft to the Peruvian highland lake where he had originally been picked up.[29]

NEWS

Nearly two years passed. Pallmann bought a property in El Salvador, Central America, a lakeside fishing and hunting lodge affording magnificent views of the surrounding scenery, including the San Vicente volcano. He began writing the manuscript of a book describing his claimed experiences with the Itibi Rayans. 'I had lost some of my diaries and it wasn't an easy job to find dates and names. Certain places and words I had synchronized by sound I could not write down in "human" language at all,' he explained. 'I simply had to use similar words and sentences so far as the dialogue with these people is concerned.'[30] The Itibi Rayans, he further explained, do not use letters or print.[31] As to use of the Egyptian word 'Ra', for example, which surprised Pallmann, this was taken as further evidence that an extraterrestrial civilization had been on Earth thousands of years ago.[32]

About two weeks after taking over the property, on 15 January 1969, he felt a burning sensation from the ring Satu had given him, the inset flashing and gleaming. Later, having taken his small motor-boat to a sandy beach near the Isla del Altar, he noticed that the normally placid waters of the lake were ruffled by several huge concentric circles. 'There could be but one explanation,' he wrote. 'Somewhere near at hand, my friends from Itibi Ra II had effected a landing.' Shortly afterwards he encountered the figure of Satu Ra, sitting motionless on a rock. 'He was inexpressibly sad. I noticed that his clothing was of a dark green, that he wore a broad instrument belt, on which was a much larger talking device than the one to which I had become accustomed when I had stayed with him.'

'Where is Xiti? Is she with you?' asked Pallmann.

'Xiti is dead,' came the shocking reply, in Satu's own language (*'Ximsi Xiti Tasat'*), followed by the confirmatory translation in French and Spanish, the languages in which he and Pallmann normally communicated. Supposedly, a disaster had befallen an expedition to another planet, killing Xiti, Mr Hua and many other crew members aboard their spacecraft. The two sat talking sadly for more than an hour.[33]

Not feeling like talking to his housekeeper or the gardener, Pallmann drove in a daze to a doctor friend at San Pedro Nonualco, where he stayed the night. The following morning he was awakened by the newspaper boys, shouting about a flying saucer having been sighted over the capital, San Salvador, and the San Jacinto Hills that surrounded the lake where he lived. When he bought the paper, there were the banner headlines: 'OVNI Vuela Sobre San Salvador'.[34] As Pallman related:

> From the reports, it seemed that shortly before [Satu Ra's] visit, the space-craft was reported over the Cerro de San Jacinto and had then continued high above San Marcos. The amazing thing is that the spacecraft had silently, and for quite a long time, stayed in an observation position directly over the extensive capital town of San Salvador, exposed to the vision of several hundred thousand people.

Pallmann reports that one of his neighbours observed the spacecraft as it came down at tremendous speed and settled, as if on an air cushion, between the isles of Los Quemados and Los Patos, exactly midway between his house and the Jiboa River outlet of the huge tropical lake.

> I do not know of any person on the lake having seen the return of the space-craft, nor do I myself know on which part of the lake Satu Ra took his speed-boat in order to be picked up. Contrary to what other space-craft observers have described, the Itibi Rayan control craft did not show any kind of illumination during darkness.[35]

AN UNFINISHED STORY

Ludwig Pallmann's book, *Cancer Planet Mission*, was published in London in 1970. There must have been some promotion, because I recall that a friend heard an interview with him on BBC Radio, and there was an article about him in the *Guardian*. The book fell into obscurity, and is known only to a few UFO researchers. A planned second volume, describing some of his experiences in more detail, was not published.

Some time afterwards, I visited the publisher in London, with the aim of tracking down Pallmann. The place was deserted and I was unable to obtain a forwarding address. Later, I learned that the company had gone into liquidation. Veteran researcher Wendelle Stevens, a former US Air Force pilot, was likewise unable to track down Pallmann, though he did come across corroboration for some of the claims.

In 1967, Stevens was delivering several Beechcraft T-34 trainer planes to the Peruvian Navy, making fuel stops at the last Colombian town, the river port of Leticia, on the Amazon. On impulse, he hired some native boatmen to take him for a trip up the river to view rare orchids in the jungle. Remarking on the lush, dense vegetation along the bank, he asked the Indians why the natives made no plantations of some of the more rare exotic tropical fruits that grew there in abundance. 'I was certain there must be a market for them,' said Stevens. 'It would only require a little organization.' The natives replied that this might be too large a project for them. Then one of them remarked that he knew of some 'Americans', three or four days up-river, who were doing just that. What was more, the native added, he knew of a white man, a German, who had gone up there to look for them some months previously, but who had not returned. Although the Indians had never seen these Americans, they had heard about them from the wilder tribes farther upstream. The native added that the Americans had aircraft at their encampment.

Further enquiries in Lima led Stevens to a somewhat inaccurate newspaper report about one 'Ludwig F. Pallimann' (*sic*), a German salesman who sold food-processing equipment and health foods to a chain of stores in Lima. This man, reported the newspaper, had gone up-river from Iquitos in the Peruvian/Brazilian border area looking for a giant arrowroot plant for possible hybridizing, seeking a greater yield by improving the strain. (This much is true: Pallmann was doing research for the Agricultural University of Lima at the time, to find an inexpensive high-protein food.) The Indians taking Pallmann up-river asked him why he did not go further up-stream, about another three days' journey, where a party of 'Americans' were doing the same thing. Intrigued, Pallmann took up the suggestion, but found that the Indians would only take him another day up-river, where they would leave him with another tribe for the remainder of the trip.

On arrival in the vicinity of the 'American' encampment, the newspaper report continues, the Indians superstitiously refused to take Pallmann any further, but put him ashore and pointed him in the right direction. Pallmann walked to the camp, consisting of plastic-like tents.

The 'Americans' were fair-skinned, dressed in toga-like garments and spoke in a strange language. Pallmann greeted them first in English, then Spanish and German, to no avail. Getting a limited response in French, he was welcomed and provided with a place to stay.

According to the Lima report, Pallmann learned that his hosts, who said they came from another planet outside our solar system, named 'Itipura', were hybridizing plants and other stock to be taken back there. These extraterrestrials were served by three streamlined disc-shaped flying machines. After a while, the report continues, Pallmann became concerned that his business associates would worry about his whereabouts. The 'Itipurans' offered to deliver him to his destination in one of their flying machines. Because of his long absence, he asked his hosts to take him to his ranch in the Dominican Republic instead of to Lima, and was transported there in 15 minutes.[36]

Stevens believes that Pallmann was covering his tracks in his interview with the Lima reporter.

He had associated the location with the Peruvian town of Iquitos because you could never get to the plantation site from Iquitos by river, and the jungle there was all but impassable. He had omitted all of the earlier contacts with the Itibians as well as what was going on in Lima and elsewhere, probably to head off possible interference for them as the operation was still going on . . .[37] Pallmann was not returned from the plantation to the Dominican Republic when he left . . . and he did not make his first contact with the extraterrestrials by river from Iquitos.[38]

'I searched for Ludwig Pallmann all over South America in 1968 and 1969, and again in 1971 and 1972,' wrote Stevens in his introduction to a reprinted edition of Pallmann's book, which he published in 1986. 'He was moving around Peru in 1968 and then disappeared. I also looked for him in West Germany in 1977 and 1978 but failed to find any productive lead.'[39] Though German by birth, Pallmann is believed to be a British citizen, having fled to England as a young man to escape the Gestapo during the Second World War.[40] My enquiries at the Passport Records Office in London drew a blank: there is no record of a British passport having been issued to a Ludwig F. Pallmann.[41] The search for him continues.

Pallmann was the first to admit that his story is unbelievable. 'As I read what I had written,' he commented ruefully, 'I came to the conclusion that all this would be in vain, because who would want to believe such a story? It's a concatenation of unlikely circumstances for which I can offer very little explanation.

'I have only tried to tell what happened, and even if it should be considered a waste of time, I felt it necessary to do so, because of the religious theme involved. It is stupid of me perhaps to expect that others should feel about this what I felt. Men will continue to be born into their present-day beliefs . . .[42]

'*Cancer Planet Mission* may seem [to be] the product of my fantasy, which I try to pass on as a true story. However, much of what I relate can be checked. Many things may not correspond to the exact date and time as it happened, simply because I did not date my diary from day to day, and because I was overwhelmed by what happened to me. I, myself, did not believe this possible for a long time.[43]

'Just to have known Satu Ra and his sister made me realize that none of us at the present time has the slightest notion of peace, *real* peace, so great was their relaxed and modest humanism, so great their contentment with "Time",' wrote Pallmann, following his initial meetings in India. 'They just seemed to live every hour, every minute, without being "Time-conscious" . . .'[44]

NOTES

1 Pallmann, Ludwig F., *Cancer Planet Mission*, The Foster Press, London, 1970, pp. 13–20.
2 Ibid., pp. 21–2.
3 Readers interested in the Vedic literature about this subject should consult *Alien Identities* by Richard L. Thompson (1993), Govardhan Hill Publishing, PO Box 1920, Alachua, Florida 32615.
4 Pallmann, op. cit., pp. 23–9.
5 Ibid., pp. 30–7.
6 Ibid., pp. 40–2.
7 Ibid., pp. 44–6.
8 Ibid., pp. 46–7.
9 Ibid., pp. 49–50.
10 Ibid., pp. 63–4.
11 Ibid., pp. 72–8.
12 Ibid., pp. 79–82.
13 Ibid., pp. 83–7.
14 Ibid., pp. 88–93.
15 Ibid., pp. 97–100.
16 Ibid., pp. 102–6.
17 Ibid., pp. 111–12.
18 Ibid., pp. 116–19.
19 Ibid., pp. 120–9.
20 Ibid., pp. 130–3.
21 Ibid., pp. 138–43.

22 Ibid., pp. 189–90.
23 Ibid., Postscript.
24 Ibid., p. 170.
25 Ibid., p. 191.
26 Ibid., pp. 175–7.
27 Ibid., pp. 198–9.
28 Ibid., pp. 156–7.
29 Ibid., pp. 141, 144.
30 Ibid., pp. 154–6.
31 Ibid., p. 199.
32 Ibid., p. 170.
33 Ibid., pp. 158–65.
34 *El Diário de Hoy*, Central American Press, San Salvador, 16 January 1969. (A facsimile of the headlines is reproduced in Pallmann, op.cit.)
35 Pallmann, op. cit., pp. 166–9.
36 Pallmann, Ludwig F., and Stevens, Wendelle C., *UFO Contact from Itibi-Ra: Cancer Planet Mission*, UFO Photo Archives, PO Box 17206, Tucson, Arizona 85710, 1986, pp. 3–5.
37 Ibid., p. 17.
38 Ibid., p. 16.
39 Ibid.
40 Pallmann, op. cit., p. 5.
41 Ibid., p. 156.
42 Letter to the author from the UK Passport Agency, Passport Records Office, Public Record Office, Hayes, Middlesex, UB3 1RF, 18 July 1997.
43 Pallmann, op. cit., Postscript.
44 Ibid., p. 48.

Chapter 16

Perplexing Trends

As I have discussed in my previous books, the UFO phenomenon has long been associated with unexplained mutilations of animals, particularly cattle. Although the first public reports came out in 1967, the mutilations began in the mid-1950s, according to Lieutenant Colonel Philip J. Corso, who from 1954–57 served on the National Security Council's Operations Coordination Board, the most sensitive executive branch of the United States Government. 'Whoever went after the animals,' says Corso, 'seemed most interested in the mammary, digestive, and reproductive organs, especially the uteruses from cows.' He continues:

> In many cases the eyes or throats were removed in a type of surgery in which the demarcation line was almost microscopically thin and the surrounding tissue showed that the incision had superheated and then blackened as it cooled[1] . . . the removal of the animal's blood – where blood had been completely drained – [was] so sophisticated that there was almost no peripheral damage to the surrounding tissue . . . We had no medical instruments that even remotely approached what the aliens could do.[2]

Seldom have we had such solid evidence linking the two phenomena, though there are a number of cases in which circumstantial evidence is provided. One such is the following.

In 1969, Wendelle Stevens delivered a C-54 transport plane to a client, Oscar Bowles, a wealthy Bolivian businessman to whom Stevens had delivered over a dozen cargo planes in the past. Bowles's ranch was at Santa Rosa, where he had a meat-processing plant. Above his property, to the west, was a sheep herder whom Bowles knew well. According to Bowles, the sheep herder and two of his Indian helpers were watching the flock one bright day when a silent disc-shaped aircraft came out of the sky, gliding down gently in a curving approach until only 50 feet above them.

The disc stopped, hovered momentarily, and flashes of white light emitted from its underside like electricity jumping a gap, striking the sheep. This happened about 30 times in quick succession, and the sheep fell to the ground. As Stevens relates:

> As it became apparent that the flock was the target of interest, the shepherd picked up a stick and, raising it, started to run toward the slowing, shiny

metallic disc. With his first threatening motion of raising the club, there was a flash of violet light and he suddenly was unable to continue the motion. It was like he was immersed in molasses. He could barely move, and then very slowly. He could roll his eyes and look around. He saw his helpers apparently in the same kind of paralysis . . . The stricken sheep did not move.

As he watched, the silver disc-shaped craft, with a shiny dome on top, descended still more, to about five or six feet above the ground, and a trap-door with a built-in stair opened down from underneath, and two human-like feet started down the stair from the center of the ship. First one and then another emerged, and as they reached the bottom of the stair and stepped off, he was sure they were normal men in strange suits.

The form-fitting suits were a bright, reflective white, of the one-piece coverall type, with a transparent dark helmet over the head. The beings wore matching white gloves and boots, and each carried what looked like a shiny silver fire-extinguisher in one hand, and a black nozzle on the end of a white hose to the bottle in the other.

These two men walked around among the flock putting the 'fire-extinguisher' nozzle to each of the fallen sheep in turn, apparently showing little interest in the shepherds. They finished their task in only three or four minutes and then walked back and boarded the ship up the stairway. The stair retracted as the big, circular ship, over 30 feet in diameter, drifted higher, to about 300 to 400 feet. From that position, there was a tremendous 'sshhoo' and the ship sped up into the blue sky at a steep angle and disappeared.

As soon as the craft had vanished, the shepherds recovered from their paralysis and rushed to their flock. All 34 sheep were dead. The shepherd and his helpers dragged the sheep to one side to bleed them – but there was no blood. 'The three men carried the carcasses to the hut and began to dress them out,' continued Stevens. 'They not only found them blood-less, they found certain organs considerably desiccated and spongy, including the brain, spleen and eyeballs.'

Although the witnesses agreed that the beings looked human, they felt certain that they did not originate from anywhere on Earth.[3]

AN UNIDENTIFIED FLYING BOAT

On the morning of 27 June 1970, Aristeu Machado, his wife Maria Nazaré, their eldest daughter Creuza, aged 23, their four younger daughters, and João Aguiar, an official of the Brazilian Federal Police, were relaxing on the veranda of the family home on the Avenida Niemeyer, overlooking the Atlantic coast to the southwest of central Rio de Janeiro. Sr Aguiar noticed what he thought was a motor-boat striking the water, throwing up spray all around it. He drew the attention of the others to the 'boat', about 700 metres from the shore. On board could be seen two 'bathers', who seemed to be signalling with their arms. They were wearing shining clothing and 'something on their heads' and seemed to be 'thickish set and quite small'.

The two persons seemed to be working on the deck of the metallic grey craft, which appeared to be between four and six metres in length, covered with a transparent cupola (see Fig. 22). While the others continued watching, Sr Aguiar ran to the nearby Mar Hotel to telephone the harbour police for assistance. No sound could be heard from the craft, nor did it bob up and down like a normal boat.

When Sr Aguiar returned about half an hour later, the 'boat' could still

Fig. 22. (*Terence Collins/ FSR Publications*)

be seen on the sea. Shortly afterwards, it took off into the air. Dr Walter Bühler, one of Brazil's most respected investigators, interviewed the witnesses on the day of the incident, having been alerted to the case by a reporter from the newspaper *Diário de Notícias*, which published a report the following day. As Dr Bühler reported:

> Sr Aguiar informed us that when the disc took off, it did not rise straight up, but skimmed along for about 300 metres on the surface of the sea, throwing out the usual sort of bow-wave, such as we see with our own fast motor-boats.

In fact, it was only when the machine had become airborne and was moving away in a low arc out to sea towards the south-east that the witnesses realized it was *not* an ordinary boat, but a flying saucer.

As the disc took off, from a point about 600 metres from the shore, Dona Maria Nazaré noticed a hexagonal-shaped object underneath it, as though retracted into the craft, and several coloured lights – green, pale yellow, red – repeatedly flashed in the same sequence. While the disc had appeared metallic grey while resting on the water, it looked transparent once airborne. She clearly saw the two occupants sitting inside the disc.

On the area of sea where the disc has rested, Dona Maria noticed a 'white, hoop-shaped object, of the size of a trunk or a chest', which sank after a while. Then it reappeared, and from it came a yellow, oval-shaped object, which began to move slowly towards the beach. A greenish 'flange' later separated from the main yellow body, and continued to follow it. About fifteen minutes later, the yellow object came to within about 120 metres of the beach, made a right-angled turn, then headed towards the beach at Gávea, contrary to the prevailing current. Dona Maria decided to go down to the beach, where she pointed it out to a group of boys, one of whom began throwing stones at it, to no effect. Ten minutes later, the object disappeared around a rocky promontory. Twenty minutes later the white 'hoop' also headed in the direction of the Gávea beach.

Not long after the disc had taken off, a police motor-launch from Copacabana Fort appeared on the scene, presumably sent out following Sr Aguiar's call. 'We do not know if its crew saw the UFO take off,' reported Dr Bühler, 'but it may be assumed that they did, for they would have had the UFO in view long before they reached the area from which it took off and where the "white hoop" remained floating.

'When they got there, the motor-launch stopped at a distance of about one kilometre from the shore. Then the witnesses saw the crew of the launch hoist aboard, with great difficulty, a cylindrical red object . . . Having done this, the motor-launch returned at high speed to its point of origin.'[4]

No further details are available regarding the recovered object.

A CALF-NAPPING

On rare occasions, animals – particularly cattle – are said to have been levitated towards a hovering UFO, prior to being mutilated. Though no UFOs were seen at the time of the following incident, and the fate of the animal is unknown, it was observed rising inexplicably into the sky. The incident is believed to have occurred at the end of October 1970, on the Palma Velha ranch, about 18 kilometres from the town of Alegrete, in the First District of Rio Grande do Sul, Brazil.

At about 16.00, Pedro Trajano Machado and his 23-year-old son, Euripides, were carrying out veterinary treatment on some cattle, and had just picked one Jersey cow which had a one-month-old calf with her, weighing about 20 kilograms. The cow was tied up but the calf was loose, about five metres away from its mother. Suddenly, the cattle became disturbed; the tethered cow began to low, constantly turning to look at her calf. As Pedro turned to look at the calf, which was also bellowing, he could not believe his eyes:

> ... the animal was hanging in the air, at about one metre above the ground, and otherwise in the normal posture (i.e. with its feet pointing downwards). [Pedro] shouted to his son to look, and both were now able to watch as ... the calf began to move away parallel to the ground, still at a height of about one metre, in the same position as before, and bellowing as it headed towards the open fields ...
>
> While the rest of the cattle were bellowing and lowing and churning about in evident fear, the calf was now moving towards the barred gate in the fence, which was open. Then it passed beneath the branches of some trees, towards the northeast, until it was now about 20 metres from its mother ... still about one metre above the ground. But now it began to move slowly upwards, still with its feet pointing down. It had stopped bellowing now. According to the two witnesses, this slow vertical ascent lasted for about three or four minutes until, while still far below the cloud-ceiling, the calf became invisible.

The calf was never found. Investigations were conducted by a team from the independent group GIPOVNI, headed by Victor Soares. They found the witnesses to be thoroughly credible; not likely to have invented such an unlikely tale. Although no UFOs were seen at the time, the Machados remarked that on a number of nights – including the date of the incident – they had seen 'red lights, coming on and going out'; 'stars' in the sky, moving about and stopping, 'doing somersaults in the sky, either separately or in groups of three'.[5]

SUSPENDED IN TIME
Among the assorted phenomena inextricably woven into the UFO mystery, that of time distortion is particularly perplexing. This phenomenon should not be confused with missing time, which, though related, I regard as a separate issue. While many researchers opt for esoteric explanations for time distortion, my feeling is that some extraterrestrials, by means of highly advanced technology, are able to manipulate space and time. Consider the following case, translated by Gordon Creighton from a prominent Brazilian magazine.

Nélson Vieira Leite, a prominent businessman and farmer from Itaperuna in the state of Rio de Janeiro, had spent the day on his farm, the Fazenda Toyota, which lay some 40 minutes by car from the town. The date: 27 May 1971. Towards sundown, while waiting for his cousin, Manoel Carlos, to pick him up, Sr Leite observed a light, pale at first but becoming increasingly bright, then blinding, as it approached him and came down in a meadow, without actually touching the ground. Leite went over to take a closer look.

A greenish object, 'resembling a soup-plate upside down', hovered less than a metre from the ground. Nervously approaching it, Leite had reached a point about 10 metres away when he suddenly found that he was no longer walking; that he had not in fact been walking for what seemed some minutes. By now, he had lost all sense of time. However much he tried, he could not walk. It was not a case of total paralysis, because he was able to wave his arms about; he was unable to walk forwards, as though an invisible barrier prevented him from doing so, and perhaps that is what it was. The craft, meanwhile, emitted a humming sound and a weaker light, though sufficient to illuminate the surrounding area.

Behind him, Sr Leite heard his cousin shouting. As Sr Carlos began running towards him, he was struck down. 'He was knocked right out,' reported the journalist, 'knocked out just as though he had been run over, or had walked into plate-glass doors.' Carlos remained unconscious for two hours.

By this point, Leite was not only frightened but also overwhelmed by depression as he pondered his inability to extricate himself. Gazing at the craft, still suspended about a metre above the ground, he noticed a brighter band of light around it, as though coming from portholes of some sort. Leite himself felt suspended – suspended in time. As the journalist described it: 'Everything was going on just as though it had been just like this for a long, long time. How much time actually did elapse he was subsequently able to estimate: about 20 minutes, at the outside. But at the time, to Nélson Leite, it was 20 years.'

> The light from the disc now began to grow stronger and the hum more piercing, until Sr Leite was obliged to put his hands over his ears. And the flying saucer took off, straight up, slowly at first, then moving so rapidly that immediately it was no bigger than a star in the sky.

At last, Leite was able to walk away from the spot. Later, when Carlos had recovered consciousness, the two examined the area over which the craft had hovered. The long grass seemed scorched.

'It must have been something to do with extraterrestrial beings, people like us, or a bit different,' Nélson Leite told the journalist. 'But nothing beyond the bounds of what is rational.'[6]

SINISTER SPACE CREATURES

encounters reflecting an increasingly sinister trend.

On the evening on 22 September, Paulo Caetano Silveira, a 27-year-old mechanic, was driving home to Itaperuna from Carangola when he noticed a low-flying object, which turned out to be a luminous disc that seemed to be following him. Frightened, he stopped in Tombos and reported the matter to the police, who did not take it seriously. Paulo set off once more, only to find the object still tailing him: it was now 19.00.

Shortly after leaving the town of Natividade, Paulo saw what at first he took to be a 'black ox' in the middle of the road, but when it turned vivid red, then brilliant white, he could see that it was the disc.

> Then a luminous beam shot out from it towards him, and he felt his engine beginning to falter. The engine died, and he found himself confronted by a craft a little bigger than the familiar Volkswagen car . . . It had small windows, just like an aircraft, and a door was open. Near this door were standing two small chubby beings about 40–50 centimetres high . . . He felt his whole body, and especially his legs, being drawn in some mysterious way towards that open door.

The creatures reminded Paulo of dwarfs, with fair complexions, slit eyes and flattish heads. They were dressed in one-piece overalls of a bright, luminous, sky-blue colour, with long sleeves to the wrists, high collars, and 'Roman helmets' with spikes on top. Moving about like automata, with rigid arms and legs, they carried objects that gave off bright red and blue beams of light.

Paulo reported that he felt overpowered: all energy and willpower drained out of him as he was 'drawn' towards the lights. He fought hard mentally, but to no avail. The creatures advanced on him and dragged him silently towards their craft. They seemed very strong for their size. Just as they got him to the door, he put up resistance, causing some injuries to his arms. Once inside the machine, he found it difficult to observe everything, owing to the dazzling white light that permeated the cabin, which he estimated at about three metres in diameter and 2.5 metres high. The fittings appeared to be very simple.

> There were now a total of seven of the creatures, and they seemed to be examining him silently, as though he were some rare species of animal. At no point did he hear any sound exchanged between them, though it was clear that they were in communication with each other.

Paulo then heard an 'infernal din' start up, and assumed that the disc was in flight or in movement, at which point he became unconscious.

> Later, after what lapse of time he does not know, he heard a strange humming sound and was aware that they were carrying him out and were laying him

and that it seemed to hang there, suspended, for a brief moment, before shooting away . . . like a flash of lightning, and was gone from sight.

AFTERMATH

Dazed, Paulo lay beside the road, waiting for help from a passing car. The first person on the scene, fortuitously, was a physician, Dr Cirley Crespo, who summoned help from the police in Itaperuna, nine kilometres away. By the time the police arrived, Paulo had flagged down another passing car, driven by Mário Alves de Brio, who commented: 'I was greatly moved. Something had obviously happened . . . The man was totally disoriented, and in urgent need of medical attention.'

On arrival at the hospital in Itaperuna, Paulo was examined by Dr Munir Bussad, who found the patient to be in a state of severe nervous shock. He had an abnormally fast pulse-rate, badly scratched and bruised arms, his eyes were badly bloodshot and he was unable to see properly. The doctor, who knew Paulo personally, stated that there was no history of mental illness. 'I do know,' he added, 'that many of the people who have had this sort of experience are not suffering from any kind of psychiatric disorder, nor do they display any signs of having any mental obsessions.'

Next, Paulo was taken to the police headquarters in Itaperuna, where he related the experience to the chief, José Luís Maron, and to an inspector, Gilberto Gomes.

When interviewed by reporters five days later, Paulo Caetano Silveira was still not able to see properly, and from time to time he fell into bouts of weeping. He told reporters that although he had no clue as to how long he had been abducted, his watch was found to have lost 15 minutes.[7]

Three days later, Benedito Miranda reported being levitated 50 metres in the air by similar entities, while driving from Itaperuna to Catagueses. After a while, though powerless to move, he felt able to use his voice and shouted to be released. When the lights of an approaching car came into view, he found himself lowered gently to the ground.[8]

The Prefect of Itaperuna, brother of Nélson Vieira Leite, clearly felt sufficiently concerned about the situation, as this news report of 1 October shows:

> The Prefect of Itaperuna, Sr Rubém Vieira Leite, is to send, next Tuesday, to the UFO Study Department of the Brazilian Air Force, and also to NASA, an extensive report on the sightings of flying saucers which have been occurring here for about two years past and which are now occurring with great frequency; two people, Paulo Caetano Silveira and Benedito Miranda, having been seized and held for some time by 'small beings'. In addition to giving a detailed account of all these happenings, the Prefect will also request that urgent measures be taken.[9]

THE DILEMMA OF OFFICIALDOM

As early as 1954, a high-ranking official of the Brazilian Air Force went on record attesting to the reality and seriousness of the UFO situation. 'There has been a staggering increase in UFO sightings since the explosions of the atomic bombs,' said Colonel Adil Oliveira, chief of the information service, Staff Headquarters. 'The Brazilian Air Force has never been unmindful regarding this mystery.'[10]

Yet what could the military, or the politicians, do about the situation in Itaperuna? Given the bizarre nature of the encounters and the rather sinister implications thereof, would it not be expedient to simply play down the situation, or ignore it altogether? 'The widespread anxiety and alarm engendered by these events,' wrote Gordon Creighton, 'may provide food for thought for those who are so ready to criticize governments for censoring or suppressing UFO reports.'[11]

REJUVENATION

Ventura Maceiras, a 73-year-old caretaker, was sipping his *maté* (the national, non-alcoholic beverage) and listening to the radio near the little wooden shack where he lived at Tres Arroyos, in the province of Buenos Aires, Argentina. Beside him were his dog, his cat and her kittens. It was 22.20 on 30 December 1972. Suddenly the radio failed and a loud humming noise could be heard, 'like the noise of angry bees, only ever so much stronger'.

As the noise grew steadily louder, Maceiras looked up and saw a powerful light, increasing in intensity and flooding the neighbourhood. Within the light could be seen an enormous object, estimated at about 20 to 25 metres in diameter, 'red-orange turning to purple' in colour, hovering almost directly above him. Protruding tubes in the lower central portion emitted sparks, while around it an enormous wheel rotated constantly. In the brilliantly illuminated upper central part could be seen a spherical cabin with two small windows. At one of these windows appeared a being dressed in dark-grey clothing 'made of rolls or cylinders joined together' (similar to the 'Michelin men' described on p. 172). From a helmet on the being's head came a tube, passing down into a box on its back.

Simultaneously, a shower of sparks shot from the underneath of the craft, directly in front of Maceiras, and the object tilted downwards and towards him, affording a better view of the interior of the cabin. Now a second, identical occupant could be seen, looking over the shoulder of the first. 'He described the eyes of both of them as slanted [and which] looked fixedly, and gave an impression of depth,' reported the principal investigator, Pedro Romaniuk, a former airline pilot and technical investigator for the Argentine Air Force's aviation accidents board. 'The mouth was but a thin line, and he remembered no details of nose or ears.'

In addition to the two small windows, there were two more windows on the further side, between which could be seen an emblem, consisting of what looked like a 'sea-horse' with signs or symbols to the right of it. Inside the cabin were what the witness described as a long panel with 'a whole lot of instruments and clocks'.

> Also simultaneously with the downward tilting movement of the object, a powerful flash of light came from the under-part, blinding the witness temporarily . . . This flash of light completely enveloped the cat and then vanished at once. Meanwhile the humming noise was growing much louder, and the colour of the object was turning to bluish-green. It began to move forward . . . and descended still lower until it was no more than from four to six metres above the ground. At this point [Maceiras] was able to see that in the upper part of the cabin there was a wheel or ring which was spinning very fast.

The object then moved off towards the northeast. A sulphur- or arnica-like smell remained for a few seconds, as the object flew away at low altitude, its colour changing slightly to reddish, then greenish-blue. The whole incident had lasted for less than half an minute.

AFTER-EFFECTS

Pedro Romaniuk emphasized that Ventura Maceiras was a poor, simple man, scarcely able to read or write, with no television set or access to films. His neighbours all vouched for his honesty. In more than 60 interrogations that ensued – by physicians including a psychiatrist, engineers, police officials, the secretary to the local government office, and others – he never changed his story and begged not to be given any publicity.

During the incident, Maceiras felt some tingling or vibration in his legs, which lasted for a couple of days. For over a week he suffered from a gradually worsening and eventually unbearable headache, and by the eighth day he was experiencing pain in the back of his neck. There were even more disturbing symptoms, as reported by Romaniuk:

> Eight or nine hours after the episode, he developed a most violent type of diarrhoea, involving about eight attacks daily. Unfortunately he did not think to check whether or not he was passing blood. The diarrhoea continued until the eighth day . . . accompanied for the first four days by nausea and vomiting.
>
> At the time of my first visit, on January 16, 1973, [Maceiras] had begun to notice that he was losing hair abnormally, for at one pull he would be losing between 170 and 200 hairs. He did this several times, in the presence of investigators, and it was evident to everybody that despite his 73 years he had abundant hair.
>
> From the 14th day on, several small red pruriginous pustules appeared on the back of his neck, so that he was constantly scratching them . . . I examined him and found some ten of these swollen pustules.

After his experience, [he] developed a marked difficulty in speaking, having trouble in moving his tongue . . . both his eyes watered constantly [and] very thin filaments, about three centimetres in length, almost capillary, also came from the eyes. This symptom finally vanished completely on the fifth day.

The tops of the many eucalyptus trees that surrounded the site were scorched or completely burnt. According to the National Atomic Power Commission, no traces of radioactivity were found, though it did not issue a written report.

Although the dog and kittens were unaffected, the mother-cat which had received the full force of the flash of light was nowhere to be found afterwards, and she did not return until 48 days later, her back showing scorch marks.

In spite of Maceiras's alarming symptoms, he made a complete recovery, though there were later remarkable developments. He found that he had regained strength to the extent that he could lift heavy weights that hitherto would not have been possible for him.[12] That was not all. One night in February 1973, Maceiras claims to have had a repeat visit. There was no craft this time, but two beings 'appeared' to him, one of whom approached, smiling, and greeted him. Between 1.75 and 1.8 metres in height, the being had small, round ears, 'slit' eyes, a small, flattened nose, and very short, very fine thin hair. Communication was telepathic.

A great deal of information and imagery was imparted to the witness. Much of it seems incomprehensible, even ludicrous, but some is interesting. The aliens communicated that they came from our galaxy, giving what seem rather silly names for their city, planet and 'empire'. Here follow some excerpts from those communications supposedly relating to the propulsion of the craft, and the purpose of the visits:

The manner in which they function is . . . as yet unknown to the people of Earth who, if they had it, would use it for making war. Their speeds are very variable, but they can attain limits that to you are inconceivable. The metal of which they are made is . . . a mineral unknown to you, which, once it has been moulded in the shape of the spacecraft, is indestructible, cannot be melted down again, and can be projected through space in the form of 'energy'. The energies of these craft are electromagnetic, and are obtained from the atmosphere (the lower and upper atmosphere) of each planet.

We are visiting you because you are causing atomic explosions which are having a great effect on the Sun, which may suffer damage as a result . . . Upon your planet there are mighty seismic movements and earthquakes coming soon, which will destroy part of the planet.

Regardless of the validity of the communications, the fact remains that Maceiras, though barely educated, began talking to the investigators about philosophical, theological and astronomical concepts well beyond his comprehension, even occasionally breaking into an unknown tongue. 'He mentions big figures in miles, and in a matter of seconds converts them to kilometres,' commented Romaniuk, adding that not all the information given out by the witness would be disclosed at once, the better to obtain possible corroboration from other sources.[13]

Researcher Jane Thomas Guma, who translated Pedro Romaniuk's reports, believes that by comparing some of the details of the story with other 'incredible' or 'crazy' contactee cases, progress might be made. 'Maybe corroborating details would be found,' she explained, 'which would help us fit together a few more pieces of this immense jigsaw puzzle.'[14]

UNDERSEA BASES OFF VENEZUELA AND ARGENTINA?

In the latter part of March 1973, thousands of witnesses reported seeing unusual objects either entering or emerging from the Caribbean Sea on the north coast of Venezuela. This led to a widespread conviction among the local populace that an alien base was located there.

At 16.00 on 22 March, two flying objects approached at high speed from the sea, veered sharply and passed over the area between Naiguata and Carenero, then came to a halt at an altitude of around 200 metres over the town of La Sabanamore, where they were witnessed by more than 3,000 people. Alerted by callers, the police observed the objects. After hovering for between five and ten minutes, the objects separated and headed off in different directions.

A few days later, a night-watchman told the press about his sighting of several orange lights, tubular in shape, 'like our rockets', which rose up out of the sea and vanished into the sky. Other, similar reports followed, many of them from the vicinity of Maiquetía International Airport, Caracas. On the night of 28 March, Armando Silva and his wife saw two bluish objects flying high above their country house, which overlooks the seashore near Carayaca. The following night, awakened by a sensation of tremendous heat, Sra Silva went to the balcony and saw:

> two long blue flying things like capsules, not very big; they were the same as we had seen the previous evening. They were of a vivid blue colour, and so close that I am absolutely sure of what I saw. When they came nearer to the shore they shone as brightly as the sun. One of them dropped into the sea, submerged, and then came out again. The second one also dropped into the sea, lay still awhile, and then moved nearer to the first one.

The sighting lasted about 15 minutes, ending when both objects took off from the water and vanished at phenomenal speed. Afterwards, the temperature in the house returned to normal. Sr Silva claimed that, while his wife watched from the balcony, he went outside and was startled to see, at one of the small windows on one of the objects, a figure like that of a five-year-old child, but with a head that looked like a gourd. He took fright and ran back to the house.[15]

SOUTHERN ARGENTINA

After years of research, the Argentine Society for the Investigation of Unusual Phenomena announced in 1973 that machines from another world had established undersea bases in the gulfs of San Matías and San Jorge, in the coastal waters of southern Argentina. By the early 1960s, flying saucer sightings had become so frequent along the coast of Patagonia that they no longer caused surprise, the Society stated. The most convincing sighting to support the hypothesis occurred on 14 August 1968, when 100 witnesses observed the trajectory, covering a distance of 700 kilometres, of five extremely luminous ellipsoid objects which rose up out of the Gulf of San Matías and then submerged again in the Gulf of San Jorge.[16]

ON THE PERUVIAN CENTRAL HIGHWAY

'I never thought I would see a flying saucer, much less photograph one,' said Lima architect Hugo Luyo Vega, following the sighting of an unknown flying machine, identical to George Adamski's 'scoutcraft', which Vega photographed on 19 October 1973 (see plates).

On the day in question, Vega had taken a client into the Lima countryside in search of a home site. They had driven about 54 miles inland along the Rimac River when they took a break near a valley surrounded by tall hills. Suddenly, Vega told reporters, 'my client, obviously excited, told me he saw a shining object in the bottom of the valley that was advancing towards us extremely slowly'.

> The car was not far away. I ran back for my camera, because in that fraction of a second I thought I, too, had seen something interesting. When I pointed my [Polaroid] camera and took the picture, the object was less than 50 yards away from us and about 20 yards off the ground. Suddenly, the object changed direction, headed toward the east and increased its speed. It rose off the ground as if trying to avoid some high-tension wires that came down from the top of one of the hills and crossed the valley, and disappeared from view.
>
> It was of the colour of burnished silver [and] shaped like an overturned soup plate with a cupola on top. At the very top of the cupola, there was a round object giving off a fixed, sky-blue light. Lower on the cupola, we could see a row of small windows like port-holes in a ship.

On the bottom of the craft was what appeared to be 'the propulsive force of the object . . . a dark red throbbing light that was aimed toward the ground from a sort of turbine in the middle of the upside-down plate. Near the turbine-like part, we could see protuberances like half-eggs.'

The architect said that only about 30 seconds elapsed from the time they spotted the object until it disappeared.[17]

'For a moment I didn't actually think the picture would come out all right, for I don't consider myself all that good a photographer, and I was greatly surprised when I saw that it had come out,' Vega continued. 'All the photo showed was the thing's shape, but at any rate this little piece of evidence is enough to prove that it was a real "UFO" and not an invention of my mind.'

It took the witnesses some 20 minutes to recover from their astonishment. Vega was reluctant to disclose the identity of his client. 'He is a wealthy man who prefers no publicity,' he explained.[18]

AN INTRUSION OVER MANZANO LABORATORY

In *Beyond Top Secret*, I reported on the 1980 series of sightings (including a reported landing) of unknown flying objects in the vicinity of the Manzano Weapons Storage Area, at Kirtland Air Force Base, Albuquerque, New Mexico. These sightings are confirmed officially in several declassified Air Force documents. Major Ernest Edwards, USAF (Retired), who had been in charge of security at Manzano, confirmed the content of these reports and pointed out to me in person where the incidents had occurred.[19] Another interesting sighting, investigated by R. C. Hecker of the Aerial Phenomena Research Organization (APRO), occurred in 1973.

At 21.45 on 6 November, an air policeman spotted a large, glowing object hovering above Plant No. 3 (nuclear weapons inspection facility) of the Manzano Laboratory area, Kirtland Air Force Base East. The object was described as shaped like an oblate spheroid, 150 feet in diameter, golden in colour, and absolutely silent as it hovered at about 100 feet. 'The nine other air policemen on duty in that area were alerted to the presence of the intruder,' reported Hecker.

> While the other air policemen moved into positions affording views of the object, a call was put through to Kirtland East for assistance. According to my informant (one of the air policemen who saw the UFO), four interceptors (F-101 Voodoos) of the 150th Fighter Group, New Mexico Air National Guard, were scrambled to intercept the object. As the interceptors grouped in the skies over Kirtland AFB West, the object began moving in an easterly direction and passed out of sight over the Manzano Mountains at treetop level (below the radar horizon). By the time the jets had arrived on the scene, the object had vanished.

'When I interviewed [my informant] approximately one week after the incident,' continued Hecker, 'he said that military officials were upset with the incident. He requested that I did not identify the source of my information due to immediate censoring of the report. He said that officially, the sighting had not occurred; there were no references to it in intelligence briefs (which he had access to) in succeeding days.'[20]

INTERCEPTION OVER BOLIVIA

Two days after the preceding incident, on the afternoon of 8 November 1973, a peculiar, top-shaped flying craft was observed by the crew of a Lloyd Bolivian Airlines plane as it approached the El Alto International Airport at La Paz, Bolivia. When first seen, the luminous object was hovering at some 12,000 metres over the snow-capped Illimani Mountain which overlooks the city. There were hundreds of witnesses.

Unable to obtain identification of, or to communicate with the strange craft, air traffic controllers contacted the commander of the Air Force Air Pursuit Group based at El Alto, Major Norberto Salomon, who at that moment was on a training flight. Heading towards the target at supersonic speed in his F-100 Super Sabre, Salomon reported that the stationary object was shaped like a top, on the sides of which could be seen what looked like small windows. When the jet approached to a distance of 2,000 metres, the object took off. 'When I had managed to approach up to that distance,' said Major Salomon, 'it began to slip away at an incredible speed.' He gave pursuit, but in a sudden manoeuvre the object reverted to its original hovering position, before disappearing vertically at fantastic speed.[21]

THE RESCUED EARTHMAN

One of the stranger stories to emerge from the plethora of incidents reported from Puerto Rico and the Dominican Republic in 1972 is the following, which appears here as it did in a Puerto Rican newspaper at the time:

It is 9.00 o'clock in the morning. On a deserted road near San Cristóbal [Dominican Republic] a man appears and flags down a car. The driver, an insurance company director whom we shall call X.X., slows down and pulls up. The man approaches. He is wearing a sort of light-green overall with a shiny glint to it. The garment covers his feet. He is wearing no shoes, no gloves, and has no pockets, no weapons, no insignia. It is noted that he is wearing a sort of watch on his left wrist. This mysterious person asks X.X.: 'Don't you know me?'

'No,' says X.X.

'My name is F—— M——,' says the man. (This is the name of a person well-known in Santo Domingo [the capital], who disappeared mysteriously in the sea about 15 years ago.) 'It was thought that I was drowned along with two other people. But I was rescued by a modern machine.'

X.X.: 'By a helicopter?'

F.M.: 'No. Supposedly a module, that is to say, what you folk call a UFO. I was rescued by those two people (he points towards two companions a certain distance away) because of my knowledge of radio techniques and my intelligence.'

The two people mentioned are over six feet high, slim, and dressed in identical fashion to F.M. Their hair is short and brown, and their skin a light colour, like that of the Chinese. They remain silent, standing there with arms crossed, observing the scene.

F.M. draws the attention of X.X. to the machine in which they have come. It is of the size of an automobile, has the shape of an American football, and its surface is nickelled.

X.X.: 'What are they doing here?'

F.M.: 'Supposedly investigating.'

X.X.: 'What sort of investigating?'

F.M.: 'Investigating.'

X.X.: 'From where do they come?'

F.M.: 'Supposedly from Venus.'

The stranger adds that they are greatly interested in the Milwaukee Depth. Then F.M. says: 'Step back. We are about to leave.'

The dialogue has lasted some five minutes. Just as the three individuals are about to depart, F.M. turns round and tells X.X. not to worry if his car won't start at once. He says everything will return to normality.[22]

I decided to make enquiries about this case. From Jorge Martín, Puerto Rico's leading investigator, I learned that the incident allegedly took place on 22 September 1972. 'X.X.' is Virgilio Gómez Contreras, a sales manager of an insurance company, while 'F.M.' is Freddie Miller, one of the pioneers of television broadcasting in the Dominican Republic.[23]

The reference to the Milwaukee Depth is interesting, for several reasons. Firstly, this is the deepest part (9,200 metres, or 30,183 feet) of the Puerto Rico Trench, which lies immediately north of the island. Secondly, many residents of Puerto Rico believe that at least one alien base exists on the island, and perhaps in the depths of the Trench, since so many observations have been made of objects entering or leaving the sea there. Thirdly, I have learned from a well-connected source that certain extraterrestrials have a number of underground and undersea bases on our planet, one of which is in the Caribbean area. Furthermore, plate tectonics (the study of Earth's geology based on the concept of moving 'plates' forming its structure) was mentioned by my source as an item of interest on the alien agenda; in particular, tectonic plates beneath our oceans. Finally, this is not the only story alleging alien liaison with Earth people.

NOTES

1 Corso, Col. Philip J., with Birnes, William J., *The Day After Roswell*, Pocket Books, New York and London, 1997, pp. 182–3.

2 Ibid., p. 181.

3 Pallmann, Ludwig F., and Stevens, Wendelle C., *UFO Contact from Itibi-*

Ra: Cancer Planet Mission, UFO Photo Archives, PO Box 17206, Tucson, Arizona 85710, 1986, pp. 6–8.

4 Bühler, Dr Walter, 'UFO on the Sea near Rio', translated by Gordon Creighton, *Flying Saucer Review*, vol. 17, no. 3, May–June 1971, pp. 3–7.

5 Bühler, Dr Walter, 'More Teleportations and Levitations', translated by Gordon Creighton, *Flying Saucer Review*, vol. 19, no. 1, January–February 1973, pp. 28–9.

6 Creighton, Gordon, 'Itaperuna Again', *Flying Saucer Review*, vol. 18, no. 2, March–April 1972, pp. 13–14, translated from *Domingo Ilustrado*, Rio de Janeiro, 17 October 1971.

7 Creighton, Gordon, 'Uproar in Brazil', *Flying Saucer Review*, vol. 17, no. 6, pp. 24–7, translated from several Brazilian newspapers, including *O Dia*, Rio de Janeiro, 10 October 1971.

8 Ibid., p. 26.

9 Ibid., p. 28, translated from *O Dia*, 1 October 1971.

10 *O Cruzeiro* (special issue on flying saucers), 1954, translated by Gordon Creighton, *Flying Saucer Review*, vol. 17, no. 3, May–June 1971, p. 7.

11 Creighton, 'Uproar in Brazil', p.24.

12 Romaniuk, Pedro, 'Rejuvenation Follows Close Encounter with UFO', translated by Jane Thomas, *Flying Saucer Review*, vol. 19, no. 4, July–August 1973, pp. 10–14.

13 Romaniuk, Pedro, 'The Extraordinary Case of Rejuvenation', translated by Gordon Creighton, *Flying Saucer Review*, vol. 19, no. 5, September–October 1973, pp. 14–15.

14 Thomas, Jane, 'The Contactee of Tres Arroyos: Some Thoughts', *Flying Saucer Review*, vol. 19, no. 5, September–October 1973, p. 16.

15 Creighton , Gordon, 'Underwater UFO Base off Venezuela?', *Flying Saucer Review*, vol. 21, no. 1, published June 1975, pp. 9–13.

16 *La Nazione*, Italy, 29 July 1973, translated by Mary Boyd, *Flying Saucer Review*, vol. 19, no. 6, November–December 1973, p. 29.

17 United Press International (UPI), Lima, 22 October 1973.

18 Creighton, Gordon, 'George Adamski Still Casts his Shadow', *Flying Saucer Review Case Histories*, supplement no. 18, February 1974, p. 12, translated from *El Comercio*, Lima, 23 October 1973.

19 Good, Timothy, *Beyond Top Secret: The Worldwide UFO Security Threat*, Sidgwick & Jackson, London, 1996, pp. 381–2, translated from *El Comercio*, Lima, 25 October 1973.

20 Hecker, R.C., 'New Mexico Reports', *The APRO Bulletin*, vol. 23, no. 2, September–October 1974, p. 5.

21 Associated Press (AP), La Paz, Bolivia, 9 November 1973, translated by Bill Armstrong from *La Nueva Provincia*, 10 November 1973, published in *Skylook*, Mutual UFO Network, no. 82, September 1974, p. 17.

22 Freixedo, Salvador, 'UFOs over the Caribbean', translated by Gordon Creighton, *Flying Saucer Review Case Histories*, supplement no. 14, April 1973, pp. 9–10, 12.

23 Interview with the author, 22 May 1997.

Chapter 17

Disquieting Developments

Australia's harsh, remote Sturt Desert, covering the borders of South Australia, New South Wales and Queensland, was reportedly the scene of a bizarre, disquieting encounter early in January 1974. The witness, 'Ben', an ornithologist from Adelaide and the author of several books on that subject, was with a lady companion, searching for the fossils that abound in that desert. Their exact location, a place called Clifton Bore, is in South Australia, about 250 miles north-northwest of Broken Hill, New South Wales.

At around 13.30, Ben was over a mile away from his station wagon. He had asked his companion, who was unfamiliar with the terrain, to keep the car in sight at all times. It was at this point that Ben felt he was being watched. Suddenly, two small beings, about a metre in height, approached him. As Kevin McNeil, author of an article on the case, wrote:

> These beings were humanoid, indeed human-looking in many respects. They appeared to be males, with short, average-cut hair styles . . . dressed in a skin-tight, silvery covering, not unlike a diver's suit. The clothing appeared to be seamless. Their faces were normal, the colouring like a very light suntan; their heads, however, were elongated at the rear . . . Their arms were considerably shorter than a comparable human's.

The beings spoke in a rapid, unintelligible language. Though alarmed, Ben did not feel threatened; when they beckoned him to follow them, he did so. About 50 feet away appeared a silver-coloured object, which he had not seen before. Shaped like a 'hot dog roll', it had no visible seams, doors, windows or protrusions. By Ben's estimate, it was about 25 feet long and four feet high. A doorway opened at the centre of the craft, and the beings motioned him to enter. What he encountered inside left him totally confounded. As he bent down and stepped inside, he found to his astonishment that the interior was vast. 'You could have fitted a full-sized battleship inside,' he told the investigator. 'How can this happen? It was 25 feet long and four feet high, I know that – I could see over the top – but inside, it was enormous . . . space meant nothing.'

Inside were 20 or so similar beings, at least four of whom were females, with longer hair. All seemed to be about the same age, with no facial ageing lines: it was as if they grew to about one metre tall, and about twenty-five years of age, then stopped growing and ageing.

REJECTION

The alien beings offered Ben a drink from a silver metallic tumbler. Though afraid to drink it, he was more afraid of the consequences if he refused. After taking the drink, he passed out. Asked later to describe the taste, he was unable to relate it to any with which he was acquainted.

On recovering consciousness, he found himself lying on the floor of the craft. The two aliens were still near him, and somehow he felt that he had been 'rejected'. He reflected that perhaps this was due to his age (he was 38 at the time) or an illness from which he was then suffering. (If this was the case, an interesting parallel can be drawn with the abduction of Alfred Burtoo in Aldershot, England, in 1983. As described in *Beyond Top Secret*, Burtoo was rejected by the small-sized aliens, who, having examined him, said he was 'too old and infirm' for their purpose.[1] Burtoo, at the time, was 40 years older than Ben.)

THE INTERIOR

Although the aliens continued talking among themselves, Ben felt, telepathically, that he 'knew' what was going on. 'He felt no hostility directed at him,' reported McNeil, 'and apart from his two "guardians", the other crew members paid him no attention.'

These crew members were walking around, talking among themselves, switching switches on and off. There were seven or eight TV screens, showing interior and exterior scenes, and one that was showing what appeared to be motors of some kind, although he could not see them physically himself. There was also a large, mirror-like screen on the wall, with dots, Catherine-wheel-shaped images (perhaps spiral galaxies?) and some strange images and symbols that he could not understand.

Together with his 'guards', Ben walked around the interior. The floor surface was of a shiny metallic substance that was neither hard nor slippery. The atmosphere inside was normal; neither he nor the aliens required breathing apparatus. Ben did not know how, nor could he understand it, but he felt that the aliens themselves were able to get from one place to another by means of thought alone.

A SHOCK

Ben was profoundly shocked by what he saw next: two human female children; one appearing to be 12 or 13 years old, the other eight or nine years old. Both were in a cage-like structure.

Ben, the father of five children, became very emotional on the tape at this point [and] was unable to describe the 'cage' properly . . . He stated that it was 'neat' and that the two girls seemed to be in some type of trance. They did not appear to comprehend what was going on around them. They had plenty of room inside this 'cage', but were not moving, just standing there. Ben did not know if they could see him or not; they made no sign. They were European (that is, white) and were dressed normally. The aliens, walking around, doing their chores aboard the craft, paid absolutely no attention to the girls.

Ben felt that the aliens planned to take the girls with them to where the aliens had come from, but he felt he had been rejected because he was of no use to them.

A PROBLEM OF CONTACT

Ben became aware of a buzzing sound, like motors of some sort, though he could not see the source of the noise. With the opened doorway now before him, he took his chance. Stepping out of the craft, he found himself back in the desert. He walked away and was not stopped.

On returning to the car and his companion, Ben learned that an hour and a half had passed since he began looking for fossils. 'He wrote down what happened to him within one hour of returning,' said McNeil. 'During his interview he was constantly referring to these notes. He was mostly disturbed about the children.

'Did this happen? I believe it is possible that it did, due to the emotion and detail supplied by Ben,' commented McNeil. 'It is, unfortunately, to be expected that the sceptics and "professional" UFO debunkers should jump up and down pouring scorn and vitriolic hyperbole on the alleged witness to close contact activity.'

> UFOs are constantly ignoring scientific protestations of their non-existence and are continually returning, from wherever they come from, to the earth we know and think we understand . . . objective, scientific investigation could enhance Man's knowledge of himself, his world, and the universe he lives in.[2]

MAYDAY!

'Mexico Centre from X-Ray Alpha Uniform. Mayday! Mayday!'

The young pilot's voice was desperate as he contacted Mexico City's Benito Juarez International Airport control tower. It was 12.15 on 3 May 1975.

'Come in, X-Ray Alpha Uniform.'

'My aircraft is out of control . . . I have three unidentified objects flying around me; one came under my aircraft and hit it. The landing gear is locked in and the controls won't release them. My position – I am on the Radial 004 from the VOR [VHF Omnidirectional Range] Tequesquitengo – I am not controlling the plane – Mexico Centre, can you hear me?'

'Taken note, X-Ray Alpha Uniform . . . We are contacting [appropriate] authorities.'

Carlos de los Santos Montiel, the 23-year-old pilot, was flying in his Piper PA-24 Comanche (registration XB-XAU) from Zihuatenajo, State of Guerrero, to Mexico City, a distance of some 180 miles, when, he claims, an object, about 10 feet in diameter, shaped like two plates joined together with a small cupola on top (see Fig. 23), positioned itself just above his starboard wing. A glance to the left revealed another object just above the port wing.

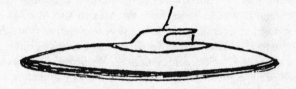

Fig. 23. A sketch of one of the three objects.

'I was petrified, after I saw a third object which seemed about to collide head-on with the windshield,' Carlos told airport officials. 'But it went beneath the aircraft and I heard a strange noise from below as though it had collided with the underside of the plane.'

The Piper's airspeed dropped from 140 to 120 nautical miles per hour. Carlos tried banking to the left in an attempt to 'bump' the object away from his plane, but the controls were frozen. He then tried lowering the landing gear, though to no avail. The Mexico control tower contacted Ignacio Silva la Mora, Carlos's uncle, an aircraft expert, who talked with Carlos on the radio to analyse the problem and assist with landing preparations.

By the time Carlos had reached the Ajusco radio beacon, he found that his aircraft had risen from 15,000 to 15,800 feet. At this point, the objects left, one by one, heading in the direction of the Popocatepetl and Ixtaccihuatl volcanoes. Control of the aircraft was regained. Meanwhile, at Mexico City International Airport, runways were closed and preparations made for an emergency landing. After 40 minutes of circling, Carlos managed to lower the undercarriage after adjustments to the control lever with a screwdriver, and the plane landed safely on a grassy area between two runways.

Immediately, Carlos was taken to the airport clinic, where he was examined and pronounced fit. Three days later the airport's chief of the Aviation Medicine Department, Dr Luis Amezcua, completing neurological, physical and psychiatric tests, inferred that Carlos might have been hallucinating as a result of low blood sugar (hypoglycaemia), because he had not eaten prior to the incident since 20.00 the previous night. It was also inferred, by a chief inspector from the Civil Aviation Directorate, that Carlos may have hallucinated because he had been flying too high without oxygen. The latter hypothesis fails to account for the radar evidence.

RADAR CONFIRMATION

According to Julio Interian Díaz, Airport Terminal Radar Controller, the blip of Carlos's plane – the only one in the sector at that time – was seen on radar when it was 43 miles from the Mexico City Airport. Radar registered another aircraft going in a different direction from Carlos's plane, and which executed a 270-degree turn in a radius of three or four miles at a speed of 450–500 nautical m.p.h.[3] 'Normally a plane moving at that speed needs eight to ten miles to make a turn like that,' air traffic controller Emilio Estanol told reporters. 'In my seventeen years as an air traffic controller I've never seen anything like that.'[4]

MEN IN BLACK

Carlos de los Santos Montiel's story received extensive coverage in the Mexican media. When Pedro Ferriz, the well-known television personality and ufologist, invited Carlos to appear on his programme, the rather retiring young pilot reluctantly agreed. As Carlos drove down the freeway on his way to the television studio, a black Ford Galaxy pulled in front of him. 'Through the rear-view mirror he could see an identical car just behind him,' reported Jerome Clark, a leading American researcher.

> Both vehicles looked brand new, almost as if they were being driven for the first time. The cars started to crowd him and soon Carlos' car had been forced over to the side of the road. Alarmed, he stopped his vehicle and was about to get out when the Galaxies also pulled over and four tall, broad-shouldered men jumped out.

One of the men put his hands on the door as if to prevent Carlos from getting out. 'Look, boy,' he said, in a rapid, rather mechanical-sounding Spanish, 'if you value your life and that of your family too, don't talk any more about this sighting of yours.' Carlos was too taken aback to respond.

The men, dressed in black suits – traditional garb since the 1950s for the so-called 'men in black' (MIBs), who have often been reported to threaten UFO witnesses – were described as 'Scandinavian' in appearance, with unusually pale skin. They returned to their cars and disappeared in the of it; that such threats by the MIBs had proved to be empty in other cases. Reluctantly, Carlos agreed to reschedule the interview.

One month later, Carlos met Dr Allen Hynek, the distinguished UFO researcher and former consultant to Project Blue Book, who was travelling in Mexico at the time. At an initial meeting, Hynek invited Carlos for breakfast the following morning to discuss details of his aerial encounter. At 06.00, Carlos left his house and drove to the Mexicana Airlines office, where he had applied for a job as a pilot, then proceeded to Hynek's hotel. As he climbed up the steps he was approached by one of the men who had threatened him a month earlier. 'You were already warned once,' the man said. 'You are not to talk about your experience.'

'All I did was accept an invitation,' said Carlos. 'Dr Hynek wants to know what I saw and I thought that maybe I could understand it better myself if I talked with him.'

'Look, I don't want you to make problems for yourself,' snapped the man, pushing Carlos back several feet. 'And why did you leave your house at six this morning? Do you work for Mexicana Airlines? Get out of here – and don't come back!' Carlos obeyed.

That was Carlos's last encounter with the mysterious men in black. 'They were very strange,' he told Jerome Clark and Richard Heiden two years later. 'They were huge, taller than Mexicans are, and they were so *white*. But the strangest thing of all is that all the while they were in my presence I never saw them blink.'[5]

Disturbing though this story is, Carlos at least lived to tell the tale. As described in *Beyond Top Secret*, another young pilot, Frederick Valentich, disappeared for ever following his encounter with an unknown flying machine during a flight from Melbourne to Tasmania, in October 1978.[6]

A REPAIR STOP IN LEEDS

Jan Siedlecki, born in Poland in 1919, came to live in England during the Second World War. He does not recall the precise date of his alleged encounter with an alien vehicle that landed near a housing estate in a suburb of Leeds, England, but believes it was during the summer of 1976.

At the time of the incident, Siedlecki was working overtime at Brydon's Garage, Cross Gates. It was about 01.00 on a hot, humid night when he put down his tools, walked the short distance home and retired to bed. Suddenly, a brilliant white glow flooded the bedroom. Thinking the garage might have caught fire, Jan jumped out of bed. What he saw was a strange craft, hovering around 10 to 15 feet off the ground, wobbling, as if trying to land. The craft was saucer-shaped, of a glistening deep-blue colour. Jan put on his shirt, trousers and slippers, and headed outside. The time was now 02.00.

fence to watch the strange machine as it hovered above a large expanse of grass, about 75 yards away, which backed on to a large housing estate. 'Jan watched very carefully as the object finally managed to rest on the

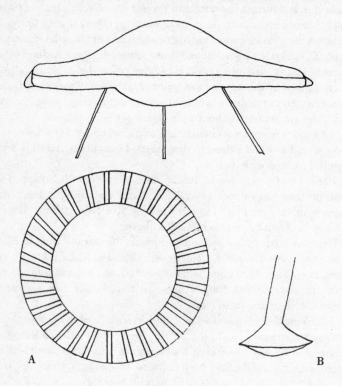

Fig. 24. Drawings by investigator Mark Birdsall, projected from Jan Siedlecki's own sketches. (A) How the 'turbine' underneath the craft appeared, and (B) one of the inverted mushroom-shaped landing legs.

ground,' reported researcher Mark Ian Birdsall, who headed an investigation by the Yorkshire UFO Society (now Quest International). 'There had been complete silence and the witness was still looking around to see if there were any other people or cars passing. He then looked carefully at the underside of the object. He could now see three "legs" which supported the object. At this point he was very frightened.'

A few seconds later, Jan observed a tube descend from the central underside of the object, which he says was a few feet wide, and which reached the ground. Although the entire object was dark, an unusual glow was pouring from the underside, which he thought strange. 'On one side there was total darkness, and on the other an intense glow; you could have cut the two with a pencil,' he remarked.

THE HUMANOIDS

Suddenly, the tube opened 'like a book'. Two humanoid figures, approximately four feet tall, walked from the tube and stood in front of the object. Noticing Jan, who had positioned himself nearer the object, they beckoned him to come closer. As he walked slowly towards them, he heard a conversation taking place, but could not understand the language. Jan stood in front of the figures, who were wearing one-piece suits of a yellow-orange colour, and a kind of helmet with a darkened visor over the face area. They wore mittens on their hands, so he could not see any fingers, and some kind of boots or shoes which appeared to be integral to the suit.

At this point, Jan noticed a panel with a series of square-shaped 'switches' and circular buttons on the chest of each humanoid (see Fig.

Fig. 25. Mark Birdsall's drawing of the humanoids, based on Jan Siedlecki's description.

25). When the humanoids began adjusting these, Jan was surprised to hear them speak recognizably. 'We are in trouble with ship, we will have to make repairs before we leave,' they said. 'We apologize for the intrusion. As soon as repaired, we go.' The tone of the voices sounded 'tinny', like a ten-year-old boy speaking. The switches and knobs were operated

only when the beings conversed with Jan, so he assumed they were translating devices.

Frightened, Jan asked the men how their ship worked. 'If you want to
come, come. We go inside now,' they replied. 'The lift will come down
for you if you want to come.' Jan watched as the doors silently closed and
the tube moved rapidly upwards. A few seconds later, the tube dropped
down once more. As he bent down to walk underneath the hull, which
was about five feet off the ground, he noticed two rows of what looked
like small rotor-blades rotating very slowly; one clockwise and the other
counter-clockwise (see Fig. 24).

INSIDE
Siedlecki was perspiring as he entered the tube, which was glowing in a
pale-blue colour. The doors closed 'like a book'; to his surprise, he could
now stand straight. The doors opened out into a metal 'cabin'. Standing
on a shiny surface, he immediately became aware of a smell as of 'rotting
grass'. The two humanoids led him up a sloping ramp that spiralled
around what he took to be the ship's inner perimeter, then into a room
they referred to as the 'cooling system'.

Around the edge of this room lay a two-foot-wide channel of flowing
water, with some kind of green grass, about two feet high, growing out of
it. As he moved with the men to another room or compartment, Jan asked
them how the ship flew. There was no reply, but he had the impression
that they were on the point of showing him the 'engine room'. 'How fast
does the ship fly?' he asked. 'B13,' came the cryptic reply. Then another
door opened. Peering into the semi-light compartment, he noticed in a
far corner four or five crouched figures, with their heads in between their
hands and knees. Unlike the other two humanoids, they were dressed in
black one-piece suits, with no helmets, and had brown hair. The figures
were gathered beside a circular pool containing a black, bubbling, oil-like
substance, from which flashes of red light darted into the air.

As reported in many other contacts, the lighting source itself could not
be discerned. 'There were no visible lighting points,' Jan explained. 'The
light inside the ship was constant; an unusual yellowish-orange light
coming from all the panels. There were no windows or visible openings.'
Gazing towards the 'ceiling', Jan saw what he thought was the central
dome of the ship. Then, a football-sized ball of orange light (which could
have been a probe) darted around the room in stops and starts. 'At this
point, he heard many footsteps, as if there was panic,' Birdsall reported.
'One of the men very politely informed Jan that he would have to leave,
explaining that they had got a "space-bug". Jan was ushered down the
spiralling staircase and into the tube. When Jan was inside, the man said,
"When you get out, run!"' The door opened and Jan ran quickly back to

his former position behind the fence.

A Noisy Departure

Now sweating profusely, Jan stared at the craft. No windows or lights could be discerned, only the dark blue of the object and the intense glow emanating from its underside. Suddenly, he heard a loud, high-pitched whistling sound.

> The front portion of the object tilted forward like a helicopter and remained stationary for a few seconds. The object was now clear of the ground. The tripods and tube lifted inside the object, the whistling intensified, then the object shot away at an angle of 45 degrees into the sky, taking only three or four seconds to reach the thick cloud cover. Jan still managed to see the object for a while after it had broken through the clouds, with red 'fire' pouring out from its underside. Just before it hit the clouds, there was a terrible noise; the whistling grew louder and louder. 'It must have been heard by the whole area,' he said.

Jan remained in position for a while, then walked carefully over to the landing site. He noticed that the grass, as well as the ambient temperature, was very warm. The whole incident lasted between 20 and 25 minutes. Arriving home, he washed, returned to the bedroom and told his wife what had happened. She did not believe him.

Discussion

Jan and the investigators remained puzzled by the fact that, although the object reportedly landed near a dual carriageway and a housing estate, it was neither seen nor heard by any neighbours; at least, none reported the incident. 'When I was running towards the ship,' Jan told the investigators, 'I was constantly looking for other people and cars.'[7] That a dark object was not observed by others is not surprising, though the noise should have been heard, but lack of supportive testimony does not itself negate the event. One explanation might be that the occupants of the craft were able to exert some sort of 'influence' over the surrounding area; an explanation given by aliens to several other contactees and abductees.

There is an interesting parallel with Jan's case in the encounters of Ben and some others. Jan estimated the diameter of the object to be about 20 feet and its height about 10 feet. He said that he walked up three levels, and on each floor was able to stand upright; since he is five feet six inches tall, one could assume the height of the craft to have been well over 10 feet. In an attempt to rationalize this inconsistency, Jan explained that the seven or so steps up to each level sloped very gradually. That still would not explain it, in my view. Jan himself admitted that the object appeared much larger on the inside than it looked when he was on the outside. 'It could quite easily accommodate 20 or 30 persons,' he added.

Grass at the landing site died soon after the incident, and Jan said it was four years later before even weeds would grow there.

Friends, neighbours and an employee testified to Siedlecki's integrity, as did the investigators. 'I believe Jan and the case to be genuine,' concluded investigator William Tree. 'The description of the inside of the object is so bizarre. Jan never altered his story, although he himself knew the data were unbelievable.'[8] In 1986, 10 years or so after the alleged incident, I was introduced to Jan by Mark Birdsall. He seemed a believable witness, still emotionally affected by his unbelievable encounter, with no hidden agenda nor ulterior motive. He died in 1992.

AN ATLANTIC ENCOUNTER

It was 23.40 on 22 December 1977. Captain Walt Hammel and co-pilot 'Slim' Dickson (pseudonyms) were flying in a Trans World Airlines (TWA) plane about 600 miles out over the Atlantic at 21,000 feet, bound for Boston's Logan International Airport. 'The weather was clear except for some widely scattered clouds beneath us,' Hammel related to investigator Donald Todd, the day after this incident. 'We had just been having a cup of coffee and Slim had set his cup on the panel next to his right elbow. As he put the cup down, he glanced out to what would be between one and two o'clock, and suddenly he grabs me by the arm.'

Reflexively I swiveled my head to look at him and caught this dazzle of twinkling lights coming at us from the starboard side, just ahead, and appearing maybe about 50 feet below. Instantly Slim and I realized that whatever the thing was, it was moving in a hurry, that it was entirely too close, and appeared to be about to cross in front of, or about to collide with us. And it was huge!

I slammed on some power, hauled the nose up and prayed we'd go over the top of that thing. Just as we started to climb, this thing swept straight up, did an impossible right-angle turn and begins to pace us. I don't see how *anything* could have executed a maneuver like that – I mean almost a simultaneous two-directional turn – up and to the right, not to mention coming to damned near a dead stop.

We couldn't detect any sound, any prop or jet wash, nor see any exhaust. It just kept flashing a lot of lights around the middle. Once we leveled off again, the thing stayed just ahead of us off to our right and we had a chance to observe it. We couldn't see any hard outline or shape to it, but you could tell it was almost circular because of the lights [which] were mostly white . . . There was a red blinking light on top of the thing with all those twinkling, silvery-white lights on around the middle . . . A little later the thing rose up and there was another red blinking light on the bottom of it . . . It looked as if about every dozen or so lights around the middle, there was a reddish-purple one, and in between them, there was a blue one. But they were all blinking off and on intermittently.

I don't mind telling you I was nervous as hell, and Slim, he was chalk white and scared stiff. It suddenly occurred to me, what if the passengers are watching that thing? What am I going to tell them? I didn't want panic back there, so I buzzed the stewardess . . . evidently none of the passengers had spotted it. The object may just have been far enough forward to be out of their sight line.

FURTHER DEVELOPMENTS

'While we were watching the UFO, suddenly this other glowing thing drops out from underneath it [that] looked like a neon-green smoke ring,' continued Hammel. 'It dropped away from the larger UFO down toward the water, and submerged.'

We saw the glowing green circle of water where it went in, and then the glow disappeared. Seconds later, two more green rings dropped out. The second one dropped away and submerged like the first one, but the third one dropped down and then shot straight ahead to disappear toward the coast.

Slim and I finally came to the conclusion that the big one must have been all of 100 feet across, or maybe even more. By comparison, the smaller ones looked like they might have been 20 feet across. I guess the big UFO paced us for about 20 minutes, then all of a sudden the lights around the middle began going out in clusters – not in banks of say, six or eight in a row, but six or eight separate, individual lights at the same time . . . The top and bottom blinking red lights went out too, with only scattered blue lights around the middle still blinking.

As our eyes became accustomed to the dark again, we could faintly see the silhouette of [what looked] like two inverted shallow soup bowls put together. Then very faintly, just above the mid-lateral line, we could see a soft subdued green glow emanating from what appeared to be trapezoidal-shaped windows . . . wider at the bottom than at the top. Just about the time our eyes got focused on the windows, the thing assumed an overall bluish corona. Then it took off straight ahead . . . leaving nothing in front of us but a blue streak in the sky. It was positively the damnedest thing I've ever seen. When we finally came in to Logan, we must have hit the runway half a dozen times . . .

'After a previous encounter reported to the FAA [Federal Aviation Administration], and upon ensuing company and other authoritative harassment,' explained Donald Todd, '[Hammel] has sworn never to report another UFO encounter. Fortunately, he has confidence in [my] discretion.'[9]

In several interesting respects, the TWA case brings to mind the Atlantic encounter reported by US Navy aircrew in 1951 (see pp. 86–9).

I must stress that the professional and personal mentality and sense of responsibility that characterizes the behaviour and attitudes of aircraft pilots and other crew members, in particular those transporting passengers, precludes the likelihood of contriving a UFO report for gaining attention, or for entertainment.

THE 'CHUPA-CHUPA' EPIDEMIC

Since 1946, Brazil has been the source of many alarming reports about alien activity, perhaps owing to its many inaccessible areas where the entities can carry out their agenda with impunity. In 1977, remote areas of the states of Pará and Maranhão were plagued by what the local people called the 'vampire light' or *luz chupa-chupa*. The Brazilian Air Force sent in teams of investigators from the First Air Zone Command (1COMAR), and at least one incident attracted an investigation by Brazilian Naval Intelligence.[10]

Bob Pratt, a respected American journalist who has travelled extensively in northeast Brazil researching the chupa-chupas and who is the author of an important book on disturbing encounters in northeast Brazil,[11] learned that UFOs were seen almost every night from April to July 1977 in the area of the town of Pinheiro. The Mayor, Manoel Paiva, told Pratt that he estimated that as many as 50,000 people had witnessed sightings. Typically, a big ball of fire would descend and then hover 300 to 400 metres above the town. Some said it hurt their eyes to look at it, others claimed it made them feel sick. Fishermen and farmers reported being chased or injured by the objects. Invariably, the 'ball of fire' seemed attracted by other light of any kind.[12]

AN ALIEN ENVIRONMENT

At 01.00 on 10 July 1977, José Benedito Bogea, a prosperous chicken farmer who lived six kilometres from Pinheiro, was walking into town to catch a bus to São Luís. The night was very dark. 'Suddenly, a bright, greenish-blue light appeared in the sky and chased me for about 200 metres,' he told Pratt.

> Then it circled over a bush in front of me and stayed there, three or four metres above the ground, for just a fraction of a second. I could see a V-shaped thing 15 to 20 metres long, with a beam of orange light going down to the ground. I raised my arm and shone my flashlight at it, and in an instant I saw a bright flash of light. It knocked me down, and I felt like I'd had an electrical shock. Then I passed out.

On recovering consciousness, Bogea says he found himself in a strange 'city', with wide avenues and beautiful gardens. 'I looked for the sun, but I didn't see it,' he said. 'I didn't see any sky at all, just empty space.' Bogea saw many people in the city, all looking very much alike; about 30 years old, five feet tall, slender, and nearly all dressed in grey and brown clothes; long gowns for the women and tunics and trousers for the men. 'They looked like us,' Bogea elaborated. 'Most were light-skinned and had eyes of different colours: blue, brown. The women were pretty and had long blond hair. All the men had short hair, beards and mustaches.'

Although the people seemed to be talking to each other, Bogea heard nothing. After having been observed for a while in a large room, he was allowed to leave, though he was followed. In one area, he encountered what looked like small transportation devices; in another, about 20 disc-shaped objects, though none like the V- or triangular-shaped craft that had brought him to the city. Eventually, Bogea was motioned to enter one of the smaller 'transporters', at which point he again became unconscious.

He next awoke at 08.30 and found himself beside a highway near the port of Itaqui, eight miles west of São Luís, more than 70 miles from where he had been abducted. Suffering from terrible pain in his lower back and his right side, he managed to hitch a ride home. For eight days, he had no appetite, and for many weeks he was obliged to use a cane to walk. Nevertheless, there was one dramatic benefit from the encounter. At the time of the abduction, he had been wearing strong eyeglasses. 'But it wasn't until I got home the next day that I realized I'd lost them. Since then I haven't needed to wear glasses anymore . . .'[13]

'CAMBURÕES'

Weird, cylindrical-shaped objects, called *camburões* by local people, typically emitting powerful beams of light that swept across terrains, were encountered frequently by farmers and fishermen during the *chupa-chupa* 'epidemic' in Brazil. In July that year, another frightening encounter was reported by João de Brito of Vila de Piriá, in the vicinity of the River Gurupi. As a friend related:

> It was 11.00 p.m., and he was sitting quietly in a hide among thick bushes, awaiting game. An animal appeared, but suddenly something flying in the sky threw down a beam of light on to the animal, which made off. João himself couldn't escape. He felt the light bearing down on his body and felt his strength being sucked out of him, and was sure he was going to die. The flying thing was shaped like a cylinder, and he could hear voices coming from it, in an unknown language. The thing went away then, but it left him powerless [and] he ended up in hospital.

By this time, few dared venture outdoors at night. 'It is important to note,' emphasized Dr Daniel Rebisso Giese, author of a seminal book on the chupa-chupa phenomenon, 'that many local folk felt sure that the UFOs came up out of the sea, because they had been seen quite often doing precisely that and coming and shining down their beams on to boats or on to villages.'[14]

THE WOMAN IN BLACK

Dr Rebisso Giese refers to fascinating stories, originating in the town of

Bragança, Pará, of encounters with the so-called 'Fish Woman', thought to be connected with the chupa-chupas, in July 1977. Described as fair-haired and fair-skinned, the young woman was believed by some to live alone on Cajueiro, a little island near Augusto Corrêa. Local people were bemused by the fact that she frequently bought large quantities of fish at the local market – usually 100 or 200 kilos a time – presumably, it was surmised, for the consumption of the 'extraterrestrials'.

Rumours prevailed. Fishermen who ventured near Cajueiro claimed to have seen the fish woman 'walking on the water', and peculiar lights were often seen near her cabin. One witness, Margarida, claims to have had a brief conversation with the woman. As Rebisso Giese reports:

> Walking one day along a lonely road, she suddenly came face to face with a pretty woman, dressed entirely in black, whom she recognized as the 'Fish Woman'. The latter was wearing a blouse with tight-fitting sleeves right down to the wrists, and her hands were encased in gloves. Enhanced by her long blonde hair, her face seemed to probe into Margarida's very soul. The woman asked Margarida what she did, and how many children she had, and . . . was she not afraid to walk alone in that place? And then, suddenly, she vanished from sight as though by magic! Shaken and perplexed, Margarida went home, with a severe headache.

Following coverage of these reports in newspapers in Belém,[15] officers of both Brazil's Air Force and Navy intelligence services arrived in the area and began making enquiries about the fish woman. Apparently, little more was ascertained, beyond the discovery of an envelope in her now abandoned cabin. The envelope, mailed from France, was simply addressed to 'Elizabeth'. 'In the opinion of many folk,' wrote Giese, 'she was connected with the UFOs, for when they vanished from the area, she vanished too.' Furthermore, according to Dr W. Cecim Carvalho, a physician working in that part of Brazil, there had been occasions when 'beautiful white women with fair hair' had been seen aboard the UFOs.[16]

THE CURTAIN OF SILENCE

The craft reported at this time appeared to witnesses in different shapes and sizes; often shaped like helicopters or ray fish. Beams of light which emitted from them were capable of penetrating solid matter, such as the roofs of houses, and of targeting and paralysing people. By October 1977, the populace was thoroughly alarmed, particularly in Vigia, as this news report confirms:

> Mayor José Ildone of Vigia will today be sending to the Regional Headquarters of both the Army and the Air Force, in Belém, an extensive account of what is happening in Vigia, Santa Antônio do Tauá and, in particular, an account of the terror being experienced by the people of

Umbituba. The decision to do this was taken after the mayor had discussed the matter with Vigia's Chief of Police, Alceu Marcílio de Souza.[17]

As sometimes perhaps exaggerated stories about the chupa-chupas proliferated, so the newspaper reports in Amazonia became correspondingly less objective. In due course, a curtain of silence descended on the matter.[18] The chupa-chupas, however, remained active in northeast Brazil.

SHOCKING ENCOUNTERS

In *Alien Liaison*, I alluded to the case of Luís Fernandes Barroso, a once prosperous businessman and rancher from Quixadá, in the state of Ceará, Brazil, who on 23 April 1976 was struck by a beam of light from a hovering aerial object. He did not recall what happened after that. Barroso suffered nausea, diarrhoea, headaches and vomiting. Psychiatrists and other physicians who examined him concluded that he had a brain lesion. His speech deteriorated, and three months later his hair turned white. Six months later, he lost all his mental faculties and his behaviour regressed to that of an infant.

The case was investigated by Bob Pratt, who provided many additional details in a 1990 report I published. Barroso had been examined first in Quixadá by Dr Antônio Moreira Meghales, who later sent him to Fortaleza, where he was examined by a dozen psychiatrists and psychologists. After a lengthy stay in hospital, he was sent home. Nothing more could be done for him. Together with Dr Meghales and the principal investigators, Reginaldo Athayde and José Jean Alençar, Bob went to visit Barroso's home, where he remained in the care of a nurse. 'He sits all day in a chair, staring, occasionally moving his eyes but apparently seeing nothing,' wrote Bob. 'He reacts to no stimuli, except that when someone takes a photo of him with a strobe light, as I did, he screams when he sees the flash.'[19]

Although we shall probably never learn what happened to Barroso, the inference is that he was abducted. He died in 1993, unrecovered from his sad state.

On 24 March 1978, a 16-year-old boy named Luís Carlos Serra disappeared in his home town of Penalva, nearly 200 kilometres southwest of São Luís, Brazil. He did not turn up until three days later, when a fisherman discovered him lying dazed in the forest, unable to stand up. He was taken to the local hospital and examined by Dr Linda Macieira, who later told Bob Pratt that the boy was completely dumb and had muscle contractions. That was not all, as Pratt reports:

Four of his teeth were missing. Two had simply been broken off, and one had been extracted completely . . . Luís also had a full head of hair before this happened, but when Dr Macieira examined him he appeared to be bald. At first she thought his head had been shaved. On closer examination, however, she discovered that his hair had been burned off. The scalp was not burned, but the top of his ears were, very slightly, like sunburn. Luís seemed to be paralysed. Dr Macieira tried to move his arms and legs but could not. She pricked his arms and legs . . . to test his reactions, but there was none at all. He went nine days without eating or drinking anything, and had to be fed intravenously [and] catheterized.

Several days later, Luís was flown to a larger hospital, in São Luís, where he was examined by more than half a dozen doctors, none of whom was able to say what had happened to him. Three days later, Luís began to come out of his paralysis. Still unable to talk, he motioned for pencil and paper and wrote down his story.

He had been gathering guava fruit just inside the forest at noon on Good Friday when he heard a loud sound like a car horn above him. Looking up, he saw a light, brighter than the sun, just above the trees, which hurt his eyes. Suddenly, something made him fall flat on his back and he was unable to move anything except his eyes. He lay there for some time then began to rise into the air, although he neither saw nor felt anything that could have levitated him. As he rose, he could see a round object above the trees, with four spheres on the bottom, one of which was illuminated.[20]

'When I got high enough,' Luís told Bob later, 'I could see a dome on top and three windows that went all the way around the dome. Only one window was open, and I just entered it head first. It was about a metre square. When I got inside, I fell down on the floor, but not hard.'[21] Inside the craft were three small beings, only a metre tall, their faces obscured by helmets and visors. They were moving around, talking in a loud, incomprehensible language, paying no attention to Luís.

Soon there came a rumbling sound and Luís felt the machine moving. He was taken to a 'strange land' with no sky or trees – just tall grass. Next, he was floated out of the machine and placed on a flat stone or table.[22] 'I was still paralysed,' he said. 'Then these little people came to me and put a tube in my nose. It didn't hurt. Then they put a transparent ball in my mouth, and the liquid just went down my throat very quickly. I fell asleep then, and I don't know what happened after that until I woke up in the jungle.'[23]

In his definitive book, *UFO Danger Zone*, subtitled 'Terror and Death in Brazil – Where Next?', as well as in the report I published, Bob Pratt relates many more disturbing, sometimes shocking encounters in north-east Brazil, including apparently UFO-related deaths. Jacques Vallée (who wrote the foreword to Pratt's book) has himself visited Brazil and spoken with some of the witnesses, whom he found credible, despite their incredible experiences.[24]

'Wherever they come from, some UFO beings *may* be kind and well-intentioned, but others definitely are not,' states Bob Pratt. 'UFOs may or may not be significant to mankind in the long run, but until we find out, we should treat them with the greatest of caution . . .'[25]

NOTES

1 Good, Timothy, *Beyond Top Secret: The Worldwide UFO Security Threat*, Sidgwick & Jackson, London, 1996, pp. 87–93.

2 McNeil, Kevin, 'The Clifton Bore Incident'. Article supplied to the author.

3 'UFOs Escort Mexican Aircraft', *The APRO Bulletin*, vol. 24, no. 2, August 1975, pp. 1–3.

4 Clark, Jerome, 'Carlos de Los Santos and the Men in Black', *Flying Saucer Review*, vol. 24, no. 4, July–August 1978, pp. 8–9.

5 Ibid.

6 Good, op. cit., pp. 168–78.

7 Private report by Mark Birdsall, William Tree and Peter Swallow, Yorkshire UFO Society, Leeds, 1984.

8 Ibid.

9 Todd, Donald R., 'Underwater UFO with "Mother Ship"', *The APRO Bulletin*, vol. 26, no. 10, April 1978, pp. 5–6.

10 Giese, Dr Daniel Rebisso, 'Extraterrestrial Vampires in the Amazon Region of Brazil', *Flying Saucer Review*, vol. 39, no. 3, autumn 1994, p. 8.

11 Pratt, Bob, *UFO Danger Zone: Terror and Death in Brazil – Where Next?*, Horus House Press, PO Box 55185, Madison, Wisconsin 53705, 1996.

12 Pratt, Bob, 'Disturbing Encounters in North-East Brazil', *The UFO Report 1991*, ed. Timothy Good, Sidgwick & Jackson, London, 1990, p. 106.

13 Pratt, *UFO Danger Zone*, pp. 100–8.

14 Giese, op. cit., p. 9.

15 *O Liberal*, Belém, 10 July 1977; *A Província Do Pará*, Belém, 11 July 1977.

16 Giese, op. cit., p. 10.

17 *A Província Do Pará*, 22 October 1977.

18 Giese, op. cit., p. 13.

19 Pratt, 'Disturbing Encounters in North-East Brazil', pp. 117–18.

20 Ibid., pp. 114–16.

21 Pratt, *UFO Danger Zone* p. 112.

22 Pratt, 'Disturbing Encounters in North-East Brazil', p. 116.

23 Pratt, *UFO Danger Zone*, p. 114.

24 Vallée, Jacques, *Confrontations: A Scientist's Search for Alien Contact*, Ballantine Books, New York, 1990.

25 Pratt, 'Disturbing Encounters in North-East Brazil', pp. 102, 124.

Chapter 18

Beyond Belief

Though encounters with the proverbial and ridicule-prone 'little green men' are rare, there are reports by reliable witnesses which warrant our attention. Such is the case in the following report from Poland, an incident that occurred in the village of Emilcin, 140 kilometres southeast of Warsaw, on the morning of 10 May 1978.

At about 08.00, Jan Wolski, a 71-year-old farmer, was passing through a forest in his horse-drawn cart when he noticed two individuals ahead, walking in the same direction, but with 'supple jumps' like 'divers on the sea bed'. When one of them approached a muddy patch, his feet seemed to 'slide' across the mud, as reported in a number of other cases. When Wolski caught up with the 'hunters', for that is what he took them to be, they walked alongside the horse and cart for a while then jumped aboard and sat down gently, one at each side, gesturing to Wolski to carry on. (The added weight caused the mare to exert extra effort.) Wolski drove on while the 'hunters' exchanged some words in an incomprehensible tongue. Shortly, as the cart approached a clearing in the forest, a strange, almost 'transparent white' object could be seen, hanging in the air about 70 metres away, emitting a faint humming sound.

A NOVEL CONTRAPTION

Insofar as I am aware, the shape of that object was like none ever reported. Described by Wolski as 'like a short bus, but with a roof like a barn', it was about five metres in length, three metres in width, and about 2.5 metres in height. It shone, as if nickel-plated. No windows were seen. As investigators reported:

> At its four corners, and half-way up, it had on the outside 'barrels' with black vertical rods running through them and carrying what looked like spirals rather reminiscent of corkscrews [see Fig. 26]. These black rods were rotating very fast. Their diameter was . . . around 30 centimetres. As for the 'barrels', their approximate dimensions were: height, about one metre; diameter, possibly 80 centimetres. The length of the black rods may have been about 1.5 metres.

The 'corkscrews' emitted a range of colours, and the 'barrels' seemed

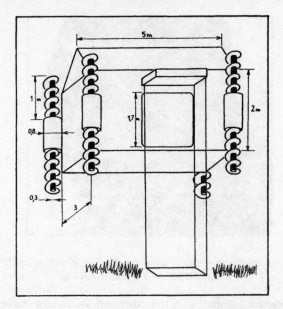

Fig. 26. The bizarre contraption in which Jan Wolksi was taken.

to have been the source of the humming. When closer to the object, Wolski said the sound was like that of bumble-bees in flight. As though cast in one piece, the craft's surface was smooth, stainless and seamless. At a height of some 50 centimetres from the ground was suspended a 'lift', held by four thin cables attached above the entrance to the craft, which had descended as the trio approached. Stepping on to the platform, one of the entities invited Wolski aboard, gesturing to him that he should grasp the cables. After rising rapidly, the 'lift' stopped in front of an opening and Wolski was motioned inside.

Stepping into the object, Wolski paused, bracing himself with his right hand against the entrance. Inside the chamber, the walls of which were almost black, were two more beings identical with the first two . . . The chamber was rectangular. There was no internal lighting other than the daylight from the open door. The walls, floor, and ceiling were a greyish black – the same colour as the overalls of the occupants. The floor was shining, 'as though polished'. The walls were smooth and hard to the touch, and made of a material resembling glass. Against the four walls there were seats, each fastened by two black cables.

No apparatus was seen inside the contraption, with the exception of two black 'tubes' that ran from one gable wall to the other and two holes, about 30 centimetres apart, into each of which one of the entities inserted alternately a smallish black 'rod'. From floor to ceiling, the height was

Fig. 27. An artist's impression of one of the alien beings encountered by Jan Wolski. (*WKPiB UFO Klub Mozaika*)

about 1.8 metres. On the floor of this cabin were about ten crows or rooks, which seemed to be paralysed, though they could move their heads and eyes.

THE BEINGS
The four identical beings, of indeterminate sex, were about 1.4 to 1.5 metres tall and had delicate, slim figures. They were dressed in tight-fitting, flexible one-piece suits of a greyish-black rubber-like material, covering the entire body except the face and hands. No pockets, belt or fasteners were seen. Their legs seemed thicker than those of normal men, and from the way these curved when the beings sat in the cart with their legs hanging down, they looked like prehensile limbs. A 'hump' was visible on the shoulders, as though something was contained under the suits. The slim, greenish-coloured hands had five fingers, between which were fine membranes, except for the gap between thumb and forefinger.

Their heads were relatively large, with faces of an olive-green or greenish-brown hue, having high cheek-bones that gave them an Asiatic appearance.

The eyes, almond-shaped, very long, were dark, and, according to [one report], had nothing corresponding to what we call the white of the eye. In the place of the nose, there was only a slight protuberance with two small vertical openings . . . but according to [another report] the nose was straight. The mouth was straight, and thin . . . They had no lips. Their teeth were white. No hair was visible on the face.

When the beings smiled, the mouth twisted to one side with the effect of a grimace. Their speech was rapid and delicate. Wolski felt no fear in their presence; indeed, their polite, gentle manner inspired his confidence.

A PHYSICAL EXAMINATION

The beings indicated to Wolski that he should take off his clothes, and one of them helped him undo his shirt buttons. Facing him less than two metres away, one of the beings held in each hand a grey disc-shaped object that seemed to be attached to the hand by something like a suction-pad. The discs were vibrating and emitting a dull humming sound. Wolski was positioned with one side facing towards the entity holding the discs, then with his back towards him, and finally with the other side. Each time, Wolski's arms were raised alternately by the entities, whose fingers were very cold. He believed that when the discs were being moved around, he was being 'photographed'. During the process, he smelled an odour similar to that of burning sulphur; a smell that lingered in his clothing for days afterwards.

Wolski's clothes were also examined by the beings. Then, after looking inside his mouth, they motioned for him to get dressed. When ready, he was shown the way out. Wolski doffed his hat, bowed, and said goodbye. The beings bowed likewise, smiling. The same lift took him back almost to ground level, so that he was obliged to jump down a short way.

CORROBORATIVE EVIDENCE

Reaching his horse and cart, Wolski turned round to look at the contraption. Two or three of the beings were watching him from the entrance. He did not see them leave. Once he was inside the cart, the mare galloped towards home. There, ten minutes later, he told his family all about the experience, then he lay down to recover for a few hours. His family and some neighbours headed for the site, but the object was not there. Traces were found; however, these included relatively long, almost rectangular footprints, uprooted stalks and broken ears of maize, twigs and small branches pulled from trees, evidence of soil samples having been taken – and black feathers.

At a farmhouse 800 metres from the scene of Wolski's encounter, a mother was preparing a meal as her children, Adas (aged six) and Agnieszka (four), played outside. Between 08.00 and 09.00, the mother heard a noise like thunder, though seeming to come from the ground. Shortly afterwards, Adas came in to say that he had seen an aircraft like 'a little house', or a 'big box', fly low over the barn, coming from (it was later determined) the direction of the landing site where Wolski had been, then vanish vertically with a sound like thunder. Adas said that the object was flying with one of its smallest 'walls' to the front, in which was a square window with rounded corners, through which he saw a pilot. On the edges of the wall, he said, were moving black rods, like the feelers of a snail.

The first investigators, Witold Wawrzonek and Dr Zbigniew Bolnar, found the witness truthful, forthright and sound mentally and physically. Yet another investigator, Dr Kietlinski, a psychologist, also found Wolski to be honest. A devout Catholic, Wolski swore to God the veracity of his testimony.

Like the British abductee, Alfred Burtoo, Jan Wolski believed that the creatures were foreigners from Earth, such as Chinese. He attributed their green skin colour to make-up, or masks![1]

Those familiar with the numerous publications, personal accounts and memories about abductions, wherein, predominantly, grey-coloured beings with huge wrap-around almond-shaped eyes, slit mouths and only nostrils for noses – the so-called 'Greys' – are described repeatedly, will have been alerted by the partial similarity of Wolski's entity to the Greys, perhaps wondering why more of these have not featured in this book. The fact of the matter is that the overwhelming majority of reports of the *typical* Greys (there are variations) did not appear in print until the late 1970s. There are earlier reports, such as the case of Carroll Watts (see Chapter 14), whose 'white or very light grey' beings had wrap-around eyes of varying colours, small slits in a nose-type bone structure, and 'a straight line with very thin lips' for a mouth, and of course the famous Barney and Betty Hill case of 1961, but such cases are the exception rather than the rule; even then, certain differences are evident.

CONTRASTING ITALIAN ENCOUNTERS

Giorgio Filiputti, a 47-year-old railway employee from San Giorgio di Nogaro, in the northeastern Italian province of Udine, was fishing in the River Corno at Melaria, at about 15.30 on 18 September 1978, when he heard a piercing whistle, like 'something scything through the air'. A few minutes later, having decided to give up fishing because of the wind, he walked up to the embankment.

No sooner had he reached the top than he was alarmed to see an unusual disc-shaped object standing on a small dry mudflat no more than 20 metres away. As Filiputti related to investigator Antonio Chiumiento:

> It was four or five metres wide and it had a cupola on top [and] was supported on three thick legs about 1.5 metres high. These latter seemed to be divided into two parts, almost cylindrical in shape, the upper part having a greater diameter than the lower . . . I had the impression that they consisted of two tubes, one sliding into the other [which] terminated in flat 'plates'. The object was totally smooth, without windows or portholes [and] seemed to be made of a metal of a brassy or yellowish colour which shone in the sunlight.

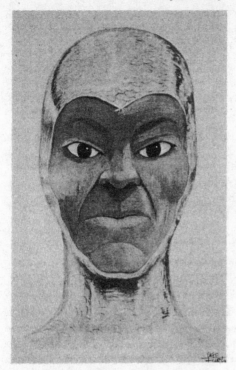

Fig. 28. The craft and its occupant encountered by Giorgio Filiputti, near San Giorgio di Nogaro, Italy, in September 1978. These illustrations were made by Ugo Furlan after lengthy discussions with the witness.

'A SORT OF ASIAN PYGMY'

'Then, almost immediately,' Filiputti continued, 'I saw someone appear, from right behind the cupola, who was walking on the rim of the disc. My first thought for a moment was [that he had] the physical appearance of the inhabitants of certain Asian countries.'

His height was maybe about 1.30 metres [and] he was wearing a completely tight-fitting overall, of the colour and brightness of silver, which flashed and sparkled vividly in the sunlight, and which left only the front part of the head, from the forehead to the chin, exposed. On his feet he had boots of the height of those worn by paratroopers and of a smoky black colour [and] his hands were clad in white gloves [see Fig. 28].

His face, dark-bronzed, had almond-shaped eyes extending back towards where his ears would be – which I did not see because that part of the head was covered by the overall. Nose and mouth were quite normal. From the moment that I observed him, particularly his eyes, I could see that these were wide open, with pupils that appeared to me to be a bit bigger and a bit more protruding than those of certain inhabitants of the Orient . . . the single-piece garment this being was wearing looked as though fashioned entirely of 'fish-scales', and he was wearing, approximately at waist-level, two containers of the same colour as his boots.

'When I caught sight of that "sort of Asian pygmy", I was overcome by a profound emotional disturbance due partly to stupefaction and partly to fear,' continued Filiputti. 'He too appeared to be gripped either by surprise and bewilderment [or] unease at seeing me, as if it had been completely unexpected for him.'

A Quick Repair Job

Becoming aware of Filiputti, the being halted temporarily then continued walking, in a rapid and agile manner, around the rim of the saucer. Finally he halted, stooped down slightly, and touched something sticking out near the base of the cupola. The gadget looked like 'a sort of half-moon or horse-shoe . . . something semi-circular'.

The individual kept on touching it with his hands for about three or four minutes, and all the while that he was doing this he kept repeatedly fixing his eyes on me . . . maybe to make sure that, while he was carrying out his task, I hadn't managed to get closer to the craft . . . in my opinion there was almost certainly something wrong with the craft, and this operation was being carried out by him to repair it . . .

Then, having finished his task, he glanced again in my direction for the umpteenth time and, following the same route as before, vanished from my sight behind the cupola and into the cabin, which no doubt was contained in the main body of the craft, though not visible to me.

A Loud Lift-Off

A few seconds later I heard a very loud rumbling noise, like a deafening clap of thunder, and then a very piercing whistle, both coming from beneath the object, which began to rise vertically. It went straight up, and slowly. As it rose vertically, I was in a position to see its under-part. It was hemispherical and its external surface looked . . . like a lozenge-patterned grid. The landing gear was withdrawn into the craft almost immediately after the take-off . . . Underneath I saw a bluish . . . tongue of flame about 60 centimetres long, of the same colour as burning kitchen gas. Then, when it had got to a height of about 10 metres, the contraption quickly turned on edge, so that for roughly a few seconds I was able to observe it in narrow profile. It went off towards the southwest at a tremendous speed . . . and very rapidly appeared like a glowing ball. It was totally out of sight in a few seconds.

Imprints from the landing gear, about 50 centimetres in diameter, three centimetres deep, and spaced between two and three metres apart, could clearly be seen at the landing site.[2]

Like Alfred Burtoo and Jan Wolski, Giorgio Filiputti was convinced that what he had seen was not extraterrestrial in origin, but a highly advanced secret aircraft, perhaps built by an Asian power.

GROTESQUE GOBLINS

It was shortly before noon on 24 November 1978. Sixty-one-year-old Angelo D'Ambros, of Gallio, in the northeastern Italian province of Vicenza, was gathering firewood near Gastagh. Suddenly, he noticed two ghastly-looking creatures watching him, suspended in the air some 40 centimetres above the ground.

One of the creatures was about 1.2 metres in height, the other a little shorter. 'They were extremely thin, and had a yellowish skin that was stretched so tightly that he could see great veins standing out on the head and hands of the bigger creature,' reported Antonio Chiumiento.

> Their heads were large and elongated, like pears, smooth and bald, with enormous ears that rose straight up and ended in a point. They had great white eyes, sunken and without eyelids, set above a nose of pronounced dimensions which almost reached down beyond the lower lip, the latter being pretty fleshy, and large mouths displaying, at their extremities, two long, pointed 'tusks'. From immediately below the knee right up to the neck, the two creatures appeared to be clad in dark, very closely-fitting overalls, which also covered the arms as far as the wrists, leaving the hands and the rest of the legs and the feet uncovered. The hands and feet were of a remarkable size and out of proportion to the rest of the body, with extremely long fingers and long nails.

As though gliding, the smaller of the creatures began moving back and forth rapidly, to either side of D'Ambros, without moving its long feet. After shouting for help at the top of his voice, the witness asked them who they were and what they wanted of him, to which the smaller creature responded unintelligibly, though it was clear what they wanted – D'Ambros's machete. A struggle ensued, with the witness fending off the taller creature, who repeatedly grabbed hold of the machete. On two occasions, D'Ambros felt an electric shock go up his arm. Managing to get hold of a large branch he had cut, the witness threatened the grotesque goblins with it, whereupon they scurried off. D'Ambros chased them to a nearby clearing, where their saucer was parked.

It was disc-shaped, estimated to be four metres wide and two metres high at the centre, with four aluminium-coloured landing legs that had apparently sunk into the ground a little. It appeared brightly coloured on the upper section, including the cupola, but blue on the lower half, with a white band around the centre. Immediately, D'Ambros saw one of the

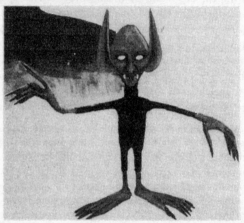

Fig. 29. The craft and creature seen at Gastagh, Italy, by Signor Angelo D'Ambros. Sketch by Ugo Furlan, based on discussions with the witness.

goblins' long hands closing a sort of trap-door, located in the dome, and a few seconds later the disc took off at an angle, silently and at phenomenal speed, emitting a burst of red 'flame'.[3]

These goblins resemble in some respects the creatures who terrorized the Sutton family near Hopkinsville, Kentucky, in 1955 (see pp. 174–5), though in that case, aggressive behaviour was not displayed. Again, it is the wide variety of entities and craft types and configurations, though of the same basic kinds, that will be surprising to first-time as well as to expert readers.

A LUFTHANSA ENCOUNTER OVER NEWFOUNDLAND

In 1994, Captain Werner Utter, for many years the chief pilot of Lufthansa, revealed that he had witnessed unknown flying craft on three occasions, the second of which took place while he was flying at 30,000 feet over the Atlantic, off the coast of Newfoundland, on 21 November 1978, at 09.55 GMT.

'It seems we have a flying saucer in sight, very bright, beaming rays, sometimes red, sometimes violet,' Utter reported to air traffic controllers. 'It looks like a spider.'

The object, estimated to be 20 to 30 metres in diameter and about a mile in front of the Boeing 747, was also reported by the crew of a Trans World Airlines plane.

A few years later, Captain Utter and his crew witnessed a gigantic cigar-shaped object, while flying over the United Kingdom.[4]

Craft with spidery appendages have been observed elsewhere: for instance, on 10 September 1976, Bill Pecha saw a domed, circular 'glowing thing', about 150 feet across, above his house near Colusa, California. On each side hung two pincer-like appendages about eight feet long and six wavy, motionless 'cables', with frayed ends, also hanging down. The craft, and some attendant objects, caused a major power blackout in the area.[5]

A RESEARCH AND STUDY MISSION

'Yes, it was true,' said Jesús Antunes Moreira to Dr Walter Bühler, 'I really did see a flying saucer and I talked with its crew.'

Dr Bühler, a physician, and one of Brazil's leading UFO researchers (who died in 1996), listened intently. It had taken him a long while to track down Moreira, who at the time of his experience, on 6 December 1978, was a security guard at the Marimbonda Hydroelectric Plant near the town of Fronteira, bordering the states of Minas Gerais and São Paulo.

'It was a bit after 8.30 p.m.,' began Moreira, 'and I was up in the guard-house, right on top of the dam, because it was raining at the time, and I was anxious to keep dry. Suddenly I noticed that something was lighting up the surface of the water in the dam. My curiosity aroused, I stepped out of the guard-house and went to see what it could be, and found myself looking at an object which was slightly above the level of the horizon, about 200 metres from me, and crossing the Rio Grande.'

It was coming in my direction, and when it got closer, I could make out that it was a space-craft about five metres wide and three metres high, white in colour, and emitting a certain amount of luminosity. It looked as though it was about to land on top of the dam, maybe right by the power-house . . . it wasn't making the slightest noise.

Getting more and more curious, I started walking along the top of the concreted part of the dam, but the craft went past where I was and then moved to the earthen part of the dam, about 1.5 metres from the foot of it. I was now able to see that its colour was not white, but a light grey. It had a door about two metres high, with a little window in the upper part of it, and it had a sort of platform running right round it.

By now it was only about seven metres from me. Then the little window opened, and in it there appeared a face in many respects very like a human face. Then the main door opened, and from it came three beings dressed in blue overalls with a metallic sheen. They were all very tall – two metres maybe – and with quite long, black, smooth hair.

The men addressed the security guard in an unknown tongue. Moreira responded that he would go and fetch someone who perhaps could speak their language. 'You see,' he explained, 'I was still thinking that maybe – who knows? – they might be some foreign engineers. When I said I would go to one of the telephones that are strung out along the 300-metre-wide top of the dam, one of them gestured to me to step back.

'At this stage, I began to get scared, and I felt for my revolver, [with] the idea of firing a warning shot should it be necessary. And indeed I did try to shoot, but the revolver jammed, and would not fire.'

At that point, one of them went inside the machine and came out with a black box, about the size of a shoe-box, and handed it to one of the others, who was the one that had the longest hair. I noticed that all of them were wearing rose-coloured gloves, which were luminous, like their blue overalls. From then on, I was able to understand perfectly what they were saying to me, in Portuguese. They asked me if I was scared, and told me to keep calm, because they said they had no intention of doing me any harm. When I asked them what they wanted, and where they came from, they said they were on a 'research and study mission', and that, if I remained calm, I would soon know all about it.

When one of the cosmonauts began gathering some stones with a kind of grab at the end of a line, Moreira objected. 'That was enough for them to put away the box and, without the slightest show of dissatisfaction or displeasure, re-enter their craft.'[6]

This is one of a number of cases I have cited in which the aliens seem to be on a research trip that involves collecting Earth's rock and soil samples. I do not find this surprising: currently, that is what NASA's Pathfinder mission to Mars is doing and also what its Apollo missions to the Moon did just over a quarter-century ago as a primary mission – to collect and analyse rock and soil samples.

AN ABDUCTION IN ARGENTINA

Thirty-four-year-old Julio Platner opened the front door of his house on the prairie ranch where he worked, some 15 kilometres from Winifreda, in the Province of La Pampa, Argentina. It was about 19.30 on 9 August 1983. Noticing a bright light approaching, he investigated. As he pulled up and got out of his van, the light he had seen suddenly came straight at him, blinding and paralysing him.[7]

'It was as if a lorry were coming towards me,' he explained to reporters the following day, still disturbed by his encounter. 'I covered my head with my arms and after that, I do not remember anything until I awoke inside a room.'[8] (In other reports, it is stated that Julio first saw some small beings, then suddenly found himself, with his van, inside a room.)[9]

> When I awoke, I found myself lying on an 'operating table' and I discovered four 'beings' around me. Two were beside me at the top of the table, one clasping my shoulder with his hand (I saw him but I did not feel his hand). In front of me, about two metres away, there were two more beings, a man and a woman. I moved my lips asking where I was and what they wanted, but no sound came out. Nevertheless, they seemed to understand me and answered mentally: 'Don't be afraid. We would not harm you. What you are experiencing now has happened to thousands of people before. You can reveal it if you wish. Some will believe you and others will not.'[10]

Platner was then undressed.[11] 'I felt very calm and comfortable,' he continued. 'The temperature inside was quite pleasant. I was not afraid. I looked around and gathered as many details as possible.'

> Once the woman came close to me and put her hand on my wrist, but I did not feel it. Then they put a strange artifact on my left arm, with some kind of tube, 30 centimetres long, half rigid and half flexible, and they took some blood. They did not use any rubber band or needle, but I could see how my blood raised inside the tube up to the middle, when they stopped. I tried to touch the being at my back, but I struck something like glass. I tried to stand up and my forehead struck [more] glass, as if I were inside a glass cube.

Platner described the room in which he found himself as spherically shaped, windowless, about three metres in diameter, with walls of a soft beige colour. Though well lighted, no source of light could be discerned. The only fittings were the examination table on which he lay and a type of glass 'bookcase' that glowed. After 15 to 20 minutes, Platner was ordered to stand up, which he managed to do without touching the 'glass'.[12]

'They took all my valuables off me, but gave them back afterwards,' said Platner. 'After that, I don't remember anything more, until I came to again and found that I had been lying asleep on top of my van, on a rural road leading to Villa Mirasol.'[13]

THE HUMANOIDS

'They were humans,' insisted Platner. 'His body had human shape from neck to feet, which were like boots. Their five-fingered hands seemed to be covered by gloves. But there were no boots . . . They had no uniform, or their own skin was their uniform, close-fitting, without holes or seams, all in one piece including face and head, like rubber.'

The head was different from ours; hairless, short nose, small mouth and ears flatter against the side of the head. The most alien feature was their eyes; circular, without eyelashes and bulging out of the face, with a small protuberance in its middle. They were about 1.65 or 1.70 metres tall . . . The 'woman' was exactly like the men, with greyish, whitish or greenish skin or uniform, but I think she was a woman because she had female shapes (breasts) and she was quite thin. They showed no emotions, made no face movements except with their lips (but without noise). Sometimes they seemed like robots.

CORROBORATION

Some employees of the Argentine National Telecommunications Board told reporters that the telegraphic link between Winifreda and the nearby town of Eduardo Castex was interrupted in precisely the same area, and at the same time, as Platner's abduction. Furthermore, two of Platner's neighbours said that the pictures on their TV sets disappeared for several minutes, also at the same time. Horses in a nearby paddock seemed unusually disturbed: the following day they were found on another part of the estate and it took a lot of effort to return them to their normal grazing area.

Platner had an excellent reputation locally, and most people believed him. His physician, Dr Adolfo Pizarro, showed the press marks on Platner's right wrist and elbow where blood had apparently been extracted, one of which was right on a vein.[14]

Julio Platner's encounter took place three nights before that of Alfred Burtoo, who was abducted in Aldershot, England. There were interesting similarities, and because we may be certain that neither witness was aware of the other's story, it is worth putting those similarities on the record. Initially, Burtoo was attracted by a brilliant light as he was fishing beside the Basingstoke Canal in the small hours of 12 August 1983. He was then approached by two small beings and escorted into a craft parked on the canal towpath. Two other, similar beings were on board. As in Platner's case, all four beings were of the same size, dressed in pale-green one-piece coveralls that covered the hands and feet, and which appeared to be moulded on to their thin bodies – 'like plastic', Burtoo told me. No facial features could be discerned, unfortunately, because these were covered by visors.[15]

Burtoo's abduction was recalled consciously, and though Platner blacked out at the moments when he was taken on board and released from the craft, he recalled most of the event without recourse to hypnosis.

MADAME X AND THE HANDSOME HUMANOIDS

In the stereotypical abduction scenario, small bug-eyed beings appear in witnesses' bedrooms prior to beaming them aboard their spacecraft. In the following case, four beings appeared suddenly in a witness's bedroom – but they were neither small nor bug-eyed; neither did they abduct the witness.

Mme X was asleep in her bedroom near Sospel, 20 kilometres north-east of Nice, France. It was 02.00 on 30 April 1983. Awakening, she saw a red football-sized object. She tried unsuccessfully to awaken her husband beside her. The light then vanished; she assumed it was 'ball-light-ning'. Mme X arose and went to an adjoining room to open the window. When she returned, she was startled by the sight of four quasi-human beings, about 1.75 metres in height. They were of a very athletic gait, with long, pale faces and long thin noses and mouths. The eyes were very elongated, with blue irises. Most surprising was the position of the pupils: rather than centred, they were close to the inner corner of the eye, giving them a 'cross-eyed' appearance. They had blond eyebrows. In spite of their peculiar features, Mme X thought they were handsome.

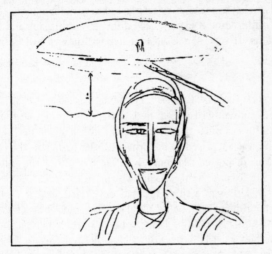

Fig. 30. Sketch by 'Madame X' of the craft and one of its quasi-human occupants. (*Lumières Dans La Nuit*)

There was another marked difference from normal human beings. 'Whereas all humans have a recess or indentation at the point where the nose is joined to the forehead, this was, she said, non-existent in them,' reported Marc Tolosano, one of the investigators. 'Neither did they possess the little vertical furrow which we all have between our nose and our mouth. They did have teeth.'

She did not see their hair, for their heads were covered by little skull-caps like those of frogmen, but with this difference: [they] were not part of the whole one-piece suit, and merely covered the entire skull and ears. [They] were wearing small bars, coloured green and yellow, 'like badges of rank' . . . Their hands, gloveless, were soft and delicate and a little larger than normal human hands.

Although Mme X was not afraid, she made another attempt to waken her husband. 'It is useless,' said one of the men, who seemed to be the leader, speaking in normal French. The German shepherd dogs, normally aggressive towards strangers, cowered under the bed. In reply to the witness's question as to whether they spoke other languages, they said they knew all the languages on Earth. Calmly, Mme X invited the strangers to be seated.

'Do you know who we are?' enquired the leader.

'You're robots!' she replied. Smiling, the leader stretched out his hand for her to feel.

'Then,' she said, noticing the soft texture of their skin, 'you are extraterrestrials', to which they responded in the affirmative.

All the while, Mme X experienced difficulty in thinking and formulating questions. It was, she said, as if her will-power had been 'slowed down'.

PROJECTED IMAGES OF EARTH'S HISTORY

The entities suddenly rose from their seats, explaining that they wanted to show her how familiar they were with Earth's history. Following them outside, she was surprised to see other, similar beings in the courtyard. The first four inspected the courtyard. 'Here it's fine,' they announced. 'We are going to make a projection for you.'

Curiously, although a thick fog had developed, and it was chilly in Sospel at that time of year, Mme X did not feel cold. Now joined by three other quasi-humans, holding black spheres in their hands, she and the others watched as predominantly sepia-coloured images, about three metres high, were projected on to the fog without any visible beams of light emitting from the spheres. The 'film', which ran from prehistory to the Second World War, retraced our wars, sometimes stopping at individual 'frames'. Mme X told them she was uninterested in wars, to which they responded that armed conflicts were all humans knew about, and that they themselves only knew this planet in that light.

FURTHER DISCUSSION – AND DEPARTURE

When the show had finished, Mme X returned to the house, accompanied by the beings. Her husband was still asleep, and the dogs remained quiet. She asked some questions on matters that she felt would be of interest to scientists: about time, distances in space, and so on. To each question they replied that she would not be able to understand. When she asked why the visitors were all men, they replied that sometimes women did accompany them, though not on this occasion.

Despite her difficulty in communicating, Mme X felt quite at ease with her guests, and invited them to eat, drink – or even to smoke, if they wished! They smiled but declined courteously, promising that they would dine with her during their next visit, six months later (a promise not kept). Meanwhile, some of the visitors glanced out of a window, as if to check that everything was all right.

At 04.00, the strangers decided it was time to leave. 'They rose, shook Mme X by the hand, and she accompanied them to the door,' reported Tolosano.

> She then perceived a long, dark-coloured oval object, about 15 metres or so in length. She could see a door in it, and inside, through the door, a diffused light. The fog was still there, everywhere around. The distance from the house to the craft was only about 30 metres. From the entrance of the craft came a gangway, and, after walking across the corrugated-iron roof of a shed, the entities went up onto this gangway . . . Mme X never saw the under-part of the craft at any moment, as it was hidden by the dense fog and by an intervening plank.

The craft then flew away, emitting a faint whistling noise. When she returned to bed, Mme X's heart began to pound violently and 'she was seized with an indescribable panic', despite having maintained her composure throughout the encounter.[16] Was this a delayed reaction to what would normally have been an alarming experience? Almost certainly. There have been a number of cases where the composure of witnesses seems induced by the aliens.

Mme X wondered whether or not she should contact the local priest and the Gendarmerie, but decided to tell no one about her experience, not even her husband, in case it was thought she had gone mad. Two weeks later, she told a friend, and the story eventually leaked out to UFO investigators. 'She has stood the experience pretty well,' commented Marc Tolosano of the UFO Research Group in Menton, who, together with investigator Claude Dufour, was favourably impressed by her character. 'And she is still amazed, thinking of their kindness, their amiability, their smiles. She thought they were handsome, and she still asks herself the question: "Why did they choose me?" '[17]

However farcical certain encounters appear to be, they should not be rejected *a priori*, solely on the grounds that they fail to conform to our preconceived notions or that they challenge our favourite hypotheses regarding alien species and their agenda. Regrettably, many investigators and writers have done just that. One must wonder how many cases have never seen publication, owing to their seemingly ludicrous content. The fact of the matter is that numerous credible witnesses have reported thoroughly incredible encounters.

THE CREATURES OF KIRGISZKAYA

'The phenomenon of UFOs is real and we should approach it seriously and study it,' stated former USSR President Mikhail Gorbachev, in reply to a question during a meeting with workers in the Urals in April 1990.[18]

That year, Russian interest in the subject surged, following numerous military and civilian reports; as it had done in 1989, when witnesses reported the landings of an alien vehicle and its occupants – three giant humanoids and a robot – at Voronezh.

It was about 21.40 on 18 May 1990, in the village of Kairma, near Frunze, Kirgiszkaya, USSR. A 10–year-old boy, Dima, came running in to tell his mother that 'space creatures' were outside. And there they were. 'I have never seen such creatures before,' she said. Just over a metre in height, the creatures wore helmets and glowing suits with stripes on their sleeves and trouser legs.

The beings had hands with three claw-like fingers. Dima reported that when they noticed him, they pulled out a box-like device from behind them and aerial-like appendages appeared on their heads. As in certain other cases described earlier, such as that of Jan Wolski, the beings appeared almost to 'slide' over the ground. When a car appeared, they jumped into a nearby stream and stayed there until it had passed by. By now frightened, Dima and his mother ran to a neighbour's house. When they returned, the creatures were gone. Throughout the encounter, the surrounding atmosphere was 'vibrating'.

At 23.30 a powerful droning sound was heard throughout Kairma, just as the local electricity supply fluctuated. Showers of sparks could be seen coming from the sub-station; later the power supply cut out altogether. After midnight, a huge red disc-shaped thing was seen over the town. Many other reports of sightings were made during the ensuing days.

On 29 May, three kilometres from where the creatures were first seen, came another such report. At 08.50, three women working at a garage heard extraordinary horn-like sounds (as reported in some other cases), increasing in volume. Through the window of the office, Ludmila Sadovskaia saw a strange creature with glowing eyes. Its face was a greenish-grey, and it had no nose and only a slit for a mouth. The women ran out of the office, shouting for help, having failed to contact anyone by radio. Two boys, living nearby, claim to have seen a 'flying saucer' appear above the garage and land nearby, at the same time. Investigators on the scene the next day reported abnormal changes in atmospheric pressure at the site.[19]

AERIAL ENCOUNTERS

Increasingly, it seems, aircraft pilots are coming forward with reports of encounters with unknown flying objects, sometimes risking their professional status by doing so. For example, a wave of sightings in Mexico during 1994 included remarkable encounters by airliners on their approach to Mexico City International Airport. On the night of 28 July, the captain of Mexicana Airlines Flight 180 radioed to the airport's control tower: 'I have an unidentified object on my right, moving very fast.' The tower responded that there was nothing to be seen on the radar.

At 21.25 that same night, many citizens of Mexico City observed an unusual bright object in the sky. Half an hour later, the crew of an AeroMexico DC-9, Flight 129, on its final approach to the airport, reported an emergency to the tower. 'My landing gear was nearly down when I felt a very hard hit,' Captain Raimundo Cervantes told investigator Britt Elders. 'I didn't know what it was.'

> I had never felt a hit as strong as this. [On landing], maintenance immediately checked the airplane and found that the shock absorber [on the nose-wheel leg] was torn off. I contacted radar control and they told me that at the moment I was making my turn, there were two UFOs. I probably crossed my path with theirs – and that's when I declared an emergency.

Television journalist Jaime Maussan reported that in early August 1994, airliners were forced to take evasive action to avoid colliding with unknown flying objects. On 8 August, Flight 304 from Acapulco, captained by Fernando Mesquita, had a near-miss with a large, silvery, metallic craft which came out of a cloud and passed just under the plane.

What also concerned pilots approaching Mexico City Airport was that digital computers on the flight deck seemed to be adversely affected when UFOs were following or buzzing airliners: the computers gave out false readings for such crucial factors as altitude and speed, forcing pilots to take manual control. Air traffic controllers remained publicly silent on the matter until August 1995, when two agreed to be interviewed for television, provided their identities were not publicized. 'We want to prove that it's happening, to prove that it's real,' they said. 'The government knows that it's happening. It could be dangerous for aircraft.'[20]

Many other areas of Mexico have seen a dramatic increase in sightings in recent years. 'Not only have there been many sightings of fly-overs, witnessed by thousands on the streets,' write my late friend Hal Starr, an American investigator who had a home near Sonora, 'but activities in the mountains to the east of our town of Alamos have been little less than

astounding.' According to one of Starr's sources, Native Americans see
UFOs so often in that region as now to be commonplace.[21] Is it possible
that an alien base exists in the Sierra Madres?

THE TUCUMCARI INCIDENT
On the night of 25 May 1995, the crew of an America West Boeing 757,
Flight 564 (Cactus 564), en route from Tampa, Florida, to Las Vegas,
Nevada, reported a large cigar-shaped craft, with a series of bright white
lights along its side and very bright white lights on each end. The initial
sighting, at 22.25 MDT, was made by Captain Eugene Tollefson and
First Officer John Waller, when the plane was at 39,000 feet near Bovina,
Texas, 60 nautical miles southeast of Tucumcari, New Mexico. The
unidentified object appeared to be moving at about 300 to 350 k.p.h. at an
altitude of between 30,000 and 35,000 feet, and came to within five miles
of the Boeing.[22] The following are extracts from radio exchanges between
Waller (564), and others, and Albuquerque Air Route Control Center
(ACC), transcribed by Graham Sheppard, an airline captain and principal associate of mine, who gave me a copy of the tapes. I have also used
some information from the definitive report by the astronomer and leading UFO investigator, Walter Webb:

> 564: Center, Cactus 564.
> ACC: Cactus 564, go ahead.
> 564: Yeah, off to our three-o-clock, there's some strobes out there. Did you
> get what it is?
> ACC: . . . I don't know what it is right now. There is a restricted area that's
> used by the military out there during the day time.
> 564: Yeah, it's pretty odd . . . did you paint that object at all on your radar?
> ACC: Cactus 564, no I don't, and talking with three or four guys around
> here, no one knows what this is . . . What's the altitude about?
> 564: I dunno, probably around 30,000 or so, and its strobe, er, is going
> counterclockwise, and the length is, er, unbelievable . . .

Albuquerque Center contacts Cannon Air Force Base (AFB), New
Mexico:

> ACC: Hey, do you guys know if there was anything like a tethered balloon
> or anything released . . . ?
> AFB: Er, no. We haven't heard nothin' about it.
> ACC: OK. Guy at 39,000 says he's seen something at 30,000; that the
> length is unbelievable and it has a strobe on it.
> AFB: Uh, huh.
> ACC: This is not good [*laughs*]. OK.
> AFB: What does that mean?
> ACC: I don't know. It's a UFO or something – it's that Roswell crap again!

Communications between Cactus 564 and Albuquerque resume:

> ACC: . . . You still see it?
>
> 564: Negative. Back when we initially spotted it, it was between the weather [*typically cumulo nimbus thunderstorm cloud*] and us, and when there's lightning you could see a dark object and, er, it was pretty eerie-looking . . . First time in 15 years I've ever seen anything like this. It's probably military . . .

Later, when another aircraft had spotted the mystery object, Albuquerque tried to obtain information from North American Aerospace Defense Command (NORAD) at its Western Air Defense Sector Headquarters (call sign 'Bigfoot'):

> ACC: . . . Around Tucumcari, New Mexico, north of Cannon, I had a couple of aircraft report something 300 to 400 feet long, cylindrical in shape with a strobe, flashing, off to the end of it.
>
> NOR: . . . Er, we don't have anything going on up there that I know of . . . You all serious about this?
>
> ACC: Yeah, he's real serious about it . . .
>
> NOR: . . . How long did he think it was?
>
> ACC: He said it was 300- to 400-foot long.
>
> NOR: Holy smoke! . . . I wonder if any of our aerostats [*tethered blimps*] got loose or something. But we don't have any aerostats there . . .

According to NORAD's Directorate of Operations and Space Control Center, no space debris re-entered the Earth's atmosphere during the period in question. Could this object have been a genuine spacecraft? In a letter to Walter Webb, NORAD explained that 'Uncorrelated Event Reports' (UERs) 'are classified Secret until downgraded by proper authority. The term "UFO" has not been used by this headquarters since the "Blue Book" was permanently closed . . .'[23]

THE PYRAMID OVER PELOTAS

Thirty-eight-year-old Haroldo Westendorff, owner of a rice-processing plant in the Brazilian state of Rio Grande do Sul, is a champion aerobatics pilot who customarily flies his Embraer EMB-712 around the state during weekends. On the morning of 5 October 1996, he took off from the airport at Pelotas, flew towards Laranjal Beach, on the outskirts of Lagôa dos Patos (Duck Lake), then headed back to Pelotas at 10.30.

Flying at 5,500 feet, the pilot was astounded by the sight of a huge craft. The silent object was turning slowly on its axis, heading towards the Atlantic coast at about 100 k.p.h. He approached for a closer look. 'It was enormous, gigantic really,' he told investigators for the Brazilian UFO research group GPCU.[24] 'It was shaped like a pyramid and had eight sides; each one of them had exactly three prominent bulges, which [presumably] constituted the windows.' (See Fig. 31.)

Fig. 31. Illustrations (A) and (B) show Haroldo Westendorff's Embraer 712 circling the huge craft, the base of which was estimated by the pilot to be 100 metres in diameter. A disc-shaped object emerged vertically, inclined at an angle (C) and shot

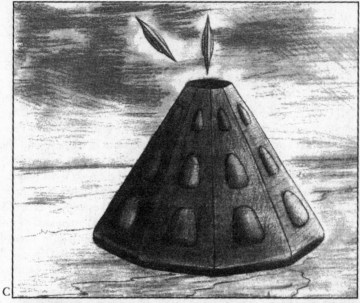

off at phenomenal speed. The large 'mother ship' then emitted beams of reddish light (D) before rotating, then taking off at an estimated speed of 12,000 k.p.h.
(GPCU)

Westendorff estimated that the base of the object was about 100 metres in diameter and its height between 50 and 60 metres. He radioed the control tower at Pelotas and asked if they were aware of the object. Airton Mendes da Silva, on duty at the time, grabbed binoculars and immediately spotted the huge craft, as did others at the tower.

Westendorff contacted the radar centre at Curitiba, operated by CINDACTA, which monitors airspace over Brazil. They confirmed that the pilot was 35 miles from the eastern sector of Pelotas, but said there were no other aircraft within a 200-kilometre radius of his position. The pilot asked for permission to be switched over to a special radio frequency so that he could give them a better description of the object, but permission was denied. He then decided to fly around the unknown craft. As he did so, his cellular phone rang. It was a friend, and Westendorff excitedly related what he was seeing. Shortly afterwards, Westendorff's son phoned. Noticing the anxiety in his father's voice, Haroldo Jr. handed the phone to his mother. Frightened by her husband's description of what he was seeing, she warned him to be careful. Westendorff daringly decided to fly around the object for a second time.

As he was circling the base of the craft, he noticed that a dome on top had opened. 'I don't know how, but I imagine that it must have retracted,' he said. Suddenly, a disc-shaped craft, about three times the size of Westendorff's Embraer, exited from the top of the giant craft and rose vertically. A moment later, the disc stopped, inclined at a 45-degree angle, then took off at a phenomenal speed towards the Atlantic Ocean. Seconds later, reddish beams of light began to come from the top of the 'mother ship', felt as waves of heat. Undaunted, the intrepid pilot decided to circle the craft yet again.

As Westendorff continued to fly around the object at a distance of about 40 metres, the enormous craft began to rotate, faster and faster. 'At this moment I began to panic,' he said. 'I thought I was going to die. I was just waiting for the explosion that would certainly destroy my plane.' He managed to retain his presence of mind, and began to go through emergency procedures in anticipation of a tremendous explosion. When the craft suddenly took off – at a speed he estimated to be 12,000 k.p.h. – Westendorff found it difficult to believe he was still alive, and his plane unaffected by the encounter.

'I know what I saw, because I was 40 metres away from that object,' declared the champion pilot. 'No one can get me to deny it. I'm certain that what I saw had nothing to do with this planet . . .'[25]

NOTES

1 'A Close Encounter with "Greenish-Faced" Creatures in Poland in 1978', *Flying Saucer Review*, vol. 36, no. 1, March 1991, pp. 1–7, prepared by the

Wroclaw Club for UFO Popularization and Exploration (WKPiB-UFO), Klub Mozaika, U1. Trzemeska 2, 53-679 Wroclaw, Poland.

2 Chiumiento, Antonio, 'The Little Oriental Airman: Another Remarkable CE-III Case in Italy', *Flying Saucer Review*, vol. 28, no. 5, September–October 1982, pp. 3–8.

3 Chiumiento, Antonio, 'An Encounter with "Rat-Faces" in Italy', *Flying Saucer Review*, vol. 28, no. 6, November–December 1982, pp. 14–19, 25.

4 *UFOs – Und Es Gibt Sie Doch* (UFOs – And They Do Exist), a documentary film by Heinz Rohde, Norddeutscher Rundfunk (NDR), 1994, translated by Dorothee Walter.

5 Sparks, Brad, 'Colusa (California) Close Encounter, 10 September 1976', *The APRO Bulletin*, vol. 25, nos. 7–10, 1977.

6 Bühler, Dr Walter, 'Conversation with Entities at Marimbonda', translated by Gordon Creighton, *Flying Saucer Review*, vol. 25, no. 3, May–June 1979, pp. 18–19.

7 *Diário Popular*, La Plata-Buenos Aires, 12 August 1983; *Tiempo Argentina*, Buenos Aires, 13 August 1983, translated by Jane Thomas, *Flying Saucer Review*, vol. 29, no. 2, December 1983, pp. 9–10.

8 *La Reforma*, Buenos Aires, 12 August 1983, translated by Luis Gonzalez.

9 *Diário Popular/Tiempo Argentina*.

10 *La Reforma*.

11 *Diário Popular/Tiempo Argentina*.

12 *La Reforma*.

13 *Diário Popular/Tiempo Argentina*.

14 Ibid.

15 Good, Timothy, *Beyond Top Secret: The Worldwide UFO Security Threat*, Sidgwick & Jackson, London, 1996, pp. 87–93.

16 Tolosano, Marc, 'They Say They Know All our Languages!: A CE-IV Case at Sospel, South-Eastern France (1983)', *Flying Saucer Review*, vol. 35, no. 2, June quarter 1990, pp. 23–4, iii, translated by Gordon Creighton from *Lumières Dans La Nuit*, no. 299, September–October 1989.

17 Ibid.

18 *Soviet Youth*, Moscow, 4 May 1990.

19 Lebedev, Nikolai, 'The Soviet Scene 1990', *The UFO Report 1992*, ed. Timothy Good, Sidgwick & Jackson, London, 1991, pp. 65–8.

20 *Voyagers of the Sixth Sun* (videotape), Genesis III, Box 25962, Munds Park, Arizona 86017, 1996.

21 Letter to the author from Hal Starr, 1 February 1997.

22 Webb, Walter N., *Final Report on the America West Airline Case*, UFO Research Coalition, July 1996, available from The Mutual UFO Network, 103 Oldtowne Road, Seguin, Texas 78155.

23 Ibid.

24 Grupo de Pesquisas Científico-Ufológicas (GPCU), Rua Barão de Azevedo Machado no. 51/301, CEP 96020-150, Pelotas, RS – Brazil.

25 Wysmierski, Michael, 'The Lagôa de los Patos Incident', *The Brazilian UFO Report*, vol. 2, no. 9, March–April 1997, pp. 3–9.

Chapter 19

Alien Base – Earth

For four decades, the United States Commonwealth of Puerto Rico has seen a proliferation of encounters with an assortment of seemingly alien species. In recent years, sightings of a demon-like creature called the *chupacabras*, or 'goat-sucker', responsible for bizarre deaths of hundreds of farm animals and pets, have received world-wide attention, though even decades ago there were reports, associated with UFO sightings, of vampire-like creatures that supposedly sucked blood out of animals. There have also been encounters with quasi-human species.

Zulma Ramírez de Pérez, whose family owned part of the land in Cabo Rojo, southwest Puerto Rico, from where have come many reports, and where it is widely believed that an alien base is located, reports that, since 1956, her entire family had observed disc-shaped domed craft coming out of or entering the waters of the Laguna Cartagena. 'You could see people or figures inside the domes,' she told Jorge Martín, Puerto Rico's leading authority on the subject. 'On some occasions, when they came out, we yelled at them and they would stop in the air in front of us.'

One night in 1964, said Sra Ramírez, her brother Quinn shouted at them that he wanted to know who they were, 'to see if they really were from outer space'. Later that night, he suddenly felt an urge to go to the lagoon, and headed down a dirt track leading there. Getting out of his jeep, he saw two figures approaching him from the lagoon.

> They were tall white men, about six to seven feet tall, with blond, long hair and dressed in one-piece, tight-fitting silvery suits. 'They were very beautiful and delicate, almost female looking,' he said. He was too nervous, and asked them not to come any closer as he couldn't take it, so they smiled sweetly and walked back into the lagoon . . . I know he had other encounters with these beings, because some nights he disappeared into the lagoon area, and he wouldn't talk about what he was doing there all night. But we knew he was with his 'friends', as he used to call them.[1]

The quasi-human extraterrestrials have been observed more recently. Jorge Martín told me that one evening in 1987, Sra Ligia Medina was fishing off Puerto Real, Cabo Rojo, together with her young grandson, when a huge disc-shaped craft came down out of the sky. She said it looked as though it was made of a transparent, crystal-like material, through which, in silhouette, could be seen human-like figures moving around and looking down at them (similarly described by Lucy McGinnis – see p. 151). The grandson shouted that he was afraid. 'Don't worry,' she told him, pretending not to be afraid herself, 'if they're going to take us away, I'll ask them to take only me.' All of a sudden, the craft flew off.[2]

RAMEY AIR FORCE BASE

The following statement, obtained by US Air Force member John Artie, was signed by Sergeant Thomas Carulli, a USAF security policeman stationed at Ramey AFB, near Aguadilla:

> [On] August 18, 1968, at approximately 2.30 a.m., off Borinquen Beach, Ramey AFB, Puerto Rico, 15 other security policemen and myself observed what was believed to be [an] unidentified flying object. Visibility was unlimited. The UFO seemed to be rising from the ocean but when first seen it was appearing to be at a 45-degree angle above the surface and rising. While also rising it seemed to yaw to its sides emitting a very bright, almost fluorescent light which was similar to that of an unblinking strobe light . . . there appeared to be struts or bars or (you could say) window panes. These struts were vertical and there were about six of them . . .
>
> When it reached its zenith, which in itself appeared to be no more than 1,500 feet off the ground, it lingered there for a few minutes; all the while it emitted this light which lit up the whole area, which before was in complete darkness. It was circular or sphere-like . . . Its size was close to a half-dollar held at arm's length. It emitted no sound . . . While in its zenith the UFO seemed to pulsate and fluxed from side to side or wobbled.
>
> Then another orb of light came from it [which] was about the size of a dime held at arm's length. It stayed by the side of the first UFO, then it too fluxed or wobbled and shot straight upwards until it was nearly invisible. After two or three minutes passed, the first UFO wobbled, turned on its side and darted upwards and outwards in a north-northwesterly direction until it disappeared. This whole spectacular phenomenon took about 12–17 minutes in its entirety.[3]

Between 1960 and 1962, William Smith (pseudonym) was assigned as a USAF air policeman to an elite corps which guarded the B-52 bombers and their cargo of nuclear weapons at Ramey AFB. This assignment called for a clean service record, as well as impeccable character. Smith told American investigator Jim Speiser that one night in late 1961 or early 1962 he was on duty guarding the bombers when, as he claims, he saw three very large objects, shaped like discs, fly over the base and position themselves precisely over the B-52s. They hovered for a short period then flew away. At least 24 other witnesses were present. The following day, two men in civilian clothing, who said they were from the Pentagon, arrived at the base. Smith was unable to determine to which department these men were attached. All witnesses were interrogated for hours and warned not to discuss the incident.

Stephen Craig, a former US Navy electronics technician who spent three years at the naval facility at Ramey AFB in the 1970s, confirmed to Jorge Martín that he had both Navy and Air Force acquaintances who had witnessed groups of UFOs flying over the flight line, as well as over the nuclear arms depot.[4]

TREMOR

On 30 May 1987, local residents near Laguna Cartagena saw a strange 'red ball of fire' descend in a controlled manner into the lagoon, making a buzzing sound. At 02.00, many witnesses sighted a huge, brilliant disc hovering over the lagoon. The following afternoon, thousands of inhabitants of southwest Puerto Rico were shaken by a powerful earth tremor and what sounded like an underground explosion. Initially, the tremor's epicentre was pinpointed at 13.5 nautical miles below the Laguna Cartagena, between the towns of Lajas and Cabo Rojo. The next day, however, the Puerto Rico Seismologic Service announced that the epicentre had been out to sea, to the west of Puerto Rico, in the Mona Passage.

On the evening of 1 June 1987, a huge unidentified cylindrical craft, with two spheres at each end, was seen by numerous witnesses in the community of Betances. 'It was really large,' said Sra Rosa Acosta. 'It was incredible. That thing came down and hovered in the air, motionless over the Laguna Cartagena. Then, about 15 minutes later, it flew up and disappeared to the south, behind the Sierra Bermeja.'

Following the explosion, several fissures, from which issued cobalt-coloured smoke, appeared in different areas of Lajas and Cabo Rojo, as well as the Laguna Cartagena. The lagoon was cordoned off, and men dressed in camouflaged fatigues, plain clothes, or white anti-contamination-type coveralls, were seen taking samples of water, mud, soil and plants. The military prevented anyone accessing the area.[5]

In late 1988 there were two occasions when pairs of US Navy jets were seen to disappear in close proximity to, or possibly to enter, huge, triangular-shaped objects. One of these incidents took place in the municipality of San Germán, the other in Betances, near the Laguna Cartagena.[6] By this time, many residents had come to accept the existence of an alien base in their vicinity.

Reports of UFOs seen by fishermen off the coast also proliferated. 'Something abnormal is going on down there,' said one witness, Arístides Medina. 'On one occasion, I was fishing late at night near Cayo

Margarita, and two of them passed under my boat, radiating a blue light. On other occasions I have seen them when they emerge from the water and fly away at great speed, and I have also seen them plunge into the water.'[7]

CREATURES GREY AND GREEN

Since 1990 I have visited Puerto Rico on many occasions, to investigate personally what to me seems a unique situation on this, the aptly named 'Enchanted Isle'. I have never seen anything really peculiar. On one occasion, though, events reportedly occurred in precisely the same area that I visited with Jorge Martín but 12 hours later. In the small hours of 31 August 1990, a number of witnesses saw five weird creatures, similar in some respects to those reported in Kirgiszkaya, USSR, three months earlier, walking down a road in the Laguna Cartagena area. Their height varied from three to five feet. When one witness, Miguel Figueroa, tried to follow the creatures in his car, they turned around, emitting a brilliant light, like a welding torch, from their eyes. 'I was blinded and scared,' he said. 'I felt, or heard something telling me not to get any closer.' But he followed and managed to get a closer look. As he told Jorge Martín:

> They were skinny, with large pear-shaped heads, long pointy ears, big slanted eyes and almost no nose . . . Their mouths were almost like a slit. They all had long arms with three fingers on each hand and three toes on their feet. At their elbows and knees they had something that looked like joints . . . I don't know if that was part of some clothing they had on, but to me they seemed to be naked. They were greyish . . .

Eventually the creatures jumped over a bridge one by one and headed along a creek in the direction of the lagoon.

The next day, Figueroa received a threatening phone call (at his unlisted number) from a man with an American accent. 'He told me not to talk or say anything to anyone about what I had seen and where the little men had gone into,' he told Jorge. 'What is happening here is real, and these beings must have a base or something underground in this area.'[8]

Other, similar species of creature appear to be based elsewhere on the island, including the Caribbean National Forest, the only tropical forest in the US Department of Agriculture's national forest system, generally referred to as El Yunque after the mountain peak near the recreation area. Many witnesses claim to have encountered beings with very large black eyes, though lacking pointed ears and usually with four, rather than three, fingers or claws.

One night in February 1991, former police officer Luís Torres and his wife, together with two of his police colleagues and their wives, were astonished by the sight of two strange little men coming down the road on Route 191, near the El Yunque tourist office. They were speaking a 'weird gibberish'. 'It was like when you listen to a tape-recording that's very fast,' Torres told investigator Magdalena Del Amo-Freixedo.

> They were maybe around four feet high, thin, dressed from head to foot in clothing [that fitted] the body very closely . . . between green and grey. It came right up to the tops of their heads, covering the skull . . . Their little arms were right down to about the knees, and their heads too were elongated, though we couldn't see them very well.
>
> Their heads were large, sort of slanted, big at the top and small down below, looking more like the shape of an egg. And a bit flattened at the top, and their faces flattened too. I saw no eyebrows on them. They had big dark eyes, blackish, protruding a bit from the face . . . Their little necks were very thin; almost no nose to be seen, and nor the mouth neither. Their skin looked to me grey or greyish-green.

The creatures walked straight past the six witnesses. 'They must have seen us,' Torres continued. 'They carried on down the road, and when they had gone about 100 feet or so past us, they turned around and started off up the road again, passing right by us once more. We tried to follow them.'

> I got out my revolver – not to harm them, but just so that they would see that we were armed, in case they should by chance try to do anything. But when I got out the revolver, it seems as though they knew it. They didn't look directly at us at any time, but they quickened their pace and, a bit further up the road, they crossed over to the left-hand side of the road and entered the thickets on the hillside.

Charmed by the creatures, Margarita Torres said she would have liked to have brought them home. 'They really were weird, and at the same time lovely,' she explained, 'because in shape they looked just like two little twins.'[9]

In the small hours of 13 August 1991, Sra Marisol Camacho encountered two similar creatures, examining a plant on her balcony with their four long, skinny fingers, in the Maguayo community next to the Laguna Cartagena. 'I don't know why, but I couldn't move,' she told Jorge Martín. 'They took leaves from the plant and left, talking among themselves in that fast mumbling gibberish.' She added that the creatures made a return visit two weeks later, but when she tried to communicate with them they made a hasty retreat, running in the direction of the lagoon. 'They didn't harm me,' she said. 'One thing is for sure: they are already here, living among us. We should prepare to face that fact . . .'[10]

Also in 1991, Ulises Pérez came across a similar creature in an irrigation canal which leads to the Laguna Cartagena. The creature appeared to have a pale, whitish skin with pinkish-red spots. When Pérez started to get away on his motorbike, the creature disappeared under the water in the canal.[11] This and other cases seem to suggest that a habitat has been established under the lagoon and perhaps other bodies of water. The webbing observed between the fingers or claws of these entities might mean that they are amphibious.

A variation on these species was encountered in July 1968. Freddie Anderson and a group of friends were visiting the El Yunque mountain when they encountered a tall creature standing in the middle of a river next to Route 191, only 12 feet away. Its height was some six feet, and it was very slim, Anderson told Martín. The hands reached almost to the knees.

> It didn't have clothes on. It was all green, and it had a large head that was wider on the bottom, on the chin, and it ended in a [conical] top. It had large, round, bulging eyes, intense green in colour, and two small holes for a nose . . . Its hands had, I think, only four fingers with some small round things on their ends, something like tree-frogs have, and some small claws that came out of those small round things on the ends.

Anderson and his friends are certain they experienced 'missing time'. 'Suddenly it's as if we came out of something,' he continued. 'We don't know what happened, but it was already night time . . . and that thing was gone.'[12]

THE SANTIAGO BASE

Another area of concentrated, unusual activity in Puerto Rico is Salinas, on the south coast, particularly in the vicinity of the grounds and airspace of the US Army Reserve Base and the Puerto Rican Army National Guard's Santiago Camp. One night in October 1987, for example, many witnesses reported a huge boomerang-shaped object hovering over the base, seeming to cause an electrical power blackout. A military officer told Jorge Martín that it looked as if it was of a whitish plastic or porcelain texture, and that it emitted a blinding light.

> It came very rapidly and stopped suddenly over a tree, suspending itself. In its central apex, or the V's point, a sort of cabin or windows with lightbulbs could be seen. After 15 or 20 seconds, it shot away sideways, and got lost at a tremendous speed toward the north. That thing was as large as a 747 jumbo jet, folded in half, that big, but much wider. The most surprising thing was that it was totally silent.

The officer also said that each time such objects were seen over the base, electrical and weapons systems failed.

On many occasions, huge flying triangles, as well as flying saucers, have also been seen over the base. Jorge Martín, whom I know as a reliable, honest investigator, himself sighted a strange flying disc over the restricted airspace of the base, one afternoon in January 1993, together with his wife and co-investigator, Marleen, and three American investigators. 'We all saw the object,' said Jorge, 'with a flat bottom and a small cupola or dome on its top part.' The disc rose upwards with a swaying or rocking motion, until it disappeared in a cloud.

Many people have reported seeing the ground 'open up' in a certain part of the base located between two mountains, revealing a 'hole' emitting brilliant reddish-orange illumination. One of the most interesting incidents occurred during exercises involving the National Guard, in 1989. As an officer informed Martín:

> We were in the field, and it was late at night. Suddenly, we all saw when one of the mountains on the north (at the back of the base) seemed to open up, and from the opening there emerged a strong light, orange-pink in colour. Suddenly, we saw many flying saucers come out of the opening – UFOs that lost themselves in the sky. After they came out, the opening closed in the mountain, and everything disappeared. We couldn't believe what had happened . . .
>
> The next day, all of us who had seen what had happened were gathered together, as some [American] officials arrived in helicopters. They told us that what we had seen was real, but that we should not worry; that the US authorities, the military and NASA were dealing with the matter. They ordered us not to speak with anyone about what we had seen, because problems could arise. At no time did they tell us that what we'd seen was of extraterrestrial origin, but something in their manner of speaking seemed to indicate as much.

In the summer of 1979, José Luís Rodríguez and his cousin also claim to have seen flying discs emerging from a flat area near the Santiago base. The actual location was in the northeastern part of the base, between Rio Juan Mountain and a sector of the old State Road No. 1, near the restricted area where ground and air artillery exercises were carried out. The incident is said to have occurred at around 14.00, as the young men were riding horseback.

'We heard a sound, and suddenly we saw that the earth was moving in front of us,' Rodríguez told Martín.

> A part of the ground arose, with dirt, rocks, underbrush, bushes – everything. They appeared like camouflaged platforms, covered with what looked like normal terrain. The platforms were raised by what appeared to be powerful hydraulic supports, and on the sides were what seemed to be nets with camouflaged strips and camouflaged underbrush, and so on.

Metallic walls extended deep into the ground. As the camouflaged platforms arose, the witnesses could see large rectangular openings, some 80 feet wide.

> Then we felt a strong humming, and that's when we saw two objects that flew out of the openings. They were flying saucers! They came out, flying slowly, and remained suspended at a height of ten storeys above the openings. They were beautiful, silvery and metallic, some 60 feet in diameter, and each had a section that had windows all around it in the centre, [from which] a strong light emitted.

What the young men claim to have seen on the centre of the flat, underneath section of the craft seems totally unbelievable: the acronym 'URSS' or 'UPSS', in low-visibility letters.[13] If the former is accurate, it was the Spanish acronym for the Union of Soviet Socialist Republics! One could speculate that the craft might have been a secret development of either the Americans or the Russians, or both, derived perhaps from alien technology. As I reported at length in *Alien Contact*, there is information that Americans have test-flown disc-shaped vehicles since the 1950s, at least. In any event, together with the hydraulically operated camouflaged platforms, the evidence suggests that the disc was of terrestrial origin.

As the witnesses continued watching the discs for two minutes, a jeep came speeding towards them from a distant white building. Assuming they had been spotted and reported to security by the pilots, the two men rode away fast.[14]

THE SPECIMEN

In 1979 or 1980, a creature resembling some of those sighted in Puerto Rico, though smaller, was allegedly killed by a youth, José Luis 'Chino' Zayas, in a cave near a mountain behind the National Guard's Santiago Camp. 'We don't really know if it's alien-related,' Jorge Martín emphasized to me. 'We only know that it came from the cave system up there, where there has been a lot of strange activity.'[15]

'I first heard about the case when [Chino] told us about it,' explained Sergeant Benjamín Morales of the Salinas Police, in an interview with Jorge Martín. He went on:

> He said that he'd been way up there by the Tetas de Cayey [twin mountain peaks] with a friend and that they'd seen a group of little animals, creatures that looked like small humans, going into a crevice within a cave. Allegedly, one of the tiny creatures attacked Chino or grabbed his leg, and he got scared. He picked up a stick, clobbered it and killed it. Later, I got to see the little man or creature myself when [Officer] Osvald Santiago [who had confiscated it from Chino] brought it to the police station . . . Chino was afraid it would decompose and took it over to Wito Morales, the owner of Monserrate

Funeral Parlor, who placed it in a glass jar with formaldehyde to preserve it.

That thing belonged to an unknown species, not a human or an animal [see colour plates]. I've been working for the Puerto Rico Police for 24 years . . . and I wouldn't say what I'm telling you unless I was sure about it . . . I'm a licensed emergency medical technician [and] those who say it was a fabrication or an ape or a fetus don't know what they're talking about, or they're lying . . . The thing had a head too large for its body and pointy little ears. Its skin was a greyish-green and [its eyes were] large and slanted. It had no nose, only two little holes, and a mouth without lips or teeth – at least, I don't recall seeing any teeth . . . The bone structure was different . . . They were already formed and hard; its crown wasn't soft, its bones weren't brittle.

Elizabeth Zayas, Chino's sister, provided further details:

It had some kind of white or blondish hair on the sides of its head, but was otherwise bald . . . It had large eyes and its pupils were like those of a cat. The eyes were strange, because they had no colour; they appeared transparent, whitish, crystalline. I don't know if it was because the little man was dead or because that's how they were. The arms were very long and thin, and its hands reached its knees or further. The hands were like forks. They had only four, clawed fingers, like the claws of a cat, and a sort of webbing in between them, like thin membranes . . .

It was real skinny, and its feet were really weird . . . They looked more like the flippers people use to swim and to skin-dive.

Regarding the 'flippers', there is an interesting correlation here with Jan Wolski's aliens (see Chapter 18).

At no time did the creature bleed from its fatal injury, though it was covered in 'a clear goo, something resembling egg whites', said Elizabeth Zayas. The creature also had a well-developed male organ and testicles. 'It was definitely neither a baby nor a monkey,' she asserted.[16]

Sr Calixto Pérez, a professor of chemistry at the University of Puerto Rico, was another who examined the creature. 'In my opinion it was something extraterrestrial,' he told Jorge. 'Its cranium was too big for the body, which was small and skinny, and its eyes were too big . . .'[17]

Eventually, the police informed the US military authorities. Not long afterwards, some men turned up at the Zayas household. 'My husband was there when the men came,' said Elizabeth, 'and told me that they had in fact shown papers, something like an order to collect the thing, along with federal identification. They said they were from NASA. They searched the house, found the [jar with the creature] and took it. Chino told me the men said they were taking the corpse to a museum in Ponce, where they had a laboratory, and then they would take it to NASA in the United States . . .'[18]

Whatever the origin of these creatures, it is unlikely they fly the saucers!

CHASING THE CHUPACABRAS

Late 1994 saw a proliferation of mutilations in Puerto Rico, involving hundreds of animals. These mutilations were attributed to the *chupacabras* – 'goat-sucker' – and what Jorge Martín calls Anomalous Biological Entities (ABEs). Typically, the animals were found with small circular holes about a quarter- to a half-inch wide, though sometimes larger, arranged in a triangular pattern and penetrating deep into the neck or lower jaws and straight into the brain (cerebellum), causing instant death. 'Whatever penetrates the animal,' says Martín, 'is at least three or four inches long, and in a few cases has been known to cauterize the wall of the wound – apparently to prevent excessive blood loss.'

> Some of the wounds of this type appear to the sides and belly of the victim. This penetration usually cuts through the stomach – down to the liver, apparently removing sections of the organ and absorbing liquid from it. Such actions would require an incision of up to five inches – a fact verified during necropsies . . . Several larger holes have been discovered [ranging] from one inch in diameter to 12 inches in length . . . located in the neck, chest, belly and anal areas . . . clean-cut openings through which certain organs are excised from the bodies. Reproductive, sexual organs, anus, eyes and other soft tissue have all been removed.

Blood and other fluids are reported to have been removed from the animals, presumably with the creature's long, snake-like tongue. According to most reports, the chupacabras appears to be a cross between a typical 'Grey' alien and the body of a bipedal, erect, dinosaur-type creature, minus tail, with short arms and three-clawed hands (see colour plates). Martín elaborates:

> Two elongated red eyes have been reported, together with small holes in the nostril area, a small slit-like mouth with fang-type teeth protruding upwards and downwards from the jaw . . . It appears to have strong, coarse hair all over its body, and while most observers claim the hair is black, it has the remarkable ability to change colours at will, almost like a chameleon, [from] black or a deep brown colour [to] green, green-grey, light brown or beige.[19]

The creature has also been reported to fly. One witness, Daniel Pérez of the Campo Rico sector of Canóvanas, to the north of El Yunque, claims to have heard a buzzing sound at around 06.45 one morning. 'At that moment, the creature descended, apparently flying,' Pérez told Martín. 'It descended on a large stone that is on my property, some 20 feet from where I stood.'

> As soon as it made contact with the stone, it took impulse again and rose into the air, and cleared the trees ahead without touching a single leaf. It's a creature [that] when it stands straight must be some five feet tall. Its hindlegs are long, its forelegs are short . . . from the top of its head all down its back it has

some type of fins that move. When it was about to take off, the fins moved in the direction it was headed . . . They're some six to eight inches long [and] he moves them so fast that he makes it seem they're hairs.[20]

In January 1997 Jorge Martín and his wife, Marleen, took me to visit Canóvanas, where we spoke to several witnesses who, like Daniel Pérez, had seen the chupacabras in broad daylight. Madelyn Tolentino and her mother showed us where they watched it walking down the street outside their home one afternoon in August 1995. As they approached the creature, it shot off into the bushes at a phenomenal speed. The witnesses, who struck me as completely down-to-earth, reported seeing a transparent membrane between the creature's arms and back, perhaps used in flight.

Jorge has learned from reliable sources that chupacabras have been captured on several occasions. He told me that on the night of 6/7 November 1995, a live creature was purportedly captured by Forestry Service personnel and taken to the continental United States. On another occasion, a live creature was taken away in a cage by federal personnel of Puerto Rico and the US Department of Agriculture. A civilian employee (name known to me) who works at a US Army base in San Juan reported that he had seen there a dead chupacabras, preserved on ice in a special reinforced-mesh cage. 'He was in an office,' Jorge told me, 'when several soldiers came by, carrying a cube-like container covered with a crystal-like mesh. They put it on top of the table in this office.

'Curious as to what was inside, he moved the cloth, and saw a very ugly creature which seemed to be dead. He remarked to the soldiers that it looked like the chupacabras. The soldiers became furious and said it was a monkey, but warned him that if he talked about what he had seen, he would get into a lot of trouble and could lose his job.'[21]

The chupacabras phenomenon was taken very seriously by the Puerto Rican authorities. On 9 November 1995, Representatives José Nuñez Gonzáles and Juan López presented a resolution at the 12th General Assembly of the House of Representatives. It states in part:

> In the last months numerous deaths and attacks on animals have been reported in Puerto Rico, which remain without any explanation . . . The number of cases has increased recently and the Puerto Rican community demands action by the Government on the alarming situation . . .
>
> It is hereby ordered that the Agricultural Commission of the House of Representatives make a profound and exhaustive investigation to clarify this unknown phenomenon and account for the damages caused by the so-called 'Chupacabras' to this country's farmers.[22]

Such official investigations are not new. In 1991, Colonel José A. Nolla, Director of the Puerto Rico State Agency of Civil Defence, who had sent a directive laying down guidelines for secret investigations into UFO sightings, stated under oath in a hearing at the Senate of Puerto Rico that he, the military and the Agency had been investigating the numerous UFO sightings and animal mutilations that had taken place over the years in the country.[23]

In a radio interview for Noti-Uno, broadcast on Christmas Day 1995, Fernando Toledo, President of the Puerto Rico Agricultural Association, expressed his conviction that the chupacabras is not an indigenous species. 'I think that if we already know it's not an ape,' he said, 'we must then be dealing with an extraterrestrial . . .'[24]

A MUTILATION WITNESSED

At about 02.15 on 12 January 1997, Piedro Viera (pseudonym), a former police officer, was driving from Caguas to Humacao, on Route 30, which adjoins the southern borders of the El Yunque National Forest, when suddenly he saw a light in the sky, coming from the direction of El Yunque, growing larger until it appeared as an object shaped like an upturned saucer. 'It came closer until it was about 200 yards away from me, then stopped,' he told me. 'So I pulled up at the side of the road, where there were some cows in a pasture.'

> It was about 150 feet in diameter, with around 12 to 15 square lights or win-dows. A blue-green conical beam of light came down from the craft and engulfed two of the cows, one of which was levitated towards the craft. The cow vanished suddenly when it was about five feet from the underside. Then the craft began to move away slowly in the direction of Humacao.

Viera tried to follow the disc in his truck. At that moment, a black 4X4 pickup came along and two men, dressed in black military-style clothes and caps, came out and ordered him to switch off his engine and remain where he was for at least 10 minutes.

'Why?' asked Viera.

'Just stay where you are and allow us to continue what we are doing,' came the reply.

The men returned to their pick-up and followed the craft. After 15 minutes, Viera proceeded on his journey. Soon, he came to a Brahman cow, lying at the side of the road, similar to the one he had seen earlier. 'It had apparently been dropped from above, because two of its legs were badly broken,' he explained. 'There were about five straight incisions along the side, one going up to the chest, and a circular hole in its rump. But there was hardly any blood . . .'

Viera impressed me as an honest witness. Even prior to publication of his story by Jorge Martín, he began to receive threats, which continued for some time. He is certain these threats came from American federal agents.[25]

A VISIT TO AN ALIEN BASE?

One night in June 1988, Carlos Mañuel Mercado claims to have been visited by three alien beings in his home in Cabo Rojo and taken by saucer to their base in the Sierra Bermeja, adjoining the Laguna Cartagena.

The beings were similar in most respects to the stereotypical 'Greys', except that they had some bumps on their facial skin. When they addressed him, he 'heard' them, though their mouths did not move. All three were dressed in tight-fitting, sandy-coloured one-piece suits. Mercado was taken outside his house and down the road to some open ground, where, he says, a domed, disc-shaped craft stood on three legs. Entering the craft via a stairway leading to an opening underneath the central section, he found himself inside what he assumed was the flight deck, with other small beings and a taller being, 'a little more human-like', the same size as Mercado (five feet nine inches), dressed in a white robe. After the taller being assured Mercado that he would come to no harm, the craft took off and Mercado says he was able to watch through the portholes as it headed towards the Sierra Bermeja, only a few miles away.[26]

'As the craft approached the El Cayúl mountain,' Mercado told me, 'I saw this brilliant light, something opened up, and the craft went in there, through a sort of tunnel and into a large cavern. We all got out of the craft. It was very brilliantly lit inside the cavern and I was given some sort of glasses to wear. Everything seemed to be made out of aluminium. "I'm going to show you how advanced we are," the taller being said to me.

'There were many barracks-like structures and hundreds of the little aliens, some wearing a type of cap, working as if on production lines, assembling machinery, and there were many crafts: saucer-shaped, like the one we went in; boomerang- or triangular- and hexagonal-shaped.'[27] The tall being then addressed Mercado:

> As you can see, we have a base here for the maintenance of our crafts' systems. We have been here for a long time and don't intend to leave. We want the Earth people to know that we mean no harm, that we don't mean to conquer you either. We want to reach out to you to establish a direct relationship which will be beneficial to both parties . . .

'Why me?' Mercado protested. 'I'm a simple man, and no one will believe me.'

'It doesn't matter,' came the response. 'People will hear you, as well as many others we are contacting and bringing here to show them the same. When [educated] people hear what you simple people – as you call yourself – are saying, they will know that you are telling the truth.'[28]

Mercado was then returned to his house.

'Are you certain that this wasn't just a dream?' I asked Mercado, during an interview at his house in Cabo Rojo in 1997.

'Sometimes I wonder if it was a dream, but it was all so very real,' he replied. 'I certainly don't remember falling asleep. I was watching TV when they came to the house, and when I returned at 3.00 a.m., I told my wife about it.'

'What do you think they are doing here?'

'My feeling is that they are investigating how they can survive here,' came his very interesting reply.

'Do you think the base is still there?'

'I feel they're probably still there. There's so much still happening.'[29]

Jorge Martín knows another resident of the area – a high-ranking military officer – who claims a similar experience. 'Everything he says corroborates the details given by Carlos Mañuel Mercado,' Jorge reports, 'especially the place where the mountain slope opens down to the purported base in El Cayúl mountain.'[30]

I have been to the Laguna Cartagena and Sierra Bermeja areas on many occasions, and my disappointment at never having seen anything unusual there is always mitigated by the beautiful scenery. I can well understand why the putative aliens might want to stay there. And thus far, the little grey men have kept the place green.

CONCATENATIONS

One of the many fascinating aspects of these reports is the suggestion of a link between the truly alien-looking species and those of a quasi-human appearance. Throughout this book I have shown repeatedly that alien species are more varied than we commonly suppose. While in most cases the various species seem to operate singly, there have been encounters where two types are seen together. A classic example of alien collaboration is the well-known case of Travis Walton who, in the presence of six human witnesses at Heber, Arizona, in November 1975, was knocked unconscious by a beam of light from an unknown craft. He was apparently 'beamed' on board and kept there for five days.

On recovering consciousness, Walton was confronted by three humanoids, a little under five feet in height, dressed in coveralls, with bald, oversized heads and almost chalk-white skin. Unlike the stereotypical 'Greys', however, who have large black eyes with no visible iris or pupils, these had large brown eyes with irises twice the size of those of humans, and very little of the white eyeball showing. Elsewhere on the craft, Walton encountered three taller, human-like men and one woman, whose only really distinguishing feature as that, as Walton wrote in his important book, *Fire in the Sky*, 'something was definitely odd' about their eyes, though he could not discern what that oddity was.

Were the smaller humanoids another race altogether? Or, as I tend to believe, biological robots? 'Who was co-operating with whom?' asks Walton. 'I saw nothing to indicate the answer to that question. In fact, I never saw the two types together in one place at the same time.' It might have helped had the aliens communicated with Walton, but all his questions drew nothing but silence from them.[31]

My earlier book *Alien Contact* devotes a chapter to the remarkable events experienced by 'Jim' [Evans], a former US Air Force security offi- cer, his family, and others, at their ranch in Colorado, beginning in 1975. Those events included cattle mutilations, sightings of a 'Bigfoot' creature and disc-shaped objects, and an eventual encounter with two quasi- human beings, who seemed to be in charge. The encounter took place late one night. Jim's description of this particular type of beings accords with those of many others, going back as far as 1920, e.g., that of Albert Coe (Chapter 2). 'They were approximately five feet six inches tall,' he reported. 'They had on tight-fitting clothing, you know, like a flight suit.'

> I noticed the clothing changing colour, from brown to silver, but I don't know how. They were very fair, had large eyes and seemed perfectly normal, com- pletely relaxed. They had blond [short] hair with something over the head . . . The thing that impressed the most was the eyes . . . Their facial features were finer. They were almost delicately effeminate [and] completely self-assured.

During the meeting, with the aid of a box-like device, the beings demonstrated their complete control over the Bigfoot creature, which, regardless of whether such creatures are indigenous to Earth or not, implies that they are used in subservient roles. Jim was convinced that an alien base was located on his ranch, which, significantly, overlooked a US Air Force base. 'I can only assume they are watching us – watching our military potential,' he told investigators.

'It's not big brothers from space who are interested in us as spiritual beings or whatever,' continued Jim. 'We're nuisances, although I think they may be more humanitarian than we are. I have no doubts that they are mutilating the cattle. They are being lifted into the air, they are being drained of blood, they are being mutilated, and they are being lowered.'

A concomitant of cattle mutilations is often the presence of military helicopters in the area where the mutilations have occurred, leading many to believe that the military is responsible. Jim disagreed. 'I figured out early in the game that the government is sending in helicopters in large numbers from several sources,' he said, 'but they are doing it to cover what is really happening.'[32]

It is significant to me that in many areas where there has been a plethora of animal mutilations – such as in Puerto Rico and Colorado – rumours abound of alien bases in the vicinity. Perhaps these rumours are not without foundation. If there is a requirement for animal blood, organs and tissue – for whatever purpose – it is reasonable that a nearby base of operations would be useful. Lieutenant Colonel Philip Corso, US Army (Retired), who served in the Army staff's Research and Development directorate at the Pentagon in the early 1960s, believes that the extraterrestrial biological entities (EBEs) are or were 'experimenting with organ harvesting, possibly for transplant into other species or for processing into some sort of nutrient package or even to create some sort of hybrid biological entity'. He claims that, although the first public reports of cattle mutilations surfaced around 1967, 'at the White House we were reading about the mutilation stories that had been kept out of the press as far back as the middle 1950s, especially in the area around Colorado'.[33]

Another state which has experienced an inordinate number of cattle mutilations is New Mexico. Over the years, I have travelled the length and breadth of this, one of my favourite American states, investigating various claims associated with the multi-faceted UFO phenomenon. Two areas where alien bases are alleged to exist, or to have existed, are Dulce (not far from the border with Colorado) and the Manzano Mountains, near Albuquerque. The latter area features in *Beyond Top Secret*, and earlier in this book, where I discuss sightings reported by security personnel in the vicinity of the Manzano Weapons Storage Area. Interestingly, I was told by a reliable source that an alien base supposedly existed in that vicinity at one time.

In March 1997, I spent some time in Dulce, on the Jicarilla Apache Indian Reservation, encompassing 840,000 acres of scenic mountains and rugged mesas, took aerial photographs of the Archuleta Mesa (where the base was said to be located) and talked with local people. I saw thousands of unmutilated cattle. Although many bizarre things had indeed happened on the reservation in the past, including unusual activity on the Mesa and numerous cattle mutilations (see *Alien Contact*), I learned that nothing much of interest seems to have happened in recent years. The only animal mutilations are those done by human hunters: deer and elk abound in the area and, in lesser numbers, black bear and mountain lion.

The bewildering profusion of alien species described in this book – and more could have been included – suggests that at least a dozen different species have visited Earth. This does not necessarily mean that they *all* come from different planets. The diverse indigenous human and animal species abundant on Earth are an indication that other planets similarly support a variety of species. Many aliens may be essentially independent of planets. Moreover, I do not discount the possibility that some might be indigenous to Earth: the stories relating to the peculiar people of Mount Shasta (Chapter 2), for example, give pause for thought.

It is hard enough for many of us to come to terms with the possibility that even one race of extraterrestrial beings may be interacting with us, let alone a possible dozen or so. However, I must emphasize that the encounters described in these pages cover a lengthy period, and I do not believe therefore that all these races have come here simultaneously; most of them probably pass through our solar system only occasionally, though the evidence does suggest periods of intensive 'multi-racial' activity.

The evidence also suggests that several extraterrestrial races have established underground and undersea bases on Earth, perhaps as a centre of operations within the solar system. Herbert Schirmer, the policeman abducted aboard an alien craft in Nebraska in 1967, was told by the crew that they had bases on Venus, Jupiter – and Earth. 'There are definitely bases in the United States,' said Schirmer. 'There is a base located beneath the ocean off the coast of Florida which is a big thing . . . used for our benefit and theirs. There is a base in [a] polar region [and] another big base right off the coast of Argentina. These bases are underground or under the water.'[34]

As to the alien agenda, I do not have all the answers, nor do I subscribe to a single hypothesis. Schirmer, for instance, was told that the aliens 'collected' samples of various types of vegetation, animals – and humans. 'He said they had a programme known as "breeding analysis" and some humans had been used in these experiments. He didn't say if humans were kidnapped and taken away . . .'[35]

It may be that one or more races are creating hybrid species. If this is the case, the reasons may not be simple. One plausibility is that the creation of such hybrid species may be due to adaptation to living on Earth – which begs profound questions.

Others based here, concerned perhaps about the potentially catastrophic combination of our war-like tendencies with nuclear, biological and chemical weaponry, as well as over-population, pollution of the environment, potentially disastrous geological disturbances, or disturbances on Earth relating to solar activity, and so on, may feel it necessary to monitor our activities – particularly if they have a vested interest in the survival of this planet. It is also likely that some of the more advanced species may feel under no obligation to declare themselves openly, preferring to remain apart from the mass of humanity, knowing that it might be centuries before we are capable of assimilating the culture shock.

The clandestine alien activity reported on the enchanted isle of Puerto Rico may be a reflection of what is happening on a wider scale on alien base Earth – this enchanting isle in the galactic ocean.

They are here, confounding our concepts of space and time. Some are malign, others benign. Contact has been established – though not necessarily on our own terms, nor according to our preconceived notions.

NOTES

1 Martín, Jorge, 'Is There an Alien Base in Puerto Rico?', *Alien Update*, ed. Timothy Good, Arrow, London, 1993, pp. 16–17.

2 Interview with the author, San Juan, 5 September 1997.

3 'UFO Phenomena in Puerto Rico', summarized from a report by Sebastian Robiou, *The APRO Bulletin*, November–December 1970, pp. 6–7.

4 Martín, Jorge, 'The Alien Presence in our Seas', *Evidencia OVNI*, no. 13, 1997, pp. 34–5, CEDICOP Inc., PO Box 29516, San Juan, Puerto Rico 00929-0516, translated by Margaret Barling.

5 Martín, 'Is There an Alien Base in Puerto Rico?', pp. 10–11, 14–18.

6 Good, Timothy, *Alien Contact: Top-Secret UFO Files Revealed*, William Morrow, New York, 1993, pp. 25–8.

7 Martín, Jorge, 'US Jets Abducted by UFOs in Puerto Rico', *The UFO Report 1991*, ed. Timothy Good, Sidgwick & Jackson, London, 1990, p. 201.

8 Martín, Jorge, 'Puerto Rico's Astounding UFO Situation', *The UFO Report 1992*, ed. Timothy Good, Sidgwick & Jackson, 1991, pp. 106–9.

9 Del Amo-Freixedo, Magdalena, 'Puerto Rico: An Area of Extraterrestrial Experimentation?', *Flying Saucer Review*, vol. 39, no. 1, spring 1994, pp. 8–9, translated by Gordon Creighton from *Espacio y Tiempo*, no. 17, July 1992, Madrid.

10 Martín, 'Is There an Alien Base in Puerto Rico?', pp. 24–5.

11 Ibid., pp. 25–6.

12 Martín, Jorge, 'Encounters with Aliens in El Yunque', *Evidencia OVNI*, no. 8, 1996, pp. 42–7, translated by Carlos L. Moreno.

13 Martín, Jorge, 'A Flying Saucer Base in the Santiago Base, Salinas, Puerto Rico?', *Evidencia OVNI*, no. 8, 1996, pp. 26–32, translated by Carlos L. Moreno.

14 Ibid., p. 8.

15 Interview with the author, 5 September 1997.

16 Martín, Jorge, 'At Last! The Truth about the Salinas "ET Corpse"', *Evidencia OVNI*, no. 11, 1995, pp. 16–25, translated by Jorge Martín.

17 Martín, Jorge, 'Reported Discovery of Alien (?) Corpse in Puerto Rico (Part I)', translated by Gordon Creighton, *Flying Saucer Review*, vol. 42, no. 1, spring 1997, pp. 15–19, from *Evidencia OVNI*, no. 3, 1994.

18 Martín, 'At Last! The Truth about the Salinas "ET Corpse"'.

19 Martín, Jorge, 'The Chupacabras Phenomenon', *UFO Magazine* (UK), March–April 1996, pp. 20–3.

20 Corrales, Scott, *The Chupacabras Diaries: An Unofficial Chronicle of Puerto Rico's Paranormal Predator*, Samizdat Press, PO Box 228, Derrick City, Pennsylvania 16727-0228, 1996, pp. 26–7.

21 Interview with the author, 5 September 1997.

22 Martín, 'The Chupacabras Phenomenon', op. cit., p. 23.

23 Martín, 'Is There an Alien Base in Puerto Rico?', op. cit., p. 31.

24 Corrales, op. cit., p. 68.

25 Interview with the author, translated by Jorge Martín, Caguas, 28 November 1997.

26 Martín, 'Is There an Alien Base in Puerto Rico?', pp. 27–31.

27 Interview with the author, translated by Carlos Moreno, Cabo Rojo, 2 September 1997.

28 Martín, 'Is There an Alien Base in Puerto Rico?'.

29 Interview with the author, 2 September 1997.

30 Martín, 'Is There an Alien Base in Puerto Rico?'.

31 Walton, Travis, *Fire in the Sky: The Walton Experience*, Marlowe & Co., New York, 1996, pp. 103, 171.

32 Good, op. cit., pp. 68–71.

33 Corso, Col. Philip J., with Birnes, William J., *The Day After Roswell*, Pocket Books, New York and London, 1997, pp. 181–2.

34 Blum, Ralph, with Blum, Judy, *Beyond Earth: Man's Contact with UFOs*, Corgi Books, London, 1974, p. 115.

35 Norman, Eric, *Gods, Demons and UFOs*, Lancer Books, New York, 1970, pp. 185–6.

Appendix

Some Recommended UFO Journals

Evidencia OVNI
CEDICOP Inc., PO Box 29516, San Juan, Puerto Rico 00929-0516
(N.B. Spanish only)

Flying Saucer Review
FSR Publications Ltd., PO Box 162,
High Wycombe, Bucks, HP13 5DZ, UK

International UFO Reporter
J. Allen Hynek Center for UFO Studies,
2457 West Peterson Avenue, Chicago, Illinois 60659, USA

Journal of UFO Studies
J. Allen Hynek Center for UFO Studies,
2457 West Peterson Avenue, Chicago, Illinois 60659, USA
(N.B. An annual journal publishing mostly scientific papers)

Lumières Dans La Nuit
BP No. 3, 77123 Le Vaudoué, France

MUFON UFO Journal
Mutual UFO Network, 103 Oldtowne Road,
Seguin, Texas 78155-4099, USA

UFO
PO Box 1053, Sunland, California 91041-1053, USA

UFO Magazine
Quest Publications International Ltd.
Lloyds Bank Chambers, West Street,
Ilkley, West Yorkshire, LS29 9DW, UK

UFO Newsclipping Service
#2 Caney Valley Drive
Plumerville, Arkansas 72127, USA

Please consult the chapter notes for details of some other UFO journals and
organizations.

Index